Green Energy
Technology, Economics and Policy

Green Energy
Technology, Economics and Policy

Editors

U. Aswathanarayana, General Editor
Mahadevan International Centre for Water Resources Management, Hyderabad, India

T. Harikrishnan, Section 3
IAEA, Vienna, Austria

K.M. Thayyib Sahini, Section 6
IAEA, Vienna, Austria

CRC Press is an imprint of the
Taylor & Francis Group, an **informa** business
A BALKEMA BOOK

CRC Press/Balkema is an imprint of the Taylor & Francis Group, an informa business

Typeset by MPS Ltd (A Macmillan Company) Chennai, India
Printed and bound in Great Britain by TJ International Ltd, Padstow, Cornwall

British Library Cataloguing in Publication Data
A catalogue record for this book is available from the British Library

Library of Congress Cataloging-in-Publication Data

Green energy : technology, economics, and policy / U. Aswathanarayana,
T. Harikrishnan, K.M. Thayyib Sahini.
p. cm.
Includes bibliographical references and index.
ISBN 978-0-415-87628-5 (hard back : alk. paper) — ISBN 978-0-203-84146-4 (e-book)
1. Renewable energy sources. 2. Renewable energy sources—Costs. I. Aswathanarayana, U. II. Harikrishnan, T. (Tulsidas) III. Thayyib Sahini, K. M. (Kadher Mohien) IV. Title.

TJ808.G693 2010
333.79'—dc22 2010022877

Published by: CRC Press/Balkema
P.O. Box 447, 2300 AK Leiden, The Netherlands
e-mail: Pub.NL@taylorandfrancis.com
www.crcpress.com – www.taylorandfrancis.co.uk – www.balkema.nl

ISBN: 978-0-415-87628-5 (Hbk)
ISBN: 978-0-203-84146-4 (eBook)

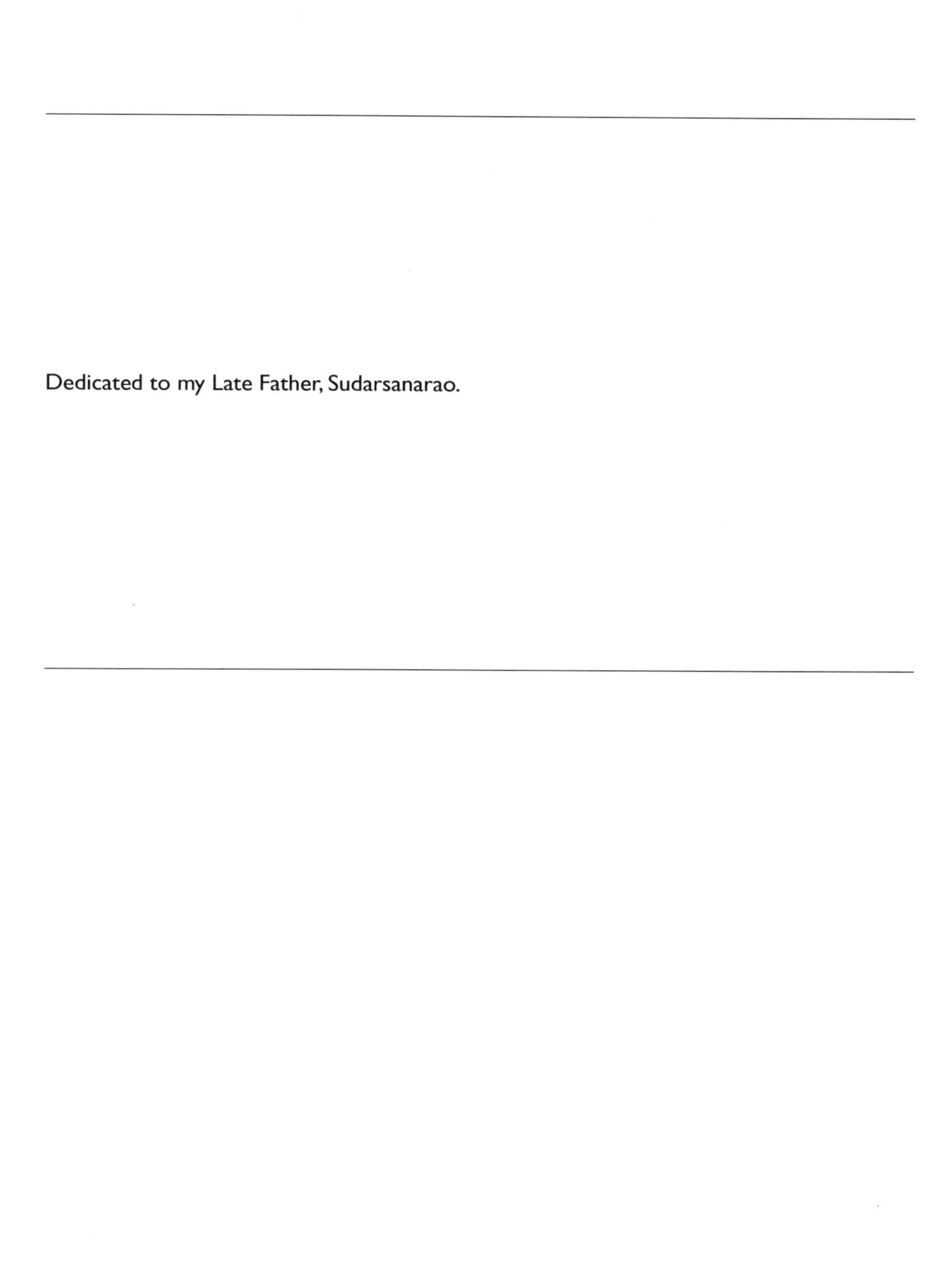

Dedicated to my Late Father, Sudarsanarao.

Contents

List of Figures

List of Tables

Preface

Lehman Brothers, a US bank with assets of over USD 600 billion, was considered to be too large to fail. But fail it did, most spectacularly – it declared bankruptcy on Sept. 15, 2008. This shook the banking industry round the world to its roots. Many banks and companies collapsed. Unemployment rose to unprecedented levels. There was a steep drop in international trade and commodity prices. Only China and India got off lightly, because of their prudent lending practices. Governments round the world came up with stimulation packages to address this catastrophic situation. Most of such packages contained promotion of green energy technologies, as a way to create employment while at the same time mitigating the adverse consequences of climate change. It has been found that renewable energy technologies employ about five times more people than the conventional energy technologies. That said, the most important challenge facing the renewables is to bring down their costs.

The issue of renewable energy has become so important that an International Renewable Energy Agency (IRENA) came to be established in Germany. Countries and institutions need a roadmap to figure out ways and means of integrating technology, economics and policy to provide clean, reliable, secure and competitive energy supply. The book addresses the need. It explains how jobs could be generated and climate change mitigated through the harnessing of the transformative power of low-carbon technologies.

The volume has seven sections. Section 1 Introduction, deals with what the book is about and how it came to be written. Section 2 gives an account of the characteristics, costs and deployment of Renewable Energy Technologies (RETs). Sections 3 and 4 deal with ways of reducing the carbon dioxide emissions and environmental impact and improving the efficiency, of Supply-side energy technologies (based on fossil fuels and CCS, nuclear power and Next Generation Green Technologies), and Demand-side energy technologies (covering industry, buildings and appliances, transport and electricity systems), respectively. The technological and R&D, improvements and economical and policy interventions needed to make the Renewable Energy Technologies (RETs) competitive in the market are covered in Section 5. Section 6 gives a vision of a Green New Deal integrating the technological, socioeconomic and policy strands. The final section 7 provides an overview and Integration.

I thank Dr. R.A. Mashelkar, President, Global Research Alliance, for his perceptive Foreword. Drs. Makarand Phadke of RIL and R. Dhanraj (formerly of AMD) kindly reviewed the manuscript, and made suggestions for improvement. The volume is meant for university students, professionals and administrators in the areas of resource engineering, energy industries, environmental science and engineering, climate change, economics, etc.

Hyderabad, India
June, 2010

U. Aswathanarayana

Foreword

Green Energy: Technology, Economics and Policy is a very timely book written by Prof. U. Aswathanarayana, a highly respected scientist with excellent record of scholarship connected with the issues of environment, ecology and socio-economics. His coauthors are Drs. Harikrishnan and Thayyib Sahini, highly accomplished professionals.

As regards green energy, there are books that deal with the individual aspects of technology or of economics or of policy, but the subtle interplay of these three is not something that has been addressed. The book fills this need admirably. The relationship between poverty reduction, protection of environment and energy security is also not at once obvious. This has been very elegantly covered in the book.

Getting into fundamentals, for our sustainable energy future, there is a common global objective that needs to be met and that is 'getting more from less for more'. And this has to be interpreted in different ways. Whereas enterprises aim at getting more (performance) at less (cost) for more (profit), they must also aim at getting more (performance) at less (cost) for more and more (people). Increasingly, affordability (more from less) and sustainability (for more and more people, both current generations and in the future) is becoming the key. In fact dealing with challenges as wide ranging as global economic meltdown to climate change, to national competitiveness, getting 'more from less for more' seems to be the only solution.

Take climate change, as an example. The answer is getting more (delivered performance) from less (carbon dioxide emissions and therefore less global warming) so that we can save the planet for more (generations to come).

This book deals both explicitly and implicitly on the technological and non-technological interventions in the form of policy, that can be made to increase the affordability and sustainability, which meet with the above objective of getting 'more from less for more'.

The book deals with the current status and the future prospects of renewable energy technology, traversing the whole range of design, development and delivery. Because the world population is growing and poor countries are becoming richer, we require fundamental changes in the way we produce and use energy. We require changes in the way we manage forests, land use, and agriculture. Greater energy efficiency, management of energy demand, and diffusion of low-carbon electricity sources such as wind, hydro, and nuclear could produce half of the required emission cuts.

The solutions that can potentially lead to 'carbon neutrality' are currently expensive. To satisfy future global energy demand will require improving the performance of low-carbon technologies and developing breakthrough technologies through not 'incremental' but 'disruptive' innovations. Our current hope rests on potential success in carbon capture and storage, second-generation biofuels, and solar photovoltaics.

The fact that a particular source of energy exists does not mean that it automatically becomes techno-economically or a socially acceptable option. Thus in the case of wind energy, despite the advances in technology, the costs need to be still brought down. Protests are heard from conscious society so social acceptability remains a challenge.

The book provides an interesting analysis of the factors that will make green energy competitive. For instance, it is brought out that the role of technology is as important as innovative energy taxation policies. The book examines both supply side and demand side energy technologies. There is a very thoughtful discussion on nuclear power economics.

This book is a 'must read' for students (both in education and research) and professionals (both producers and managers of sustainable energy solutions) and policy planners (dealing with energy, environment and economics).

Provision of energy supply with the characteristics of cleanliness, reliability, security and competitiveness is a challenge facing the 21st century world. This book is a valuable resource, which guides us towards an analysis of critical factors that will help us in achieving this path. I offer my congratulations to the authors for this timely and valuable contribution.

Pune, India
June, 2010

R.A. Mashelkar
Bhatnagar Fellow &
President, Global Research Alliance

List of Contributors

Dr. T. Harikrishnan
Nuclear Technology Specialist
Division of Nuclear Fuel Cycle & Waste Technology
International Atomic Energy Agency
Vienna, A-1400, Austria
Email: T.Harikrishnan@iaea.org

Dr. Takashi Ohsumi
Civil Engineering Research Laboratory,
Central Research Institute of Electric Power Industry,
1646, Abiko, Abiko-shi, Chiba 270-1194, Japan
Email: ohsumitk@ohsumi-jp.com

Dr. K.M. Thayyib Sahini
Division for Africa, Dept. of Technical Cooperation
International Atomic Energy Agency
P.O. Box 100
Wagramer Strasse 5
A-1400 Vienna, Austria
Email: T.Kadher-Mohien @iaea.org

Associated as a researcher with:
UNESCO Chair of Philosophy for Peace
Universitat Jaume I, Castello de plana, Spain
University of Vienna, Austria

Dr. Jayaraj Manepalli
Institut für Politikwissenschaft
Universität Wien

Albert Schweitzer Haus,
Garnisongasse 14-16/604
A1090, Vienna
Austria
Email: jayarajmanepalli@gmail.com

Sabil Francis (author C.12)
Research Academy Leipzig
Graduate Centre Humanities and Social Sciences
University of Leipzig
GC Humanities
Emil-Fuchs-Str. 1
04105 Leipzig
Germany
Email: francis@uni-leipzig.de

Prof. U. Aswathanarayana
Honorary Director, Mahadevan International Centre
for Water Resources Management
B-16 Shanti Shikhara Apts.
Somajiguda, Hyderabad - 500 082
India
Email: uaswathanarayana@yahoo.com

Units, Abbreviations and Acronyms, Definitions, Conversion Constants

Units

bbl barrel
bcm billion cubic metres
boe barrels of oil equivalent (1 BOE = 159 L)
GJ gigajoule = 10^9 joules
Gt gigatonnes = 10^9 tonnes
Gtpa gigatonnes per annum
GW gigawatt = 10^9 watts
GWh gigawatt hour
kW kilowatt = 10^3 watts
kW_{el} kilowatt electric capacity
kWh kilowatt hour
kW_{th} kilowatt thermal capacity
lge litre gasoline equivalent
mb million barrels
mbd million barrels per day
Mbtu million British thermal units
Mha million hectares
MJ megajoule = 10^6 joules
Mpg miles per gallon
Mt megatonne = 10^6 tonnes
Mtce million tonne of coal equivalent
Mtoe million tonnes of oil equivalent
Mtpa million tonnes per annum
MW megawatt = 10^9 watts

MW_e	megawatt electrical
MWh	megawatt hour
PJ	petajoule = 10^{15} joules
EJ	Exajoule = 10^{18} joules
ZJ	Zettajoule = 10^{21} joules
toe	tonne of oil equivalent
TW	terawatt = 10^{12} watts
TWh	terawatt hour
Wp	watt-peak

Abbreviations and Acronyms

(source: IEA's *Energy Technology Perspectives*, 2008. p. 602–613)

AFC	Alkaline Fuel Cell
API	American Petroleum Institute
APU	Auxiliary Power Unit
ASU	Air Separation Unit
ATR	Auto Thermal Reforming
B2B	Business-to-Business
B2C	Business-to-Consumer
B2G	Business-to-Government
BEMS	Building Energy Management System
BFB	Bubbling Fluidised Bed
BIGCC	Biomass Integrated Gasification with Combined Cycle
BtL	Biomass to Liquids
CAES	Compressed Air Energy Storage System
CAT	Carbon Abatement Technologies
CBM	Coal-Bed Methane
CCS	CO_2 Capture and Storage
CDM	Clean Development Mechanism
CdTe	Cadmium Telluride
CFB	Circulating Fluidised Beds
CFL	Compact Fluorescent Light-bulb
CHP	Combined Heat and Power
CIS	Copper-Indium-Diselenide
CIGS	Gallium-doped Copper – Indium–Diselenide
CNG	Compressed Natural Gas
CSP	Concentrating Solar Power
CTL	Coal To Liquids
DME	Dimethyl Ether
EGR	Enhanced Gas Recovery
EIA	Environment Impact Assessment
FBC	Fluidised Bed Combustion
FDI	Foreign Direct Investment
FGD	Flue Gas Desulphurisation
HTGR	High Temperature Gas Cooled Reactor

IAEA	International Atomic Energy Agency
IEA	International Energy Agency
IET	International Emissions Trading
IGCC	Integrated Gasification Combined Cycle
IGFC	Integrated Gasification Fuel cell combined Cycle
ITER	International Thermonuclear Experimental Reactor
LED	Light Emitting Diode
LNG	Liquified Natural Gas
LPG	Liquid Petroleum Gases
NEA	Nuclear Energy Agency
NGL	Natural Gas Liquids
NSG	Nuclear Suppliers Group
O&M	Operating and Maintenance
OECD	Organisation for Economic Cooperation and Development
OPEC	Organization of Petroleum Exporting Countries
PFBC	Pressurised Fluidised Bed Combustion
PM-10	Particulate matter of less than ten microns in diameter
PPP	Purchasing Power Parity
P&T	Partitioning and Transmutation
PV	Photovoltaics
PWR	Pressurised water Reactor
RDD&D	Research, development, demonstration and deployment
RETs	Renewable Energy Technologies
SACS	Saline Aquifer CO_2 Storage
SCSC	Supercritical Steam Cycle
SMR	Small and Medium-sized Reactor
T&D	Transmission and Distribution
USCSC	Ultra Super Critical Steam Cycle
VHTR	Very High Temperature Reactor

Definitions

Source A: *Energy Technology Perspectives*, 2008, p. 601–605
Source B: *International Energy Markets*. 2004. Carol A. Dahl, Pennwell Corporation, p. 475–533.

Ad valorem tax: A tax that is a percentage of the price of a good or a service (B)
Amortization: Allocating the cost of intangible assets over their legal life as specified in the tax code (B).
API Gravity: Specific gravity measured in degrees on the American Petroleum Institute scale. The higher the number, the lower the density. Twentyfive degrees API equals 0.904 kg/m^3. Forty-two degrees API equals 0.815 kg/m^3. (A)
Avoided cost: The amount avoided for the incremental purchase or the production of a good (B).
Benefits of Pollution: Any costs that you forego by being able to pollute rather than to abate. Benefits of pollution are then equal to the costs of abatement. (B)

Biodiesel: Biodiesel is a diesel-equivalent, processed fuel made from the transesterification (a chemical process which removes the glycerine from the oil) of vegetable oils or animal fats. (A)

Biogas: A mixture of methane and carbon dioxide produced by bacterial degradation of organic matter and used as fuel. (A)

Blackouts: A non-isolated power loss over an extended period of time due to capacity shortage. It may result from peak loads higher than available capacity or from equipment failure (B).

Black liquor: A by-product from chemical pulping processes which consists of lignin residue combined with water and the chemicals used for the extraction of lignin. (A)

Breakeven Pricing: Charging a price for which revenues exactly equal all costs including opportunity costs. (B)

Brent Forward Market: The over-the-counter market for buying Brent Crude oil at some future date. (B)

Clean Coal Technologies (CCT): Technologies designed to enhance the efficiency and the environmental acceptability of coal extraction, separation and use. (A)

Clearinghouse: An institution that is a part of an organized exchange that guarantees each transaction and matches buyers to sellers when contracts come due. (B)

Coal: Lignite (with gross calorific value of less than 4165 kcal/kg), sub-bituminous coal (4165–5700 kcal/kg) and hard coal (greater than 5700 kcal/kg, on an ash-free but moist basis). Clean Coal Technologies (CCTs) are designed to enhance the efficiency and the environmental acceptability of coal extraction, preparation and use. Coal-bed methane is methane found in coal seams, and is a source of unconventional natural gas (A).

Coases Theorem on Externalities: In the absence of transaction costs and market power, that private markets will arrive at an optimal allocation in the presence of market externalities no matter how property rights are originally distributed. (B)

Condensates: Condensates are liquid hydrocarbon mixtures recovered from non-associated gas reservoirs. They are composed of C4 and higher carbon number hydrocarbons and normally have an API between 50° and 85°. (A)

Cross Price Elasticity: The percentage change in quantity of one good that results from the percentage change in price of another good. (B)

Data Mining: Techniques for extracting information from large databases. (B)

Deregulation: Removing government regulations. (B)

Discounted Cash Flow (DCF): The present value of future flows of income. (B)

Discount Rate: The interest rate for converting or discounting future cash values to present values. (B)

Electricity Generation: Total amount of electricity generated by power plants. It includes its own use, and transmission and distribution losses. (A)

Energy Futures: A standardized contract offered and guaranteed on an organized exchange to buy or sell an energy product in the future. (B)

Enhanced Coal-bed Methane Recovery (ECBM): A technology for the recovery of methane through CO_2 injection into uneconomic coal seams. (A)

Enhanced Gas Recovery (EGR): A speculative technology in which CO_2 is injected into a gas reservoir in order to increase the pressure in the reservoir, so that more gas can be extracted. (A)

Ethanol: Ethanol is an alcohol made by fermenting any biomass high in carbohydrates (such as, starches and sugars). Emerging technologies will allow ethanol to be produced from cellulose and hemicellulose fibres that make up the bulk of the most plants. (A)
Financial Derivatives: Financial assets that derive their value from an underlying asset upon which they are based. (B)
Gas: Gas includes natural gas (both associated and non-associated with petroleum deposits but excluding natural gas liquids) and gas-works gas. (A)
Gas-to-Liquids: Fischer-Tropsch technology is used to convert natural gas into synthetic gas (syngas) and then, through catalytic reforming or synthesis, into very clean conventional oil products, such as diesel. (A)
Hydrocracking: Refinery process that heats heavy oil products under pressure in the presence of hydrogen to remove sulphur and increase lighter product yields. (B)
Hydropower: Hydropower refers to the energy content of the electricity produced in hydropower plants, assuming 100% efficiency. It excludes output from pumped storage plants. (A)
Marginal Production Cost: The cost of the last unit of production. (B)
Marketable Permits: Permits to pollute that can be bought and sold in the market place (B).
Metcalfe's Law: A network's value increases as the square of the number of connections (B).
Multivariate Time Series: A statistical forecasting technique in which a variable is forecast by using historical values of itself and other related variables (B).
Natural Gas Liquids (NGLs): They are the liquid or liquefied hydrocarbons produced in the manufacture, purification and stabilization of natural gas. These are those portions of natural gas which are recovered as liquids in separators, field facilities, or gas-forming plants. NGLs include but not limited to ethane, propane, butane, pentane, natural gasoline and condensates. (A)
Negative Externalities: An externality is an effect from an economic activity that involves some one not directly involved in the economic activity. (B)
Nuclear: Nuclear refers to the primary heat-equivalent of the electricity produced by a nuclear plant with an average thermal efficiency of 33%. (A)
Oil: Oil includes crude oil, condensates, natural gas liquids, refinery feedstocks, and additives and other hydrocarbons (including emulsified oils, synthetic crude oil, mineral oils extracted from bituminous minerals such as oil shale, bituminous sand, and oils from coal liquefaction) and petroleum products (refinery gas, ethane, LPG, aviation gasoline, motor gasoline, jet fuels, kerosene, gas/diesel oil, heavy fuel oil, naphtha, white spirit, lubricants, bitumen, paraffin waxes and petroleum coke). (A)
Opportunity Cost: What you forego by undertaking an economic activity. (B)
Outage: A temporary loss of power from isolated electricity transmission, generation or distribution failure. (B)
Peak load pricing: Charging higher prices during peak hours than off-peak hours. (B)
Pollution tax: A payment of tax to the government for the right to pollute. (B)
Power Generation: refers to fuel use in electricity plants, heat plants, and combined heat and power (CHP) plants. Both main activity producer plants and small plants that produce fuel for their own use (autoproducers) are included. (A)

Price Elasticity of Demand: Percentage change in the quantity demanded of a good divided by the percentage change in its own price (B).
Public good: A good that no one is excluded from using. (B)
Purchasing Power Parity (PPP): The rate of currency conversion that equalizes the purchasing power of different currencies, by making allowances for the differences in price levels and spending patterns between different countries. (A)
Research, development and Demonstration (RD&D)
Renewables: includes biomass and waste, geothermal, solar PV, solar thermal, wind, tide, and wave energy for electricity and heat generation. (A).
Straight-line Depreciation: Using an annual depreciation charge for tax purposes equal to the value of the asset divided by its allowed depreciable life. (B)
Strike Price: The price at which a put or a call entitles you to sell or buy the underlying asset (B).
Total Final Consumption (TFC): It is the sum of the consumption by the different end-use sectors: industry (including manufacturing and mining), transport, other (including residential, commercial and public services, agriculture/forestry and fishing), non-energy use (including petrochemical feedstocks), and non-specified. (A)
Total Primary Energy Demand: Total Primary Energy Demand represents domestic demand only, including power generation, other energy sector, and total final consumption. It excludes international marine bunkers, except for world energy demand where it is included. (A)

Conversion Constants

Source: "*World Energy Outlook 2007*", International Energy Agency, Paris, 2007, p. 633–641.

General Conversion factors for Energy

To	TJ	Gcal	Mtoe	MBtu	GWh
From	Multiply by				
TJ	1	238.8	2.388×10^{-5}	947.8	0.2778
Gcal	4.1868×10^{-3}	1	10^{-7}	3.968	1.163×10^{-3}
Mtoe	4.1868×10^{4}	10^{7}	1	3.968×10^{7}	11 630
MBtu	1.0551×10^{-3}	0.252	2.52×10^{-8}	1	2.931×10^{-4}
GWh	3.6	860	8.6×10^{-5}	3.412	1

TJ = Tera Joules; Gcal = Gigacalories ; Mtoe = Million tonnes of oil equivalent
MBtu = Million British Thermal Units; GWh = Gigawatt hours
1 million tonnes of oil equivalent = 1.9814 million tonnes of coal
= 0.0209 million barrels of oil/day
= 1.2117 billion cubic metres of gas

Conversion factors for mass

To	kg	t	lt	st	lb
From:	multiply by				
kilogramme (kg)	1	0.001	9.84×10^{-4}	1.102×10^{-3}	2.2046
tonne (t)	1 000	1	0.984	1.1023	2 204.6
long ton (lt)	1 016	1.016	1	1.120	2 240.0
short ton (st)	907.2	0.9072	0.893	1	2 000.0
pound (lb)	0.454	4.54×10^{-4}	4.46×10^{-4}	5.0×10^{-4}	1

Conversion factors for volume

To	gal U.S.	gal U.K.	bbl	ft^3	L	m^3
From:	multiply by					
U.S. gallon (gal)	1	0.8327	0.02381	0.1337	3.785	0.0038
U.K. gallon (gal)	1.201	1	0.02859	0.1605	4.546	0.0045
Barrel (bbl)	42.0	34.97	1	5.615	159.0	0.159
Cubic foot (ft^3)	7.48	6.229	0.1781	1	28.3	0.0283
Litre (L)	0.2642	0.220	0.0063	0.0353	1	0.001
Cubic metre (m^3)	264.2	220.0	6.289	35.3147	1 000.0	1

Section I

Introduction

U. Aswathanarayana

Each country has to figure out its own energy portfolio, consistent with its endowment of energy resources, and employing technologies, which are economically viable and socially equitable, and have minimal adverse impacts. The unacceptable degradation of environment through the use of fossil fuels could only be mitigated by the decoupling of economic growth from energy demand, and reduction in the use of fossil fuels. Improvements in the energy economy are sought to be accomplished through greater energy efficiency, greater use of renewable and nuclear power, CO_2 Capture and Storage (CCS) on a massive scale, and development of carbon-free transport. The author's book, "*Energy Portfolios*" (Taylor & Francis, 2009) provided the knowledge base to facilitate countries making informed choices in regard to energy sourcing and energy technologies to achieve a job-led, low-carbon economic growth.

As this volume was being processed for printing, the world experienced a catastrophic economic meltdown in the second half of 2008. The reckless and unsustainable lending processes of the US banks triggered the subprime mortgage crisis. On Sept. 15, 2008, Lehman Brothers with assets of over USD 600 billion filed for bankruptcy – the largest bankruptcy filing in the US history. This had a profound domino effect. Investment and commercial banks suffered huge losses, and many went bankrupt. There were reverberations round the world. There was a sharp drop in international trade, rising unemployment and slumping of commodity prices.

All the countries, rich and poor, were urgently in need of finding ways and means of getting out of recession. The adoption of Renewable Energy Technologies (RETs) constituted a win-win situation, as renewables are not only green and job-generating, but most of them do not get depleted when used. The author's latest book, "*Green Energy: Technology, Economics and Policy*" (Taylor & Francis, 2010 July) is useful to the countries to decide upon the actual mix of RETs, and timing of the policy incentives, depending upon the local biophysical and socioeconomic situations. All RETs are evolving rapidly in response to technology improvements and market penetration. The deployment of RETs has two concurrent goals : (i) exploit the "low-hanging" fruit of abundant of RETs which are closest to market competitiveness, and (ii) developing cost-effective ways for a low-carbon future.

Renewable fuels, such as wind, solar, biomass, tides, and geothermal, are inexhaustible, indigenous and are often free as a resource. They just need to be captured efficiently and transformed into electricity, hydrogen or clean transportation fuels. In effect, the development of renewal energy invests in people, by substituting labour for fuel. The renewables have hardly any carbon footprint, and do not require environmentally-damaging mining and transport. For these reasons, "green" energy should be considered as the energy of the future. That said, the most important challenge facing the renewables is to bring down their costs.

Though the Copenhagen Accord did reiterate adherence to the goal of limiting global temperature rise to 2°C by 2050, the assessment of UNFCCC is that because of the inadequacy of the mitigation efforts, the temperature rise is likely to be 3°C by 2050. This is a serious matter. M.S. Swaminathan drew attention to the implications for India of a 1°C rise in global temperature. There would be an annual loss of about 6 million tonnes of wheat production, worth USD 1.5 billion, and gross loss of about USD 20 billion to the farmers.

The volume is addressed to the delineation of the ways and means of using technology, economics and policy to provide clean, reliable, secure and competitive energy supply. China and to a lesser extent India, have shown how exactly this could be done. China has emerged as the largest maker of wind turbines and the largest manufacturer of solar panels in the world. It is building the most efficient types of coal power plants. China today has 9 GW of nuclear power. It is preparing to build three times more nuclear power plants in the coming decade as the rest of the world put together. Renewable energy industries have created 1.12 million new jobs in 2008. China's top leadership is intensely focused on energy policy – it created a National Energy Commission which is a kind of "superministry" headed by Prime Minister Wen Jiabao himself. *New York Times* makes a prescient observation that just as the West has been dependent on Middle East for oil, it is destined to be dependent on China for renewable energy technologies.

The purpose of the book is to elucidate Green New Deal models, whereby the twin objectives of job generation as a way of promoting economic development, and mitigation of climate change impacts through low-carbon technologies, are achieved through the harnessing of the transformative power of technology. Energy science and technology are sought to be linked with energy economics and markets and energy policy and planning. This is sought to be accomplished through public – private partnership in the prosecution of Innovation Chain (Basic Research → Applied Research & Development → Demonstration → Deployment → Commercialization).

Denmark shows the way for developing a low-carbon economy through innovative application of energy technologies. Denmark gets bulk of its electricity from coal. During the last twenty years, the greenhouse gas emissions were reduced by 14%. While the energy consumption remained unchanged, GDP went up by 40%. Denmark is the most energy efficient country in Europe. Energy tax revenues are ploughed back to industry to subsidize environmental innovation. Renewable resources, particularly wind power and biomass, provide 30% of Denmark's electricity. A country of just five million people has some of the leading wind, biofuel, heating and cooling and efficiency companies in the world. Energy technologies constitute 11% of Denmark's export.

Climate change could also be mitigated "by pushing the envelope of the possible". The Weizmann Institute of Science, Israel, is working on a win-win paradigm in this regard. Instead of CO_2 emitted to coal-fired power plants being a nuisance, it could be harvested by "super-algae" farms located near the plants to yield 30 times more "green" crude. "Solar" nanostructures "painted" on the exterior of the buildings and cars can help harness sunlight and destroy pollutants in the air.

An event took place in San Jose, Calif., USA, on Feb. 24, 2010 (WE) which is likely to have a profound effect on decentralized electricity production all over the world when K.R. Sridhar, an Indian American, unveiled a "Power plant in a box". The "Bloom Box" is a fuel-cell device, consisting of a stack of ceramic disks with secret green and black "inks". These disks are separated by cheap metal plates. The "Bloom Box" can covert air and nearly any renewable and fossil fuel (e.g. natural gas, biogas, gas from paddy hisk or agricultural waste, coal gas) into electricity by electrochemical process. Since no combustion is involved, there will be no emissions, vibrations or smell. Unlike solar or wind energy, which are intermittent, Bloom technology would be able to provide electricity 24 × 7. Though the presently available Bloom Energy Server has a capacity of 100 kW, Bloom Box of 1 kW capacity, costing about USD 3 000, would be available in 5–10 years to provide clean, reliable and affordable electricity to individual households.

The volume covers the following five themes: (i) Renewable Energy Technologies (RETs), (ii) Ways and means of making RETS competitive, (iii) Reduction in the CO_2 emissions and improving the efficiency of supply-side energy technologies, (iv) Reduction in the CO_2 emissions and improving the efficiency of Demand-side energy technologies, and (v) A Green New Deal – Ways of mitigating the adverse consequences of climate change, while alleviating poverty. Quo vadis? deals with where do we go from here.

Section 2

Renewable energy technologies

U. Aswathanarayana (India)

POINTS TO PONDER

"The only source of sustainable growth is technology"

– Anonymous

"If renewables are to succeed, they must succeed in a competitive market"

– (Paul Komor, 2004)

"Listen to technology. Find out what it is telling you"

– Carver Mead (Caltech)

"Tracking vectors in technology over time to judge when an intriguing innovation is ready for the market place. Technical progress, affordable pricing and consumer demand – all must jell to produce a blockbuster product"

– Steve Jobs of *Apple Inc.*

"Real innovation in technology involves a leap ahead, anticipating needs that no one really knew they had, and then delivering capabilities that define product categories"

– David Yoffe of Harvard

"Technology changes, economic laws do not"

– Hal Varian (Berkeley, USA)

"Market structure is an important determinant of how firms behave"

– Carol A. Dahl (2004)

An external cost exists when ... an activity by one agent causes a loss of welfare to another agent, and the loss of welfare is uncompensated

– David W. Pearce & R. Kerry Turner (1990)

The parties should protect the climate system for the benefit of the present and future generations of humankind, on the basis of equality and in accordance with the common but differentiated responsibilities and respective capabilities

– Global Climate Convention, Rio de Janeiro, 1992.

"Drill for oil ? You mean drill in the ground to try to find oil ? You're crazy !"

– Derisive comment of the drillers whom Edwin L. Drake tried to persuade them to work in his project to drill for oil in Titusville, Pennsylvania, USA in 1859.
http://www.freemaninstitute.com/quotes.htm

"Just as it is impossible to make out whether a fish which lives in the water sometimes drinks the water, it is difficult to make out whether a Treasury official who deals with money, may sometimes help himself to it"

– *Arthasastra* by Chanakya, 300 B.C.E.

Chapter 1

Renewables and climate change

U. Aswathanarayana

There is little doubt that the Renewables are the energy resources of the future, for the simple reason that, unlike the fossil fuels, they do not get depleted when used. Most are related to sun in some way. Sunlight produces solar energy directly. It indirectly produces hydropower (through the movement of rain water), biomass (through photosynthesis), and tidal power (through tides caused by moon and sun).

In the Baseline scenario (i.e. business-as-usual), CO_2 concentrations would rise from 27 Gt in 2005 to 62 Gt in 2050, corresponding to rise of 385 ppm today to 550 ppm of CO_2 in 2050.

In the ACT Map scenario, CO_2 emissions would peak around 34 Gt, and drop to today's levels by 2050, corresponding to the rise of CO_2 concentration to 485 ppm by 2050. This would finally result in the stabilization of concentrations around 520 ppm.

In the BLUE Map scenario, which is most ambitious, CO_2 concentrations would stabilize at 450 ppm by 2050. Only the BLUE Map scenario is consistent with the long term stabilization of CO_2 at 450 ppm.

IPCC recommends the stabilization of CO_2 levels at 450 ppm, in order to limit the temperature rise to 2 degrees Celsius above pre-industrial levels.

The above data is depicted in Fig. 1.1 (source: ETP, 2008, p. 51, © OECD-IEA).

1.1 PROJECTED GROWTH OF RENEWABLES

The share of the total renewables in the world primary energy supply (TPES) in 2005, was 12.7%. Coal: 25.3%, Oil: 35.0%, Natural gas: 20.6%; Non-renewable waste: 0.2%, Nuclear: 6.3%, Hydro: 2.6%, Renewable combustibles and wastes: 9.9%, others: 0.57%.

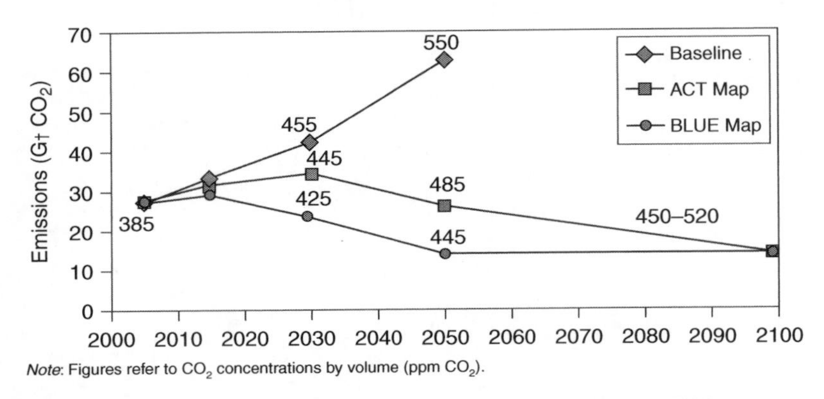

Figure 1.1 CO_2 concentration profiles for the Baseline, ACT and BLUE Map scenarios (Source: ETP, 2008, p. 51, © OECD-IEA)

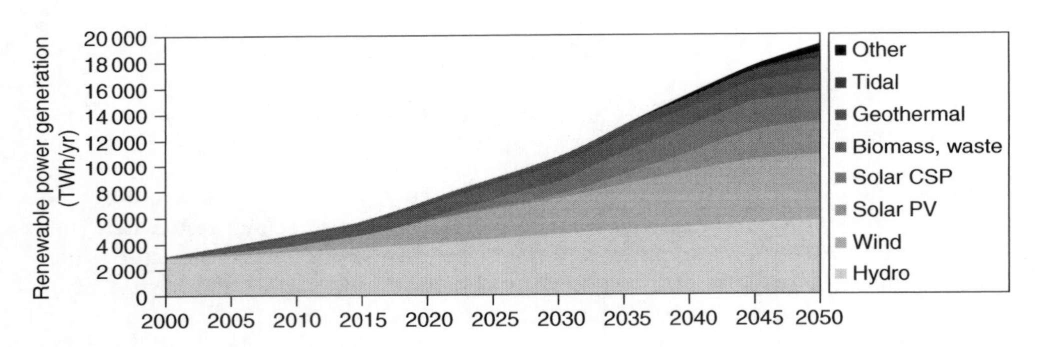

Figure 1.2 Growth of renewable power generation in the BLUE Map scenario, 2000–2050

Product shares in the world renewable energy supply, 2005: Renewables combustibles and waste: 78.6% (comprising liquid biomass: 1.6%, renewable municipal waste: 0.7%, solid biomass/charcoal: 75.6%, gas from biomass: 0.9%); Wind: 0.6%, hydro: 17.4%, solar/tide: 0.3%, geothermal: 3.2%.

The contribution of renewables to electricity generation increases from 18% in 2005 to 35% in 2050 in the ACT Map scenario, and 46% in the BLUE Map scenario. In the BLUE Map scenario, electricity generation from renewables (wind, photovoltaics and marine) is projected to rise to 20.6% (about 3 500 GW) by 2050.

Up to 2020, bulk of renewable energy production will come from biomass and wind. After 2020, solar power production will become significant. Hydro will grow continuously up to 2050, but this growth will achieve a plateau around 2030 to 2050, because of the constraints of finding suitable sites. The contribution of hydro, wind and solar will be roughly equivalent in 2050.

About two-thirds of solar power will be provided by solar PV, with the balance one-third coming from Concentrating Solar Power (CSP). As the capacity factor of CSP is higher than PV, CSP may account for 40% of the solar power generation.

The intermittency of solar power is not a problem as its peak coincides with the demand for air-conditioning. Electricity storage capacity is sought to be increased from 100 GW today to 500 GW by 2050 (in the form of pumped hydro storage, underground compressed air energy, etc.) to cover the variability in the case of systems like wind.

The BLUE Map scenario envisages a strong growth of renewables to achieve the target of 450 ppm CO_2 (Fig. 1.2; source: ETP, 2008, p. 88, © OECD-IEA).

Currently about 50% of the global population lives in urban areas, and this trend is likely to continue in the future. Consequently, urban authorities have to figure out ways of providing renewable energy services to the urban residents. Cities located on the coast could tap the offshore wind energy and ocean energy. Building-integrated solar PV (such as, solar shingles) would be most suitable to cities in low latitudes, with good sunshine. Geothermal power could be developed for the use of cities located near high heat-flow areas. Bioenergy is not usually suitable for the cities, except those, which have forests nearby.

Chapter 2

Wind power

U. Aswathanarayana

2.1 INTRODUCTION

Wind energy is believed to be the most advanced of the "new" renewable energy technologies. Since 2001, wind power has been growing at a phenomenal rate of 20% to 30% per annum. Wind power (2 016 GW) is expected to provide 12% of the global electricity by 2050, thereby avoiding annually 2.8 gigatonnes of emissions of CO_2 equivalent. This would need an investment of USD 3.2 trillion during 2010–2050. Atmospheric scientists are developing highly localized weather forecasts to enable the utility companies to know when to power up the wind turbines.

Wind turbines do not need gusty winds; they need only moderate but steady winds. Wind turbines start producing electricity when the wind speed reaches 18–25 km/hr (5 to 7 m/s), reaching their rated output when the wind speed reaches about 47 km/hr (13 m/s). So any area where the wind speeds are greater than about 18 km/hr (5 m/s) is suitable for generating wind electricity, and such areas are plentiful. When the wind speeds exceed 22 to 26 m/s, the turbine is shut off to avoid damage to the structure.

Availability of wind turbine is defined as the proportion of the time that it is ready for use. Operation and maintenance costs are determined by this factor. Availability varies from 97% onshore to 80–95% offshore.

Improved turbine design is aimed at extracting more energy from the wind, more of the time, and over longer period of time. Affordable materials with higher strength-to-mass ratio are needed for the purpose. More power is captured by having a larger area through which the turbine can extract energy (the swept area of the rotor), and installing the rotor at a greater height (to take advantage of the rapidly moving air).

Table 2.1 Cost structure of wind energy

	Onshore wind	*Offshore wind*
Investment cost	USD 1.6–2.6 M/MW	USD 3.1–4.7 M/MW
Operation & Maintenance	USD 8–22/MWh	USD 21–48/MWh
Life-cycle cost	USD 70–130/MWh	USD 110–131/MWh

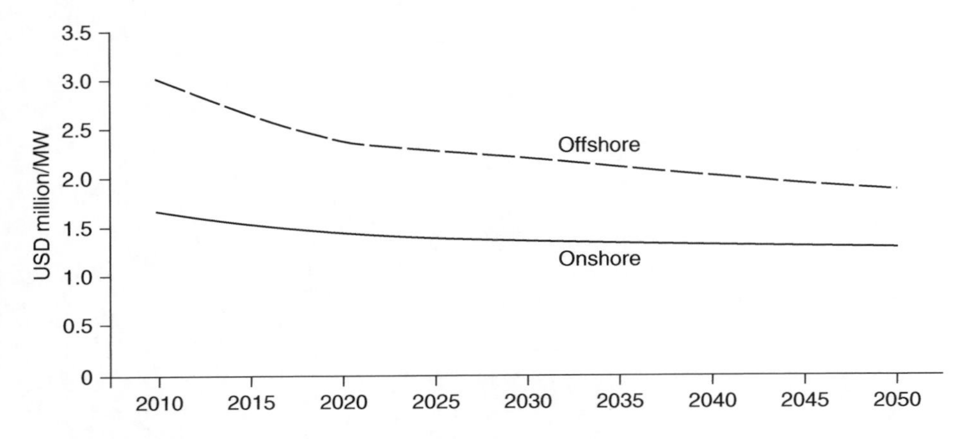

Figure 2.1 Investment costs for the development of onshore and offshore wind (Source: "*Technology Roadmap: Wind Energy*", 2009, p. 17, © OECD – IEA)

A typical 2 MW wind turbine has two or three blades, each about 40 m long, and made of fiberglass or composite material. The nacelle, which is the housing on the top of the tower, contains the generator and gearbox to convert the rotational energy into electricity. The tower height is ~80 m. The largest wind turbines presently in operation in the world, has a capacity of 5–6 MW each, with rotor diameter up to 126 m.

Table 2.1 (source: *Technology Roadmap: Wind Energy*, 2009, p. 12) gives the cost structure of wind energy.

The investment costs of onshore and offshore wind energy are depicted in Fig. 2.1

2.2 ENVIRONMENTAL FACTORS

The plus point of the wind power is that it has no carbon dioxide emissions. Wind power has three environmental impacts: visual impact, noise, risk of bird collisions and disruption of wild life. Wind power growth has to reckon with two impediments: siting and intermittency. Improved designs of wind turbines have reduced the noise pollution from wind farms. The best places for siting wind farms are tops of hills, bluffs along the open ocean and areas, which are not obstructed by topography. But these happen to be the very places, which people cherish for their scenic beauty.

Two kinds of noises are associated with wind turbines: aerodynamic noise from the blades, and mechanical noise from the rotating machinery. Design improvements are

Table 2.2 Cost estimates of wind power

Type of cost	*Capital Cost*	*Cost (US cents/kWh)*
Turbine	USD 650–700/kW	2.5–2.7*
Rest of the plant	USD 270/kW	1.0
Operating		0.5–0.9
Total	USD 920–970/kW	4.0–4.6

*Assuming capacity factor of 28%, discount rate of 7%, a lifetime of 20 years, and no decommissioning costs.

bringing about a sharp reduction in the noise. The risk to migratory birds could be avoided by siting the wind farms where the routes of the migratory birds do not cross.

2.3 COSTS

The cost of the turbine constitutes 74–82% of the capital costs of the medium-sized onshore power stations (i.e. 850 kW to 1 500 kW). Other costs are as follows: Foundations: 1–6%; Electric installation: 1–9%; Grid connection: 2–9%; Consultancy: 1–3%; Land: 1–3%; Financial costs: 1–5%; Road construction: 1–5%.

The onshore wind power production costs depend upon the wind conditions.

Operation & Maintenance costs average 20–25% of the total cost per kWh produced. They tend to be low (10–15%) in the early years of the turbine, and may rise to 20–35% in the later years. To bring down O&M costs, manufacturers are developing new turbine designs that have less down time and require fewer service visits. Experience in Europe suggests O&M costs of US cents 1.5/kWh to 1.9/kWh of the produced wind power over the lifetime of the turbine.

Wind power is capital intensive with capital costs accounting for 75–80% of the production costs – the corresponding figure for fossil fuel power stations is 40–60%. The onshore wind power production costs depend upon the wind conditions – they are low in areas of high wind speeds (such as, coastal areas) and high in areas of low wind speeds (such as, inland areas).

The cost estimates are given in Table 2.2 (source: Komor, 2004, p. 37).

That the calculated levelized price of US cents 4.0–4.6/kWh may not be off the market price is indicated by the fact that in 2001, California signed for a contract for 1 800 MW of wind power at an average price of US cents 4.5/kWh. Earlier, in 1998, U.K. contracted for 368 MW of wind capacity at an average price of US cents 4.2/kWh.

2.4 WIND POWER MARKETS

There are wind farms in about 40 countries in the world, with thirteen of them having a capacity of 1 000 MW of installed capacity. The top ten countries in the world in terms of installed wind power capacity are listed in Table 2.3.

In 1980, Denmark and California were virtually the only markets in the world for wind turbines. The market collapsed in California when the financial incentives were

Table 2.3 Top ten countries in installed wind power capacity

Country	*MW*	*%*
Germany	22 247	23.6
United States	16818	17.9
Spain	15 145	16.1
India	8 000	8.5
China	6 050	6.4
Denmark	3 125	3.3
Italy	2 726	2.9
France	2 454	2.6
United Kingdom	2 389	2.5
Portugal	2 150	2.3
Rest of the world	13018	13.8
Total top ten	81 104	86.2
Global total	**94 122**	

(Source: *Energy Technology Perspectives*, 2008, p. 342)

Table 2.4 Global top ten wind-turbine manufacturers

Manufacturer	*Capacity supplied in 2006 (MW)*	*Market share (%)*
VESTAS (Denmark)*	4 329	28.2
GAMESA (Spain)	2 346	15.6
GE WIND (USA)	2 326	15.5
ENERCON (Germany)	2 316	15.4
SUZLON (India)	1 157	7.7
SIEMENS (Denmark)	1 103	7.3
NORDEX (Germany)	505	3.4
REPOWER (Germany)	480	3.2
ACCIONA (Spain)	426	2.8
GOLDWIND (China)	416	2.8
Others	689	2.6
Total	**16 003**	

*Country designation refers to the corporate base

(Source: BTM Consult, 2007)

withdrawn. Denmark survived by falling on the stable domestic market. In mid-1990's, Germany entered the market, followed by Spain. There has been a great boom in the wind power industry. Six leading turbine manufacturers account for 90% of the global market. The global top ten wind turbine manufacturers are listed in Table 2.4.

Three parameters determine the amount of power from wind turbine: wind conditions, turbine height, and efficiency of the turbine. While the wind regime of a site is a given, higher output of power can be realized by making the turbines larger and taller. Germany and Denmark increased the productivity of their prime sites by replacing the earlier-installed smaller and shorter turbines by larger and taller turbines. The efficiency of energy production is measured on the basis of annual energy production per unit of swept rotor area (kWh/m^2). The same parameter determines the manufacturing costs. The trend is therefore towards larger and taller and more efficient wind turbines. The

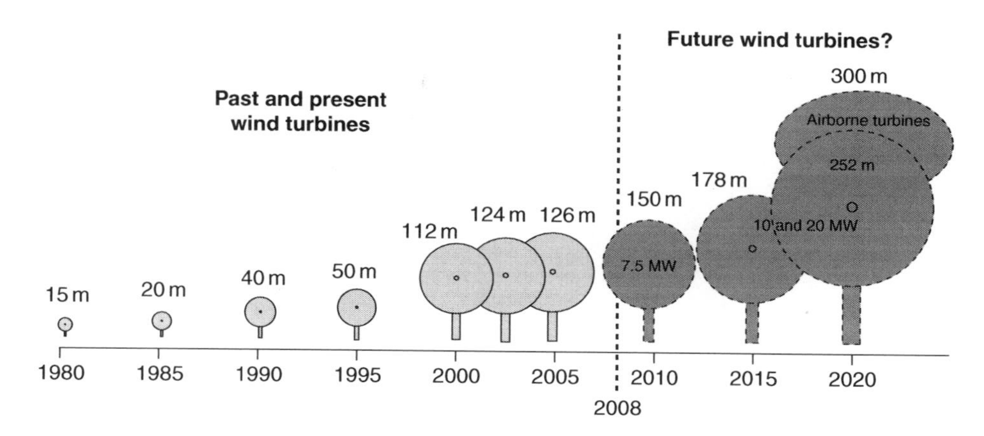

Figure 2.2 Growth in the size of the wind turbines
(Source: *Technology Roadmap: Wind Power*, p. 22, © OECD – IEA)

efficiency of the wind power sector has increased by 2–3% annually during the last 15 years through better turbine siting, more efficient equipment and higher hub heights.

2.5 PROJECTED GROWTH OF WIND POWER

ACT scenario

From its current capacity of 94 GW, the global wind power capacity is projected to grow to 1 360 GW by 2050 under the ACT scenario. Electricity production from wind power is projected to contribute 2 712 TWh/yr by 2030 and 3 607 TWh/yr by 2050.

BLUE scenario

The BLUE scenario assumes profound technoeconomic improvements, in the form of higher CO_2 incentives, greater cost reductions, extensive offshore wind power development and improvements in innovative storage, grid design and management.

It envisages the installation of 700 000 turbines of 4 MW size by 2050. Wind power installed capacity will go up to 2 010 GW by 2050, with wind electricity generation of 2 663 TWh/yr in 2030, and 5 174 TWh/yr in 2050. Wind power contribution to global energy production will reach 12% by 2050, thereby reducing the CO_2 emissions by 2.14 Gt CO_2 /yr. By 2050, China will be the world leader in wind power, with electricity from wind power accounting for 31% of the world production

2.6 OFFSHORE WIND POWER

General considerations Till now, offshore wind turbine designs have been essentially "marinised" forms of onshore turbines. It is realized that future designs of offshore

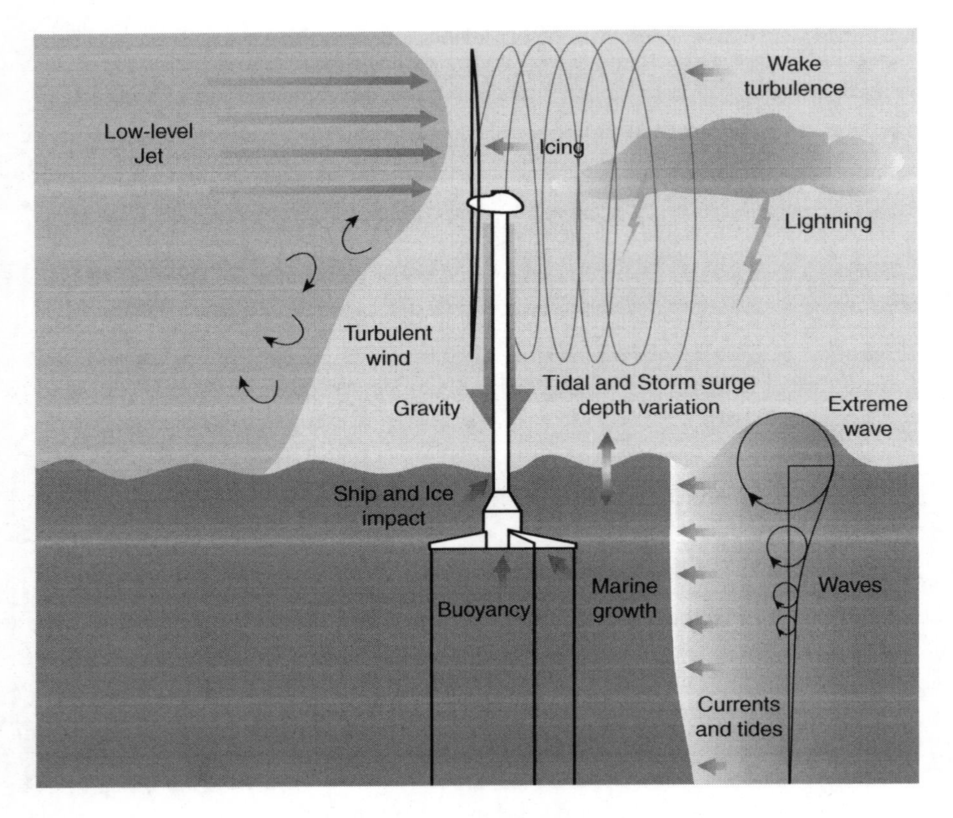

Figure 2.3 Offshore operating conditions

wind turbines should take into account the special characteristics of the marine environment (vide Fig. 2.3, Offshore operating conditions, Source: *Technology Roadmap: Wind Energy*, 2009, p. 24, ©OECD – IEA).

New designs of offshore wind turbines will have two blades rotating downwind of the tower, with a direct-drive generator. There will be no gearbox. The rotor will be 150 m in diameter. The turbine capacity could be 10 MW. It will have a self-diagnostic system, which is capable of taking care of any operational problems on its own. Such an arrangement will reduce the requirement of maintenance visits to the minimum.

Foundations will be a major area of technological development. Instead of the current monopile foundations which account for 25% of the installation cost, new types of foundations based on improved knowledge of the geotechnical characteristics of the subsurface, are being developed to reduce costs. Currently, offshore wind farms operate at depths of less than 30 m. New designs of tripod, lattice, gravity-based and suction bucket technologies are being developed for use in depths of 40 m. Technologies used by offshore oil and gas industry are being adapted by Italy and Norway to develop floating designs for offshore wind turbines.

Offshore turbines are the next future. Europe expects to obtain 30% of the energy from offshore wind.

Presently, most of the wind power is generated by land-based wind turbines. Offshore wind installations are 50% more expensive than land-based wind installations. Still companies are going in for offshore wind power because the output of offshore installations is 50% more than onshore installations, due to better wind conditions. Offshore wind power installations have to operate "under harsh conditions, shortage of installation vessels, competition with other marine users, environmental impacts and grid interconnection" (*Energy Technology Perspectives*, 2008, p. 352).

Five countries (Denmark, Ireland, Netherlands, Sweden, and U.K) have established offshore wind power stations with a total capacity of 1 100 MW. Most of these installations (typically 2 MW capacity) are sited in relatively shallow water (<20 m deep) and close to the coast (<20 km). U.K. is establishing a large (~1 000 MW) facility situated more than 20 km offshore. When completed, it will be capable of providing power to one-quarter of the households in London.

Denmark which made extensive studies on the behavioral response of the marine mammals and birds to offshore wind farms, has developed guide-lines for minimizing impact of offshore wind farms on marine biota. These could be applied to estuarine and open sea sites of offshore wind stations.

Investment costs

As should be expected, the capital costs of the offshore power stations are dependent upon wind speeds, water depth, wave conditions and distance from the coast. The experience in U.K. is that the costs range from USD 2 225–2 970/kW. The higher capital cost of the offshore wind installations is partly offset by the lower costs of production of the offshore wind electricity. This is so because the offshore installations are exposed to higher wind speeds for longer periods (i.e., 3 000–3 300 full load-hours per year, or ~34% capacity factor) relative to the onshore installations (2 000–2 300 full load hours per year, or ~25% capacity factor). Danish wind farms have recorded high load hours of 3 500–4 000 hours per year.

Investment costs vary from USD 1.5 million to 3.4 million/MW, depending upon water depth and distance from the coast. Foundations and grid connections account for the difference in costs between onshore and offshore wind power.

Water depth and distance from the coast determine the offshore wind power costs. United Kingdom established a 90 MW offshore wind turbines in 2006. The costs ranged from USD 2 226/kW to USD 2 969/kW. Offshore turbines cost about 20% more than the onshore turbines. Also, offshore towers and foundations cost 2.5 times more than similar structures on land.

The breakdown in the offshore wind power investments costs is given in Table 2.5.

Annual Operation & Maintenance costs are in the region of USD 20/MWh, averaged over the lifetime of the turbine, normal operating conditions and discount rate of 7.5%.

Steel which is used for the construction of the turbine, accounts for 90 % of the cost of the turbine. Turbine fabrication costs are being brought down by replacing steel with lighter and more reliable material, and by improving the fatigue resistance of the gear boxes. During the last five years, there has been a phenomenal growth in the use of rare-earth elements in the energy industries. Tiny quantities of dysprosium can make magnets in electrical motors lighter by 90%, thereby allowing larger and more powerful wind turbines to be mounted. Use of terbium can help cut the electricity use of

Table 2.5 Offshore wind power investment costs

	Investment costs USD 1 000/MW	Share%
Turbines, ex-works, including transport and erection	1 020	49
Transformer station and main cable to coast	340	16
Internal grid between turbines	105	5
Foundations	440	21
Design, project management	125	6
Environmental analysis	75	3
Miscellaneous	12	<1
Total	**2 117**	

(Source: Lemming et al., 2007)

Table 2.6 Estimated offshore wind turbine costs during 2006–2050

Year	Average investment costs (million USD/MW)	O&M (USD/MWh)	Capacity factor (%)
2006	2.6	20	37.5
2015	2.3	16	37.5
2020	2.0	15	37.5
2030	1.8	15	37.5
2050	1.7	15	37.5

lights by 80%. Dysprosium prices have gone up sevenfold since 2003, with the current market price being USD 116/kg. Terbium prices have quadrupled during the period 2003–2008 to USD 895/kg. The recession brought down the price to USD 451/kg.

Estimated offshore wind turbine costs during 2006–2050 are given in Table 2.6 (source: Lemming et al., 2007).

Offshore wind power will be progressively cheaper in future.

Further technology development

The European Union has launched an impressive wind energy R&D initiative, code named *Up Wind*, aimed at developing very large turbines (8 to 10 MW) and large wind farms of several hundred megawatt capacity. The programme would involve better understanding of wind conditions, development of materials with high strength to mass ratios, and improved control and measuring systems.

Some innovative approaches in this regard are described below:

Superconducting generators: Denmark Technical University, Ris?, is developing a 10 MW generator which achieves 50–60% reduction in weight through the use of high-temperature, superconducting materials. By making direct drive possible, it avoids the use of gear-boxes, and thus brings down O&M costs.

Compressed air energy storage (CAES): When the demand is low, wind electricity is used to compress air, which is then stored in a geological formation, say, salt domes. When the demand rises, the flow is reversed. The compressed gas is fed into natural gas-fired turbine, thereby enhancing its efficiency by more than 60%.

Table 2.7 Activities, milestones and actors to develop wind power

Activity	*Milestones and Actors*
Resource	
1. Refine and set standards for wind resource modelling techniques, and site-based data measurement with remote sensing technology; improve understanding of complex terrain, offshore conditions and icy climates.	Ongoing. Complete by 2015. Wind industry and research institutions, climate and meteorological institutions.
2. Develop publicly accessible database of onshore and offshore wind resources and conditions, with the greatest possible coverage taking into account commercial sensitivities.	Complete by 2015. Industry and research institutions.
3. Develop more accurate, longer-horizon forecast models, for use in power system operation.	Ongoing. Complete by 2015. Industry, research institutions and system operators.
Technology	
4. Develop stronger, lighter materials to enable larger rotors, lighter nacelles, and to reduce dependence on steel for towers; develop super-conductor technology for lighter, more electrically efficient generators; deepen understanding of behaviour of very large, more flexible rotors.	Ongoing. Continue over 2010–2050 period. Industry and research institutions.
5. Build shared database of offshore operating experiences, taking into account commercial sensitivity issues; target increase of availability of offshore turbines to current best-in-class of 95%.	Complete by 2015. Wind power plant developers, owners and operators, industry associations.
6. Develop competitive, alternative foundation types for use in water depths up to 40 m.	Ongoing. Complete by 2015. Industry and research institutions
7. Fundamentally design new generation of turbines for offshore application, with minimum O&M requirement.	Commercial scale prototypes by 2020. Industry and research institutions.
8. Develop deep-water foundations/sub-surface structures for use in depths up to 200 m.	Ongoing. Complete by 2025. Industry and research institutions.
Supply Chains	
9. Accelerate automated, localised, large-scale manufacturing for economies of scale, with an increased number of recyclable components.	Ongoing. Continue over 2010–2050 period. Industry.
10. For offshore deployment, make available sufficient purpose-designed vessels; improve installation strategies to minimise work at sea; make available sufficient and suitably equipped large harbour space.	Sufficient capacity by 2015. Wind industry, shipping industry, and local governments.
Environment	
11. Improve techniques for assessing, minimising and mitigating social and environmental impacts and risks.	Complete by 2015. Industry, research institutions, governments, and NGOs.

Floating platforms: As in the case of oil industry, platforms are built on land and towed to sea. Wind turbines are mounted on these platforms.

Hybrid systems: In the Poseidon's Organ arrangement, a floating offshore wave power plant also serves as a foundation for wind turbine.

System aspects: Traditionally, electricity grids and power markets link large-scale electricity producers with consumers. Wind power does not conform to this pattern. In 2006, wind energy provided 17% of Denmark's electricity demand. This was possible because the Danish wind power could export surplus power to, and import electricity from, the Nordic Power market. The moral of the story is that dispersed wind power plants needs to be aggregated, and linked to power markets in such a way that short falls cost the least.

2.7 PROGNOSIS

The actions needed to reduce the life cycle cost of wind energy production are listed in Table 2.7 (source: *Technology Roadmap: Wind Energy*, 2009, p. 42).

Atmospheric scientists are developing highly detailed, localised weather forecasts to enable utility companies when to power up the wind turbine and when to power down the fossil fuel electricity generation.

The juggernaut of wind power is unstoppable. China has emerged as the largest market for wind energy. It is now building six wind farms with a capacity of 10 000 to 20 000 MW each. By 2020, Britain is planning to obtain a quarter of its electricity needs from offshore wind energy system. US plans to meet 20% of electricity demand through onshore and offshore wind power by 2030 (as against 2% today), by building 100 000 wind turbines at a cost of USD 100 billion. This would create 140 000 new jobs, and reduce the CO_2 emissions by 800 million tonnes.

Chapter 3

Solar energy

U. Aswathanarayana

3.1 INTRODUCTION

Solar PV has many strong points – (i) PVs will work any where the sun shines – the sunnier it is, the more the electricity produced, and the lower the per kWh cost. Some electricity is produced even when it is cloudy, (ii) unlike the wind turbines which needs to have winds with speeds of more than 18 km/hr to start producing power, PVs have no such constraints – they work any where the sun shines, (iii) it is quiet, has no moving parts, can be installed easily, and can be sized at any scale, ranging from a single bulb, to powering the entire community. It has two serious drawbacks : (i) it is expensive – its levelized cost (US cents 20–40/kWh) is several times more than that of electricity from the fossil fuels (US cents 3 to 5/kWh). The PV costs are coming down all the time, but they are a long way from being competitive. (ii) It is intermittent – no power is generated during nights when there is no sun.

Solar energy can be used in the following ways: (i) Direct supply of solar heat to buildings and industrial processes – provision of heat accounts for about 40% of the global energy needs, (ii) electricity can be produced through the photovoltaic cells, or through steam turbines by the concentration of solar rays, and (iii) production of hydrogen which can be used as fuel. Among all the energy systems, solar energy is projected to grow the fastest. Between now and 2050, solar energy is expected to grow thousand-fold, to 2 319 TWh/yr in the ACT scenario, and 4 754 TWh/yr in the BLUE scenario. It is assumed that during the next ten years, there will be sustained support to the solar energy sector to enable it to become competitive. Under both ACT and BLUE scenarios, major growth is likely to occur after 2030. PV is expected to grow fast in the solar-rich OECD countries (e.g. North America) and the emerging economies of

China and India. CSP (Concentrated Solar Power) will grow strongly not only in these countries, but also in the sun belts of Africa and Latin America.

A photovoltaic cell (PV cell) is a semiconductor device that is capable of converting solar energy into direct current (DC) electrical energy. A PV cell is typically a low voltage (~0.5 V) and high current (~3A) device. PV modules are built by combining a number of PV cells in series. A commercial module with an area of 0.4 m^2 to 1.0 m^2 can produce peak power of 50 Wp (peak Watts) to 150 Wp. By linking together appropriately large number of PV cells, it is possible to build huge power units with a capacity of tens of MWs.

In mid-1990s, stand-alone, off-grid PV systems were common, as they are most economically viable for use in rural areas. Water pumping and rural electrification continue to account for about 10% of the PV market. Subsequently, grid-connected systems, particularly those that are integrated into building design, have come into vogue in a big way. Since 2000, the total cumulative PV capacity in the world has grown eight-fold to 6.6 GW, 90% of which is composed of grid-connected systems. Distributed generation in buildings accounts for 93% of the grid-connected systems.

Germany, Japan and USA account for 63% of the global PV production. China, India, Australia, Korea and Spain are expanding their PV installed capacity and manufacturing capability. China already accounts for 15% of the global production of PV cells. Japanese companies – Sharp (17.1%), Kyocera (7.1%), Sanyo (6.1%) – lead the world in PV manufacturing capacity, followed by Q-Cells (10%) of Germany and Suntech (6.3%) of China. The shortage of purified silicon is expected to ease soon.

Several plants are fabricating hundreds of megawatts of PV modules yearly. Japan is planning to build one GW manufacturing plant.

By 2010, ACT map scenario projects a market of 6 GW/yr and the BLUE map scenario projects a market of 10 GW/yr. The industry envisages a much higher annual production of 23 GW/yr of PV cells/modules by 2011. By 2050, the annual power generation from PV is expected to reach 1 383 TWh/yr as per ACT scenario and 2 584 as per BLUE scenario (the latter figure would correspond to 6% of the global production of electricity in 2050).

3.2 PV TECHNOLOGY

Wafer-based crystalline silicon (c-Si) is the basic material for the fabrication of most (~90%) of the PV modules. After oxygen, silicon is the most abundant element in the earth's crust, but because of its great affinity for oxygen, silicon always occurs as silica (SiO_2). The value of one kg. of quartzite gets increased 65 million times when it is made into computer chips, through the following steps:

1. Quartzite ($0.02/kg) to metallic silicon ($2/kg) – 100 times increase in value
2. Metallic silica ($2/kg) to polysilicon ($40/kg) – 20 times increase in value
3. Polysilicon ($40/kg) to silicon wafer ($1500/kg) – 37 times increase in value
4. Cutting the wafer ($1500/kg) into chips ($1.3 million) – 860 times increase in value

Ingots of silicon (which are made from silica) are sliced to make solar cells, which are then electrically inter-connected. A module is fabricated by encapsulating strings of cells. The module built of single crystalline silica (sc-Si) tends to have a higher conversion efficiency, which is presently 15% now but is expected to increase to 25–28% by 2050. A module can be made from multi-crystalline silica (mc-Si), but such a module, though cheaper than sc-Si, has lower conversion efficiency. Ribbon technologies have conversion efficiencies similar to those of mc-Si, but make use of silicon feedstock more efficiently.

About 40% of the Indians have no access to electricity – they use kerosene wick lamps for lighting. These lamps give poor quality light, emit unhealthy fumes and constitute a fire hazard in thatched houses in which poor people live. As a substitute for the kerosene lamp, D.T. Barki's NEST (Nobal Energy Solar Technologies), Hyderbad, India, has successfully developed and is marketing in developing countries, an inexpensive (about USD 30, which can be paid for in 16 monthly installments of USD 2 each), portable and sturdy solar lantern which gives three hours of good light. The battery has a life of three years, and solar panel has a life of ten years. The solar lantern is a good example of globalization – it is fabricated in China, with silicon feedstock from Japan, and marketed from India.

3.3 THIN FILMS

Thin film technology is rapidly emerging as a viable alternative to silicon wafer technology. A thin layer of photosensitive material is deposited on a low-cost backing, such as, glass, stainless steel or plastic. Initially, amorphous silicon (α-Si) was used, but now-a-days, Cadmium Telluride (CdTe) or Copper-Indium-Diselenide (CIS) are used instead. The efficiency of CIS gets improved when it is doped with gallium, to produce a CIGS module. The thickness of the thin film may range from 40–60 μm in the case of c-Si to less than 10 μm in the case of CdTe.

Thin films use smaller quantities of feedstock, and are amenable to automation. They can be integrated into buildings more readily and have better appearance. Their efficiencies are, however, lower than c-Si modules. Recent improvements in CIS modules have allowed them to have efficiencies of the order of 11%, which figure is comparable to the efficiency of mc-Si modules. An efficiency of 22% is projected for CIS modules by 2030. But the availability of Cd and Te may prove to be a constraint. Thin films are likely to increase their market share by 2020. After that, hybrid systems which combine crystalline and thin-film technologies, may dominate the market. These hybrid systems have the best of both the worlds- higher efficiencies of the order of 18%, lower material consumption and amenability to automation.

The module efficiencies of different PV systems are summarized in Table 3.1.

The consensus in the PV industry is that after 2020, the market share of c-Si PV systems will decrease, and that thin-film technology will dominate the market. Two types of Third generation PV devices are expected to come up during 2020-2030 :

(i) Ultra-low cost, low to medium efficiency cells and modules, such as dye-sensitized nanocrystalline solar cells (DSC) which could attain an efficiency of 10%, if not 15%, by 2030. Organic solar cells with efficiencies of the order of 2% are being developed. It is too early to speculate on their economic viability. They may figure in applications where space is not a problem.

Table 3.1 Present module efficiencies for different PV technologies

	Wafer-based c-Si		*Thin Films*		
	Sc-Si	*mc-Si*	*α-Si, α-Si/mc-Si*	*CdTe*	*CIS/CIGS*
Commercial module efficiency (%)	13–15	12–14	6–8	8–10	10–11
Maximum recorded module efficiency (%)	22.7	15.3	–	10.5	12.1
Maximum recorded laboratory efficiency (%)	24.7	19.8	12.7	16.0	18.2

(Source: Frankl, Manichetti and Raugei, 2008)

(ii) Ultra-high efficiency cells and modules, based on advanced solid-state physics principles, such as, hot electrons, multiple quantum wells, intermediate band gap structures and nanostructures. It is difficult at this stage to predict their efficiency levels, but some experts predict that these devices may attain efficiencies of 30–60%.

Most likely, several types of PV devices may coexist in 2050. They are listed in Table 3.2.

3.4 COSTS

As PV systems have no moving parts, Operation and Maintenance costs are minimal (mainly to wash the modules of dust and dirt) at around 0.5% of the capital investment per year.

Presently, PV modules cost about 60% of the total PV system costs. Costs of mounting structures, inverters, cabling, etc. account for the rest of the 40%. PV costs are characterized by a high learning rate of 15–20% (learning rate means reduction in cost per each doubling of cumulative installed capacity). The cost of total PV systems was USD 6.25/W in 2006. A sustained high learning rate, and increased integration in buildings are expected to bring down the total PV investment costs to USD 2.2/W in 2030, and USD 1.24/W by 2050 under the ACT scenario. Under the BLUE map scenario, the corresponding figures would be USD 1.9/W in 2030 and USD 1.07/W in 2050.

Dow Company has unveiled solar shingles. The solar shingle can be handled like any other shingle – it can be dropped from roof, or trod on. It can offset 40% to 80% of the home electricity consumption. The solar shingle is expected to have a market of USD 5 billion by 2015.

The cost of electricity generated from PV systems depends upon the total solar irradiation, system lifetime (typically, 35 years) and the discount rate assumed (typically, 10%). It is expected to be in the range of US Cents 5/kWh to US Cents 7/kWh under conditions of good irradiation (>1 600 kWh/kWp*yr).

3.5 RESEARCH & DEVELOPMENT NEEDED

c-Si module technology has been successful as it is reliable, takes advantage of the electronics industry, with ready availability of feedstock. Further advances that are

Table 3.2 Technology and market characterization of different PV technologies in 2050

	Wafer-based c-Si		*Thin Films*		*New Concept Devices*	
	Cz, Fz	*mc, ribbon*	*CIS, Cd-Te, α-Si/μc-Si thin Si films*	*Pin-Asi and ASI-THRU*	*Ultra-high efficiency (3rd. Generation, quantum wells nanostructures concentrators)*	*Ultra-low cost (dye-sensitized cells, organic cells)*
Module efficiency (%)	**24%–28%**	**20%–25%**	**CIS: 22–25%, Si: 20%**	**6%–8%**	**>40%**	**10%–17%**
Module lifetime (years)	40–50 years	40–50 years	30–35 years	30 years	>25 years	10–15 years
Provided service	High pressure at premium price	Cost-effective Power applications	Additional solutions for cost-effective power applications	Low-cost, low-efficiency "solar electricity glass"	High power supply	Colour to PV Low-material cost option
Market segment	Niche markets, space	Mass market ("The PV work horse")	Mass market	Mass market	Niche market/mass market	Mass market
Applications	All applications with surface constraints (e.g. specific BIPV), ground-mounted Very large-scale PV	All	All Special added value in BIPV (semi-transparency, screen-printing, etc.)	Consumer products Special applications Large surface buildings	All applications with surface constraints; Ground-mounted, very large-scale PV	All

(Source: *Energy Technology Perspectives*, 2008, p. 374)

needed in order for PV systems to reach large production volumes and low target cost of USD 1.25/W, are summarized below (PV-TRAC, 2005; EUPVPLATF (2007):

c-Si technology

(i) Materials: Epitaxial deposition; substitution of silver (because of cost) and lead (because of its health impact).
(ii) Equipment: Manufacturing processes (including ribbons) that use less silicon and less energy per watt. High degree of automation.
(iii) Device concepts and processes: Design of new modules which can be assembled more easily, low-cost and longer life span (25–40 years).

Thin Film technologies

(i) Materials and devices: Increase of module efficiencies from the current 5% to 15% or more; development of new multi-junction structures; reduced materials consumption; alternative module concepts, such as, new substrates and encapsulation modalities; enhancement of the module life for 20 to 30 years, with less than 10% reduction in efficiency.
(ii) Processes and equipment: Techniques of ensuring uniformity of film properties over large areas, and bridge the efficiency gap between laboratory modules and large-scale industrial modules; reduction in the pay-back time of the modules from the present 1.5 years to 0.5 years.

3.6 NEW CONCEPT PV DEVICES

Technical advances in thin-film production and "building-integrated PVs" (BIPV), such as roofing tiles, will continue to keep PV industry in the forefront of renewables.

The consensus in the PV industry is that after 2020, the market share of c-Si PV systems will decrease, and that thin-film technology will dominate the market.

Fundamental research in physics, chemistry and materials is needed to develop and operationalise these devices. The organically sensitized cells and modules need to have their stability enhanced to about ten years, with improvements in efficiency from 5% to 10%. The inorganically sensitized cells which are characterized by very low efficiencies, need to have their efficiency increased to 5–10%. The efficiency of very low-cost, nanostructured devices needs to be improved to 5–10%. Two improvements are needed in the case of polymer and molecular solar cells: improvement in the efficiency from the present 3–5% to about 10%, and greater stability period (unto, say, ten years). Devices based on concepts for super-high efficiency and full-spectrum utilization, are being developed.

3.7 CONCENTRATED SOLAR POWER (CSP)

General considerations

Concentrated solar power needs cheap water, cheap land and plenty of sun. The water need not be fresh water – brackish water or sea water will do.

Concentrated Solar Power (CSP) systems concentrate direct sunlight to reach high temperatures. This heat can then be used to power a steam turbine which drives a generator. CSP is best suited to areas with high direct solar radiation, the minimum requirement being 2 000 kWh/m^2.

Unlike c-Si systems in which solar energy is directly converted to electrical energy, CSP system has thermal energy as an intermediate phase. In other words, it can store heat in various forms, and release power as and when needed. CSP can be used for continuous solar-only generation. Also it can burn fuel in hybrid plants and use the traditional steam turbines to generate power.

A strong plus point in favour of CSP is that it reaches peak production (at noon) exactly at the time when the electricity demand is at its highest (say, for air-conditioners) in tropical, arid and semi-arid areas. In these areas, CSP electricity is much cheaper than PV electricity, though it is costlier than fossil fuels and wind power.

As CSP plants are invariably large (typically several hundred megawatts), they have to be linked to transmission networks. For instance, it is possible to export CSP electricity from North Africa to Europe at the cost of USD 30/MWh – this figure is les than the cost difference in solar electricity between the two regions (DLR, 2006). Barring some special situations, space should not be a constraint for CSP plants. It has been estimated that the electricity requirements of the whole of USA, could be generated with CSP plants occupying an area of a hundred square miles. CSP could be combined with conventional devices, such as, steam turbines, and it is possible to scale up CSP plants to several hundred megawatts, using well-established technologies.

Apart from electricity generation, CSP plants can be used to heat/cool buildings, to desalinize water, and to produce fuels like hydrogen. In areas where water is scarce, dry coolers may be used. In some arid countries, cogeneration of heat for desalinization and power may turn out to be an attractive proposition.

The constraint of intermittency of solar energy can be got over either by storing heat and conversion to electricity when solar energy is unavailable, or to have a fossil fuel backup that uses the same steam cycle as the CSP plant (this needs only an additional burner).

Description of CSP Technology

The solar flux is concentrated in three ways, with different flux concentration ratios: troughs (30–100), towers (500–1 000) and dishes (1 000–10 000).

Troughs are parabolic, trough-shaped mirror reflectors. They concentrate sunlight onto receiver tubes, thereby heating a thermal transfer fluid, such as, mineral oils, molten salts and water (direct steam generation). Molten salts are often used for heat storage, but may in future, be replaced by phase-change materials. Integrated Solar Combined cycle plants use solar heat (350–400°C) and fossil fuels. Trough plants offer a maximum concentration of 200 suns, maximum temperatures of 400°C, solar-to-thermal efficiency of 60% and solar-to-electric efficiency of 12%. In the case of Fresnel collectors, the absorber is fixed in the space above the mirror field. Fresnel collectors are cheaper, but less efficient. Compact Fresnel linear collectors which are more efficient, are being developed. A 354 MW trough plant has been in operation in California. Algeria and Spain are building trough plants.

Towers: consist of flat, double-axis tracking heliostats. There are several designs, depending upon the heat transfer fluid, which may be molten salts and saturated steam. Italy, Spain, France, Ukraine, Japan and USA have built tower plants of capacity of few tens of MWs. South Africa is developing a large tower project (100 MW in a single tower).

Dishes: Parabolic dish-shaped reflectors concentrate sunlight in two dimensions and run a small Stirling engine (about 10 kW capacity) or turbine at the focal point.

Costs

Investment costs for trough plants are in the range of USD 4–9/W, depending upon the local solar conditions and construction costs. Capital costs for 10 MW tower plant are in the range of USD 9/W, but will be cheaper for large plants. Capital costs for dishes are over USD10/W, but the prices are going down.

Current costs for electricity production are in the range of USD 125–225/MWh, depending on the location of the plant. Both the capital and O & M costs can be brought down, by using large turbines (which allow better conversion rates) and smaller mirror surface.

Increased volume production, plant scale-up and technological advances would bring down costs. If in the next 10–15 years, CSP capacities of 5 000 MW are built, the electricity generation costs may come down to USD 43–62/MWh for trough plants, and USD 35–55/MWh for tower plants.

Projected Research & Development efforts

R&D in the following areas will improve efficiencies and bring down costs:

(i) Direct steam generation for trough plants. Fundamental studies on flow patterns and heat transfer in horizontal tubes will help in identifying cheaper substitutes for mineral oil with water, and more efficient ways of using superheated steam.

(ii) Towers using pressurized air with solar hybrid gas turbine. The gas turbine in the French Pegase project has achieved higher power conversion efficiencies by using high-temperature solar heat, further heated by fossil fuel burning.

(iii) CSP plants in arid areas are being increasingly used to desalinize brackish water and seawater, while producing power. If the solar plant delivers exhaust steam from the turbine at a temperature of 70°C, the heat can be used for desalination. Thus a 100 MW plant can produce 21 000 m^3 of fresh water per day (DLR, 2007).

(iv) Solar energy can be used for the production of hydrogen and metals through solar thermolysis. Production of hydrogen from water is an endothermic process. Concentrated solar radiation provides the high-temperature process heat needed for the purpose. Steinfeld (2005) describes the various routes for hydrogen production from solar energy. Solar-assisted fossil fuel steam reforming is in the advanced stages of development. CO_2 emissions in the extractive metallurgical industries can be reduced through solar thermal, carbothermal and electrothermal reductions of metal oxides.

Chapter 4

Biomass

U. Aswathanarayana

4.1 INTRODUCTION

Biomass is biologically-produced matter, and includes agricultural and forestry residues, municipal solid wastes and industrial wastes, renewable landfill gases, etc. Biodiesel is produced from leftover food products, such as animal fats and vegetable oils and crops grown solely for energy purposes. Algae which are aquatic organisms that range from pond scum to seaweeds, have great potential for producing liquid transportation fuels (Exxon Mobil has just announced funding of USD 600 million for algal biofuels). A strain of *Escherichia coli* has been engineered to produce biodiesel fuel directly from biomass.

Further details of Biomass gasification (11.2) and algal biofuels (Chap. 12) have been given elsewhere.

A strong point of biomass power is that it is dispatchable – it can be turned on and off, in contrast with wind and solar power which are non-dispatchable. Biomass burning does produce particulates, but emissions of SO_x and NO_x are much less than from fossil fuel power stations. Some biomass fuels, like sewage sludge and animal wastes, have foul odours. The purpose of some biomass plants in countries like Holland and Belgium is not electricity production, but the disposal of the large quantities of animal wastes arising from the large bovine populations in the countries.

4.2 TECHNOLOGY

The various renewable combustibles and biofuel pathways are summarized as follows:

Renewable combustibles

Arable annual crops: oilseed rape, wheat, maize, sugarbeet, potatoes
Herbaceous perennials: Miscanthus, switch grass, Reed canary grass

Woody perennials: Short rotation coppice, pine/spruce
Residues and wastes: waste fats and oils, forestry residues, straw, organic municipal wastes.

Conversion technologies

Pressing/esterification, hydrolysis/fermentation, gasification, pyrolysis, digestion.

Fuels

Biodiesel, ethanol, methanol, DME, FT Diesel, hydrogen, bio-oil and bio-methane.

Presently, biomass is the principal energy source for cooking and heating for 1.6 billion poor people in the developing countries. Biomass burning can lead to undesirable emissions and particulates. About 668 million people in India continue to use animal dung, agricultural waste and fuelwood as fuel for cooking. Because of the low fuel efficiency of the cook stoves (~8%), the particulate matter in the Indian households burning biomass is 2 000 $\mu g/m^3$ (as against the allowable 150 $\mu g/m^3$), leading to 400 000 premature deaths.

Biomass constitutes a vast energy resource, but only a fraction of it is *actually* available for electricity generation because of environmental, financial, political and other concerns. Biomass issues are best understood in terms of two aspects: fuel and electricity conversion technology.

Table 4.1 (source: ETP, 2008, p. 332) gives the typology of liquid biofuels.

The government and local policies have considerable bearing on the production and availability of biomass feedstocks: Land-use, and change in land use; biodiversity; reclamation of degraded lands (through the cultivation of Jatropha, for instance), genetically modified crops, water use and quality, treatment of wastewater and solid wastes, support for rural industries, and provision of low-cost energy to stimulate economic growth, etc. There is the other side of the coin. The demand for biofuels has some times led to deforestation, and the deterioration of wetlands and peat soils, and thereby increased the CO_2 emissions and food prices (Fargione et al, 2008).

Increased demand for palm oil to produce biodiesel is leading to deforestation in Malaysia and Indonesia. Though Malaysia claims that the higher production of palm oil in the country is attributable to superior gene stock which replaced older trees, Indonesia is planning to convert 1 to 2 million hectares of tropical forests into palm oil plantations.

In USA, increased use of corn for the production of biodiesel, led to increased prices for corn. Consequently, the US farmers started cultivating corn in preference to soybean, thus necessitating the import of soybean. Thus producing biofuels at the expense of food and fibre would not be beneficial to the world, and is hence not a sustainable proposition. One way out is to cultivate plants like Jatropha (from which biodiesel is produced) and bitter cassava (which is inedible and which would not be pilfered by pigs and monkeys) from which ethanol is produced, in degraded and non-arable land.

The technoeconomic characteristics of bioenergy conversion plant technologies are given in Table 4.2 (source: ETP, 2008, p. 312);

Table 4.1 Typology of liquid biofuels

Fuel	*Feedstock**	*Regions where currently mainly produced*	*GHG reduction impacts, vs. petroleum fuel use*	*Costs*	*Biofuel yield per* hectare of land	*Land types*
1st. generation ethanol	Grains (wheat, maize)	US, Europe, China	Low-moderate	Moderate-high	Moderate	Croplands
	Sugar cane	Brazil, India Thailand	High	Low-moderate	High	Croplands
2nd. Generation ethanol	Biomass (cellulose)	None used, but widely available	High	High	Medium-high	Croplands, Pasture lands, Forests
1st. generation biodiesel (FAME)	Oil seeds, (oil seed rape, soybean)	US, Europe, Brazil	Moderate	Moderate – high	Moderate-Low	Croplands
	Palm oil	Southeast Asia	Moderate	Low-moderate	Moderate – high	Coastal lands, Forests
2nd. generation biodiesel	Any biomass (via F-T**)	None used Commercially	High	High	Medium-high	Croplands, Pasture lands, Forests

*Other crop feedstocks that can be used are sugar beet, cassava, Jatropha, sunflower oil and sorghum, as well as purpose-grown vegetative grasses such as *Miscanthus* and reed canary grass, and short rotation forest crops, such as *Salix* and *Eucalyptus*.
**Fischer – Tropsch process converts gasified biomass or coal to liquid fuels.

(Source: ETP, 2008, p. 332) gives the typology of liquid biofuels

Table 4.2 Bioenergy conversion plant technologies

Conversion type	*Typical capacity*	*Net efficiency*	*Investment costs (in USD)*
Anaerobic digestion	<10 MW	10–15% electrical, 60–70% heat	
Landfill gas	<200 kW to 2 MW	10–15% electrical	
Combustion for heat	5–50 kW_{th} residential 1– 5 MW_{th} industrial	10–20% open fires, 40–50% stoves, 70–90% furnaces	$\sim$23/kW_{th} stoves; 370–990/kW_{th} furnaces
Combustion for power	10–100 MW	20–40%	1 975–3 085/kW
Combustion for CHP	0.1–1 MW 1–50 MW	60–90% overall 80–100% overall	3 333–4 320/kW 3 085–3 700/kW
Cofiring with coal	5–100 MW existing. >100 MW new plant	30–40%	123–1 235/kW + power station costs
Gasification for heat	50–500 kW_e	80–90%	864–980/kW_e
BIGCC* for power	5–10 MW demos 30–200 MW future	40–50% plus	4 320–6 170/kW 1 235–2 470/kW future
Gasification for CHP using gas engines	0.1–1 MW	60–80% overall	1 235–3 700/kW
Pyrolysis for bio-oil	10t/hr demo	60–70%	864/kW_{th}

BIGCC – Biomass Integrated Gasification with combined cycle

4.3 ALGAL BIOFUELS

Further details about algal biofuels have been given in Chap. 12.

Photosynthesis which is a fundamental biological process, needs sunlight, carbon dioxide and water. Cyanobacteria can convert up to 10% of the sun's energy into biomass. This rate is 5% for algae, and 1% for corn and sugar cane. The fossil fuels that we use now, were produced by cyanobacteria in the geological past. Single-cell algae are capable of producing a chemical "mix" that contains extractable fuel usable in cars and trucks. The "green" crude thus produced is chemically identical to crude oil, but it is carbon-neutral, non-toxic and sulphur-free. Sapphire Energy of California claims has reportedly succeeded in producing the green "crude", and expects to produce one million gallons (3.8 M. liters) of biodiesel and jet fuel per year starting from 2011 (vide image on the cover page – courtesy: Sapphire Energy).

Algae can be cultivated in tanks and ponds. As oil-producing algal strains are not fast, there is always a danger of faster-growing, non-oil producing algae invading such tanks (on the analogy of weeds). This can be prevented by covering a pond with a greenhouse, or using tubes which allow sunlight, and allow the circulation of nutrients and CO_2 to enhance the productivity ("bioreactors"). Closed systems are generally more expensive to build and operate, but their costs can be brought down by locating the algal plants near sources of CO_2. Firm cost information is not available in this regard.

Second generation biofuels which are expected to be produced from a range of ligno-cellulosic feedstocks through the use of thermochemical or biochemical pathways, will be superior to first generation biofuels in that they will be high yielding, they will be characterized by low emission of GHGs, and most importantly, they can be commercially produced sustainably. Second generation biofuels may initially complement first generation biofuels from grains and oil seeds, but may later supersede them.

Since propulsion of LDVs will in future be relying more on electricity and fuel cells than on internal combustion engine, the greatest demand for biofuels will be from heavy goods vehicles, marine and aviation transport, since aviation fuels are fully substitutable by BtL fuels.

On the analogy of oil refining to produce a variety of petroleum products, "bio-refining" of biomass could serve several purposes. Thus, second generation biofuels may be produced in conjunction with other value-added products, such as bio-chemicals and bio-materials, besides energy and heat.

A number of steps need to be taken to realize the potential of the second-generation biofuels:

(i) A global inventory of ligno-cellulosic biomass resources that could be used for biofuel and biorefinery purposes, optimal growing areas, costs, environmental and socioeconomic implications, potential co-benefits, such as energy, rural employment, and mitigation of local air pollution, and so on,
(ii) Basic research, applied R&D, Demonstration and Deployment through public-private cooperation, Development of policy whereby bioenergy development is harmonized with rural employment and agricultural development.
(iii) Preparation of "score cards" for individual bioenergy ensembles, in terms of energy balance, greenhouse gas emissions, water requirements and ecosystem impacts, etc. to facilitate decision-making.

IEA has made a projection of costs and potential market penetration of two principal second-generation conversion technologies, namely, enzymatic hydrolysis of cellulosic materials, and gasification/F-T liquefaction, making use of a variety of biomass materials. Production costs will decrease depending upon feedstock prices and economies of scale, and learning experience. Fig. 4.1 (source: ETP, 2008, p. 335, © OECD-IEA) gives the cost projections about second generation biofuels to 2050 (BtL = Biomass to Liquids; LC = Lignocellulose).

Land requirements of biofuels

There is much variation in the biofuel productivity of feedstocks. For instance, ethanol production from sugar cane in Brazil has higher yield, relative to the production of biodiesel from soybean and oil seeds in USA and Europe. Land use conflicts arise when the land which has high biofuel productivity, is also most suitable for food crop production.

It has been estimated that 375 to 750 Mha of land is required to produce the amount of biomass that the world needs. This constitutes 3 to 4% of the 6 billion hectares of the agricultural area in the world. Biofuel production may not be uniformly distributed around the world, but may be concentrated in a few favourable areas.

Land requirements for various biofuels are listed in Table 4.3 (source: ETP, 2008, p. 337).

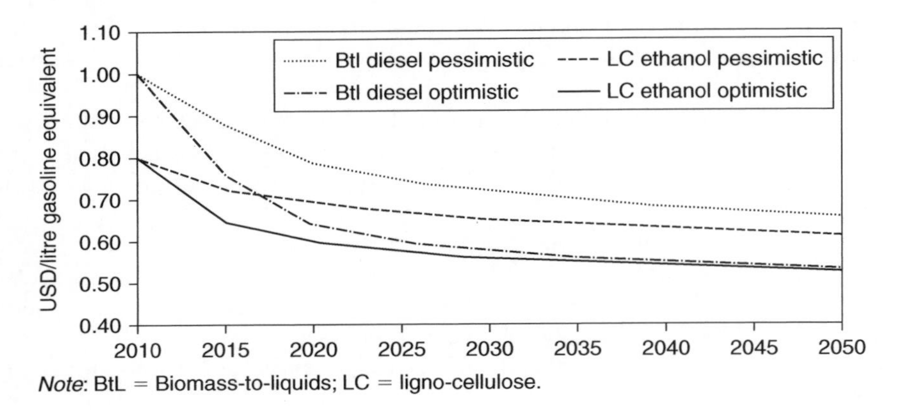

Figure 4.1 Cost projections of second-generation biofuels
(Source: ETP, 2008, p. 335, © OECD – IEA)

Table 4.3 Land requirements of various biofuels

		Yields, 2005 (t/ha)			
Region-Biofuel	*Feedstocks*	*Nominal*	*Gasoline equivalent*	*Average*	*Resulting yields in 2050, lge/ha*
Europe-ethanol	Wheat	2 500	1 650	0.7%	2 260
Europe-ethanol	Sugar beet	5 000	3 300	0.7%	4 520
Europe- FAME Biodiesel	Oilseed rape	1 200	1 080	0.7%	1 480
US/Canada-ethanol	Corn	3 000	1 980	0.7%	2 710
US/Canada-FAME-biodiesel	Soybean/Oilseed rape	800	720	0.7%	990
Brazil-ethanol	Sugar cane	6 800	4 490	0.7%	6 140
Brazil – FAME biodiesel	Soybean	700	630	1.0%	990
Rest of the world- ethanol	Sugar cane	5 500	3 630	1.0%	5 680
Rest of the world-ethanol	Grain	2 000	1 320	1.0%	2 070
Rest of the world-biodiesel	Oil palm	2 500	2 250	1.0%	3 520
Rest of the world-biodiesel	Soybean/oil seed rape	1 000	900	1.0%	1 410
Second Generation					
World-ethanol	Ligno-cellulose	4 300	2 840	1.3%	5 080
World – BtL biodiesel	Biomass	3 000	3 000	1.5%	5 360

FAME – Fatty acid methyl esters; lge/ha – litres gasoline equivalent per hectare; l/ha – Litres per hectare; Ethanol converted gasoline equivalent = ethanol 67% the energy content of gasoline; Biodiesel converted to diesel equivalent = biodiesel 90% the energy content of diesel, except BtL, biodiesel with 100% the energy content of petroleum diesel.

4.4 BIOMASS WASTES

Some biomass wastes, such as dung from farm animals and straw, are better used to fertilize the soil than for producing energy. Since people consume corn as food, there is a strong movement against using corn to produce ethanol to run cars. Some biomass residues are in the form of bits and pieces left over in lumber processing and pulp and paper mills. It is expensive to transport them to the electricity generating facilities. Also, they occupy a lot of room. Burning them in open air causes lot of pollution, and land filling is expensive.

Electricity conversion technology: Biomass fuel, like coal, is directly burnt in boilers to produce steam. Steam is used to run a turbine to produce electricity, and also as process heat (*cogeneration*). For instance, electricity and process heat that lumber mills produce by burning their wood waste are used within the plant, and any excess electricity is sold to the grid. As the moisture content and fuel value of the biomass materials are highly variable, the efficiency of biomass boilers is low (14–18%), roughly half of that of coal-fired boilers. The low efficiency increases costs and makes biomass less competitive. The efficiency is sought to be improved by drying the biomass and compressing it, and co-firing with coal (replacing 5–10% of the coal with biomass).

A completely new approach for biomass use is *gasification* – converting biomass into gas, and using the gas to drive the turbine. The interaction of biomass with steam and oxygen, at high temperature (a few hundred to a thousand degrees Celsius) and pressure (more than one atmosphere to 30 atmospheres), generates *producer gas* . This gas is composed of combustible constituents, such as, carbon monoxide, hydrogen, and higher hydrocarbons, and has fuel value of only 3–5 MJ m^{-3}, which is one-tenth that of natural gas. Small experimental generators with a capacity of 10–30 MW have been in operation, but the financial viability of large scale gasification plants, has not yet been established.

In India, 40% of the population has no access to electricity. In Bihar, India, rice husk waste is being used to generate electricity. Gases generated when the rice husk is heated, are used to drive the generator which produces electricity. The plant is operated 6–12 hours/day, and produces 35 kW to 100 kW of electricity. Sixteen plants of this type are in currently in operation. Villagers welcome this, as they save the cost of kerosene. The waste left after the rice husk is burned, makes a good manure. It is rich in silica, and could be used in cement manufacture. Husk Power Systems plans to develop 50 to 70 plants by 2010. By 2014, the company plans to provide electricity to 10 million people in 4000 villages.

Gasification continues to attract interest because of its potential in respect of three low-carbon, biomass –based systems (Boyle, 2004, p.131):

- Electricity from integrated biomass-fuelled gas turbine plant,
- Liquid fuels as substitutes for petroleum products,
- Hydrogen or other fuel for fuel cells.

4.5 COSTS

Cost of biomass-sourced electricity: Where biomass is a waste (i.e. available without payment) and can be burnt at the site where it is produced (e.g. lumber mills), the costs will be low. On the other hand where the biomass fuel has to be purchased and/or has to be transported, the costs will be high. Experience has shown that transport costs outside a radius of 80–120 km. become prohibitively high. Consequently, biomass

Table 4.4 Ethanol yields

Raw material	Litres per tonne	Litres per hectare per year
Sugar cane (harvested stalks)	70	400–12 000
Maize (grain)	360	250–2 000
Cassava (roots)	180	500–4 000
Sweet potato (roots)	120	1 000–4 500
Wood	160	160–4 000

electricity plants have to use biomass available nearby, even if that means low capacity, and high costs. In order to have the advantage of economics of scale, a biomass power plants need to have a capacity of more than 100 MW.

In general, prices of electricity from biomass power stations vary widely depending upon specific fuels, technologies and contracts. They are generally higher than that fossil fuel based power generation. The US Department of Energy estimates that electricity produced by biomass plants costs about US cents 9/kWh. California contacted 3.4 MW of landfill gas at US cents 6.1/kWh, and 26 MW of biogas from dairies at an average price of US cents 7.5/kWh. UK contracted 800 MW of biomass power at US cents 3.4–3.8/kWh.

4.6 ETHANOL

Ethanol (C_2H_2OH) is produced by the fermentation of sugars. Table 4.4 (source: Boyle, 2004, p. 135) gives the ethanol yields from various raw materials.

Brazil's production of ethanol from sugar residues is the largest commercial biomass system in the world. Its impetus came from high oil prices and low sugar prices. The production reached the maximum of about 15 billion litres/yr. Ethanol is blended with gasoline up to 26% (gasohol). Brazil saved about USD 40 billion in foreign exchange by the use of gasohol to run cars.

4.7 LANDFILL GAS

A tonne of largely organic waste in a landfill should be able to produce during its lifetime 150–300 m^3 of gas, with methane constituting 50–60% of the gas by volume. Though the nominal heat energy is 5–6 GJ/tonne of refuse, the actual heat energy output is less than 2 GJ/t. Methane has 21 times more global warming potential than CO_2, per tonne. So when methane is burnt to form CO_2, its climate forcing effect gets drastically reduced.

Methane is collected by vacuum suction through perforated pipes which are sunk into the landfill to depths of 8 to 22 m. In a large landfill, there may be several kilometers of pipes, producing as much as 1000 m^3 of gas per hour. Fig. 4.2 shows the changing gas composition in a landfill site (source: Boyle, 2004, p. 121 © Open University).

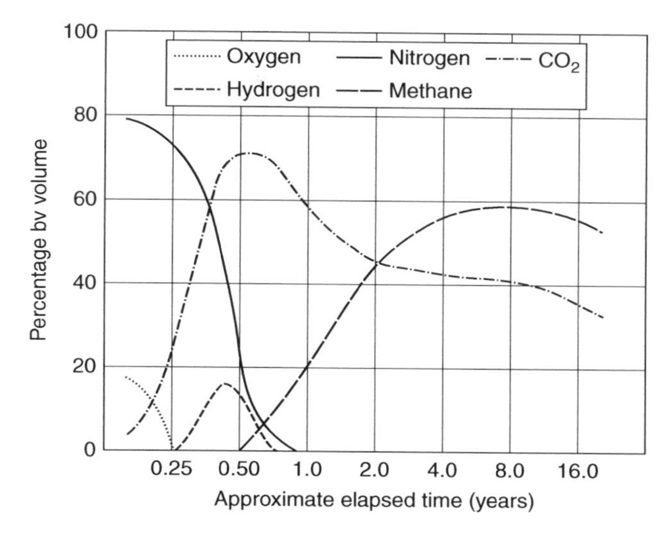

Figure 4.2 Changing gas composition in a landfill site, © Boyle, 2004, p. 121

After removing condensates and various other impurities, the landfill gas is compressed. The compressed gas may be used in three ways:

- Electricity generation via engines and turbines: About 70% of the gas is used this way. Internal combustion engines are preferred as they are of low cost, have reasonable efficiencies and are characterized by a proven technology. A disadvantage of this application is that NO_x impurities in the gas corrode the engine. The engine sizes are typically 250 kW to 1 MW, and clusters of engines could be set up if large capacities are needed. Gas turbines can be used for capacities of 3 to 4 MW, and steam turbines are used for capacities of more than 8 MW.
- Burning in a boiler to produce hot water or steam for industrial purposes.
- Injection into natural gas pipeline (this requires CO_2 removal and high compression).

In order for a landfill to be able to produce enough landfill gas to generate electricity, it should fulfill the following conditions: it should contain more than one million tonnes of waste, and should be either in operation or should have been closed recently; the waste should be organic, rather than construction debris or industrial waste; the rainfall in the area should be more than 500 mm.

Typical costs for electricity from landfill methane are US cents 6 to 9/kWh, depending upon the productivity of the landfill, its proximity to a nearby market for steam or hot water, and availability of municipal financing (which will be cheaper than bank financing). The larger the size of the plant, the less the production cost of electricity.

In USA, federal law requires that a landfill site which has a capacity of more than 2.75 million tonnes, should install landfill gas recovery system.

The landfill gas (LFG) from a million tonne landfill can support a 2 MW plant over 15–20 year generating time, with 88% load factor. Assuming capital cost of USD 1 200/kW, O & M costs of US cents 1.6/kWh, and discount rate of 8%, the levelized cost of landfill gas electricity comes to US cents 3.5/kWh. LFG plants in UK have been commercially successful, with an installed capacity of 400 MW, and annual electricity output of 2 700 GWh in 2002.

4.8 PROGNOSIS

Bioenergy is not only the largest contributor to renewable energy presently but also has the highest technical potential among the renewable energy resources. Biomass is used inefficiently for traditional domestic cooking and space heating in developing countries. According to a projection by IEA, biomass use would increase four-fold by 2050 (150 EJ/yr or 3 604 Mtoe/yr). This would involve the delivery of 15 000 Mt of biomass to the production plants annually, with half of this coming from crop and forest residues, and the rest from purpose-grown energy crops. The projected use of bioenergy would be as follows: about 700 Mtoe/yr for transport biofuels, about 700 Mtoe/yr to generate electricity (2 450 TWh/yr), and 2 200 Mtoe to produce biofuels, heating, and cooking (including production of dimethyl ether), and industry (including process steam from CHP (Combined Heat and Power) and black liquor (used for the extraction of lignin). By co-firing biomass with coal in steam turbines of about 100 MW capacity, electricity could be produced at USD 60–80/MWh. Biofuel from algae has great potential – much research is being devoted to cultivating specific algal strains suitable for biodiesel production.

Chapter 5

Hydropower

U. Aswathanarayana

5.1 INTRODUCTION

Presently, hydropower accounts for 90% of the renewable power generation. Some countries, notably Sweden, Switzerland and Norway, get virtually all their electricity supplies from hydropower. The total hydropower potential in the world is estimated to be 2 000 GW (Taylor, 2007). IEA estimates that by 2050, hydropower capacity could go up by 1 700 GW, producing 5 000–5 500 TWh/yr. Though the world's technically feasible hydropower is 14 000 TWh/yr, the realistic potential is 6 000 TWh/yr. Around 808 GW of hydropower is in operation or under construction. Hydropower is one of the cheapest ways of producing electricity. Most plants have been built many years ago, and their capital costs have already been amortized. The capital costs of new plants vary from USD1 000/kW in developing countries to USD 2 400/kW in OECD countries. Generating costs vary from USD 0.02–0.06/kWh.

Table 5.1 (source: Boyle, 2004, p. 153) provides the hydro potential and output of different regions in the world.

Hydropower output (in TWh/yr) of some important countries are: Canada – 345; Brazil – 288; USA – 264; China – 231; Russia – 167; Norway: 129; Japan – 91; India – 76; Sweden – 74; France – 74; Venezuela – 61; Italy – 51; Austria – 42; Switzerland – 40; Spain – 35, etc.

The investment and production costs of hydropower are given below, assuming 10% discount rate.

Assuming a 50-year lifetime, 45% capacity factor, operation and maintenance cost of US cents 0.8/kWh, the total levelized cost of hydroelectricity comes to US cents 3.9/kWh.

Table 5.1 Hydro potential

Region	*Technical potential (TWh/yr)*	*Annual output (TWh/yr)*	*Output as percentage of technical potential*
Asia	5 093	572	11%
South America	2 792	507	18%
Europe	2 706	729	27%
Africa	1 888	80	4.2%
North America	1 668	665	40%
Oceania	232	40	17%
World	14 379	2 593	18%

Table 5.2 Investment and production costs of hydropower

	Investment cost (USD/kW)			*Production cost (USD/kW)*		
	2005	*2030*	*2050*	*2005*	*2030*	*2050*
Large hydro	1 000–5 500	1 000–5 400	1 000–5 100	30–120	30–115	30–110
Small hydro	2 500–7 000	2 200–6 500	2 000–6 000	56–140	52–130	49–120

(Source: *Energy Technology Perspectives*, 2008, p. 400)

Making use of the force of falling water as a source of mechanical power has been in vogue since the ancient times (we are all familiar with the poem of the jolly miller of the Dee, who makes his living from the flour mill operated by the water impounded by the weir on the river). There are three kinds of hydropower facilities: (i) "Storage "projects (Aswathanarayana, 2001, p. 191–195), (ii) "Pumped Storage" projects, and (iii) "Run-of-river" projects.

5.2 "STORAGE" PROJECTS

Large surface reservoirs are constructed to increase the availability of water during the periods of low flow or dry years. Though the construction of water reservoirs has been going on since ancient times, all the large reservoirs with a total volume of more than 50 km^3 have been constructed during the second half of this century. According to Shiklomanov (1998), the total volume of the reservoirs of the world is about 6 000 km^3, and their total surface area is about 500 000 km^2. The volume of water in the reservoirs corresponds to about 5% of the annual rainfall on land or 15% of the annual runoff. They trap 2–5 Gt y^{-1} of sediments, which is about a quarter of the total sediment yield of the land. Thus, the reservoirs have a profound impact on the circulation of water on the land surface, discharge of sediments and nutrients to the sea. The evaporation from the surfaces of the reservoirs is considerable, and thus the reservoirs are one of the greatest users of freshwater. The map by Vörösmarty et al. (1997) shows the location of the major reservoirs in the world.

Large-scale hydropower projects often become controversial, as they tend to affect water availability downstream, inundate historic sites and ecosystems, and have serious socioeconomic impacts because of the need to rehabilitate displaced populations.

The water in the reservoir becomes warmer and more stratified. Dams make it almost impossible for fish to swim through, and this problem is most acute for salmon. California is so concerned about the environmental consequences of dams that it excluded new hydropower from its renewable portfolio standard. High dams, such as the Sardar Sarovar dam on the Narmada River in western India, and the Three Gorges Project on the Yangtze river in China, have drawn much criticism and agitations particularly on account of the hardship caused to the people who have been or will be displaced, on account of the submergence.

An issue of concern is how climate change will affect the hydropower potential and economics, because of depleted river flows. As a dam may have a lifetime of about 100 years, the depletion of water supply due to climate change or the reduction in the storage capacity of the reservoirs due to sedimentation, have to be kept in mind in working out the economics. Some flow is needed for ensuring the life and reproduction of the indigenous fish. Similarly, some flow is needed for development. There is no simple way to determine the minimum flow in a river which will satisfy the environmentalists and developers at the same time, and therein lies the problem.

The following discussion about the high dams takes into account the lucid exposition about high dams by P.V. Indiresan in an article in "*The Hindu*" (Madras, India) of Dec. 1, 1999:

(i) Other things being equal, doubling the height of the dam, increases the water stored *eight* times and the power potential *sixteen* times. Also, for a given amount of storage, the higher the dam the smaller the area that will be submerged. *It therefore follows that the construction of the highest dam that is technically feasible is the best way to minimize submergence and hence the number of people to be displaced.* If, instead of a high dam, it is proposed to have a number of low dams, the aggregate of the areas submerged by them will be considerably more, and the amount of hydroelectricity that could be generated will be minimal. The amount of water available for drinking and irrigation may be unaffected, however.

(ii) In countries with monsoon climate (e.g. India), the rainfall is restricted to a few months of the year. It is also erratic. Instances are known where the annual precipitation in an area occurred in a matter of few days of intense downpour. Consequently, flash floods are common, and very few rivers are perennial. Under these circumstances, harvesting and storage of surface runoff in reservoirs is the only way to provide irrigation and drinking water round the year. The runoff in a watershed can be harvested both through a high dam on a river *plus* a series of small check dams over small streams, and the recharge of groundwater. This is not a case of *either/or* but *and*. However, there may be situations where an area is so arid (e.g. Sinai desert in Egypt or Thar desert in India) that check dams are of no avail. Thus, Egypt is irrigating some areas in the Sinai desert by the transfer of Nile water. But for the water from Bhakra high dam in Punjab, the greening of desert areas in Jaisalmer, Rajasthan, India, would not have been possible.

(iii) In tropical countries, there is intense evaporation. In India, it is about 1.2 m. In other words, irrespective of the depth of the water body, the top 1.2 m layer of water in a reservoir will be lost due to evaporation. If the reservoir is (say) 100 m deep (because of a high dam), the loss due to evaporation would be only 1%. On the other hand, if the reservoir is shallow, (say, about 10 m), the loss due to evaporation would be 10% of the total storage.

Thus, in the ultimate analysis, a country or a community may have to make the choice, one way or the other. Indiresan is critical of the agitators whose sole aim is to stop the construction of high dams. This is a negative approach. Such agitators rarely, if ever, come up with techno-socio-economically viable alternatives, which can lead to benefits similar to the high dam. Indiresan suggests some practical steps to ease the problems arising from high dams : (i) Engineers should confine themselves only to building dams, and agronomists (not engineers) should manage water allocation, as is the practice in USA, (ii) The resettlement process should be handled by specialists, (iii) There should be complete transparency about all aspects of the dam, reservoir, submergence, etc., (iv) The displaced persons should be offered a package of benefits (such as retraining for the new jobs that will be created), and not just monetary compensation.

5.3 PUMPED STORAGE HYDROELECTRICITY

When there is more generation of electricity than the load available to absorb it (as during nights), excess generation capacity may be used to pump water into a reservoir at a higher elevation. When the electricity demand is high (as during daytime or due to the failure of a component of the grid), water is released back into the lower reservoir through the turbine, generating electricity. Also, Francis turbines, which are reversible turbine/generator assemblies, are capable of acting as both a pump and a turbine, could be used, as needed. One m^3 of water atop a 100 m tower, has the potential energy of about 0.272 kWh. Pumped storage is thus an effective way of storing energy through water. The two reservoirs may be man-made or natural.

Pumped storage is a high capacity form of grid energy storage presently available. It can be used to flatten out load variations on the power grid, which may be linked to coal-fired plants, nuclear plants or renewable energy power plants. In the context of the back-up provided by pumped storage, these plants could continue to operate at peak efficiency (Base Load Power plants), while reducing the need for "peaking" power plants that use costly fuels. Thermal plants are less able to respond to sudden changes in the electricity demands, and may cause voltage and frequency instabilities. In contrast, pumped storage plants, like the normal hydropower plants, can respond to load changes almost instantly (less than 60 seconds). Pumped storage is expected to become important for load balancing in tandem with large capacity solar and wind mill plants.

A recent technological development is the variable speed turbines which can generate electricity in synchronization with the network frequency.

Pumped storage system involves the evaporation losses from the exposed water surface and conversion losses. Approximately 70–85% of the electrical energy used to pump the water to the elevated reservoir is regained, depending upon the capital costs and geographic setting. This loss of about 20% of electricity is more than compensated by selling more electricity during periods of *peak* demand, when the electricity prices are highest.

Pumped storage capacity worldwide now is about 100 GW, which is about 2% of the hydropower generation capacity. Pumped storage capacity has the potential of 1 000 GW, which is roughly half of the global hydropower potential (Taylor, 2007).

5.4 "IN-RIVER" HYDROELECTRIC PROJECTS

Small-scale hydropower projects are designed to run in-river. These are environment-friendly energy conversion options as they do not affect the river flows significantly. They can be stand-alone applications to serve rural communities or as replacements to diesel generators.

Technology needs of small power stations are as follows (source: *Energy Technology Perspectives*, 2008, p. 391):

Equipment: Turbines with less impact on fish populations; low-head technologies; in-stream flow technologies.
O&M practices: Package plants which require only limited O&M.
Hybrid systems: Wind-hydro systems; Hydrogen-assisted hydrosystems.

Chapter 6

Geothermal energy

U. Aswathanarayana

6.1 INTRODUCTION

"Geothermal energy" covers both the direct use of geothermal power for space, heating, water heating and industrial processes, which are more common, and the generation of geothermal electricity, which are rarer. Geothermal electricity plants of more than 100 MW installed capacity are listed below, country-wise (MW installed capacity in 2000): USA – 2 228; Philippines – 1 909; Italy – 785; Mexico – 755; Indonesia – 590; Japan – 547; New Zealand – 437; Iceland – 170; El Salvador – 161; Costa Rica – 143. The total capacity of geothermal power plants in the world is 10 GW in 2007, generating 56 TWh/yr of electricity.

Geothermal energy has several advantages: (i) It is non-polluting and has no carbon footprint, (ii) It is of large magnitude – the heat stored in the earth is estimated to be about 5 billion EJ , which is 100 000 times more than the world's annual energy use, (iii) It is available all the year round, and production costs are low. There are, however, some drawbacks: (i) Air pollution may sometimes be caused by H_2S, CO_2, NH_3, Rn, etc. gases vented into the air, (ii) Low magnitude earthquakes may be triggered and land subsidences may take place due to changes in the reservoir pressure, (iii) The overall efficiency of geothermal power production (15%) is less than half of the coal-fired plants, (iv) Drilling costs are high (USD 150 000–250 000 per well).

Compared with wind electricity and solar PV electricity, which are intermittent, geothermal electricity can be generated round the clock, and could therefore serve as baseload electricity. This factor is reflected in the capacity factor which is defined as the actual plant output as a percentage of the maximum output of the plant operated at full capacity. Geothermal plants have a capacity factor of 90%, compared to 25 to 30 %in the case of wind electricity.

Aswathanarayana (1985, p. 159–162) summarized the geological and economic aspects of geothermal energy. The vertical temperature gradient in the earth's crust has an average of 30° C/km. It varies from 10–20°C/km in the Precambrian shield areas to 30–50°C beneath tectonically active areas. There are areas where the gradient is as high as 150°C/km. Areas of high heat flow (more than 2 HFU – Heat Flow Units) on the continents are characterized by hot springs and products of Tertiary volcanic activity. Lardarello (Italy), Geysers, Casa Diablo, Niland (USA), Wairakei and Waistapu (New Zealand) Hvergardi (Iceland), Pauzhetsk (Russia), Otake and Matsukawa (Japan) are some of the areas where geothermal power is being tapped economically.

6.2 TECHNOLOGY

High-temperature geothermal energy sources can be used to generate electricity. Lower temperature geothermal sources are best used for space heating (90% of all homes in Reykjavik, Iceland, are heated this way), domestic and industrial refrigeration, heating of green houses and animal shelters, crop drying, dehydration, etc. Freshwater is a highly valuable by-product of tapping geothermal sources. When brackish water is desalinated by geothermal energy, useful chemicals are obtained as a bonus.

Among the geothermal regions, fault block terrains with Quaternary volcanism (like those of the East African Rift system) have the highest average reservoir temperature (~250°C). In order to be economic, a geothermal well should be able to produce more than 20 tonnes/hr of steam. Geological criteria (such as, age, structure, thermal manifestations), geochemical criteria (like the dissolved silica content, Na/K ratios of surface and spring waters), and geophysical studies (deep resistivity surveys, heat flow measurements) are used for prospecting for and evaluation of, geothermal energy sources.

While potential sites for geothermal resources could be identified on the basis of geological considerations, technoeconomic evaluation can only be made on the basis of drilling. Even after this study, it is not always possible to project how long the resource will last. For instance, the production of electricity from the famous Geysers complex in California, has dropped sharply because of depletion.

The geothermal electricity potential of western USA has been estimated to be 20 GW. How much of it can be tapped would be determined by energy prices.

6.3 RESOURCES

The total capacity of geothermal power plants in the world is 10 GW in 2007, generating 56 TWh/yr of electricity. There are three kinds of commercial geothermal plants, depending upon the temperature of water:

(i) Dry steam plants, which use direct steam resources at temperatures of about 250°C,
(ii) Flash-steam power plants which make use of hot, pressurized water at temperatures hotter than 175°C. In these types of plants, pressure is lowered when the high temperature, high pressure fluids enter the plant, thereby making them boil or flash. The steam is used to run the turbine, and water is injected back into the reservoir.

(iii) Binary plants which use geothermal resources at temperatures of about 85°C. The heat contained in the hot water is exchanged through the use of a fluid that vaporizes at lower temperatures. This vapour drives a turbine which generates power. Hot water in the reservoir fluid generally contains dissolved salts, but since it is a closed system, the dissolved salts do not affect the environment. The fluids with the dissolved salts are injected back into the reservoir. As the system is environmentally benign, the binary power plants have become popular.

Large scale geothermal plants are currently possible in high heat flow areas such as, plate boundaries, rift zones, mantle plumes and hot spots, that are found around the "Ring of fire" (Indonesia, The Philippines, Japan, New Zealand, Central America, the west coast of USA) and the rift zones (East Africa, Iceland).

The geothermal electricity potential of western USA has been estimated to be 20 GW. How much of it can be tapped would be determined by energy prices.

A geothermal field need not have a surface manifestation in the form of a hot spring. In fact, fields of dry, hot rock are the most promising sources of geothermal energy, though technology for their exploitation is yet to be commercially developed. On the basis of abnormally high thermal gradients (ten times the normal value of 20°C/km), David Blackwell found at Marysvale, Montana, USA, a 31 sq.km. area underlain by hot rock (at temperature of over 400°C) at a depth of 1 km, which is accessible to drilling. It has been estimated that this field alone could provide a supply of one-tenth of America's electricity needs for 30 years.

6.4 COSTS

Geothermal electricity costs may be estimated in two ways:

(i) Summation of component technology costs: The initial costs of geothermal plants depend upon the depth of the well, the temperature of the geothermal fluid, the length of the piping, the level of contaminants and access to transmission lines. Komor (2004, p. 58) estimates the initial cost of the flashed-steam geothermal power plant system at USD 1 500–2 000/kW for a 5+ MW plant, with the costs roughly split equally between the power plant and the infrastructure (well construction, piping, water treatment, and so on). Binary plants are more expensive (USD 2 000–2 500/kW). On the basis of the above costs, assuming 7.5% discount rate, and 89% plant capacity, the levelized energy cost comes to US cents 5.0/kWh for flashed steam plant, and US cents 5.8/kWh for binary plants. In an ideal situation (very hot water or steam close to the surface, power plant close to the well, and proximity to transmission lines, etc.), the cost of electricity could be less. For instance, Geysers plant sells power at US cents 3.5/kWh.

(ii) Market conditions: In 2001, California Power Authority signed letters of intent for purchasing power at US cents 6/kWh. The price could be different under a different set of market conditions. In any event, geothermal electricity commands a premium over wind or solar electricity because of its being baseload power.

In the case of geothermal electricity, well drilling accounts for half of the capital cost. Efforts are being made to bring down these costs. The capital costs vary from USD

Table 6.1 Investment and production costs of geothermal energy

	Investment cost (USD/kW)			*Production cost (USD/kW)*		
	2005	*2030*	*2050*	*2005*	*2030*	*2050*
Hydrothermal	1 700–5 700	1 500–5 000	1 400–4 900	33–97	30–87	29–84
Hot dry rock	5 000–15 000	4 000–10 000	3 000–7 500	150–300	80–200	60–150

(Source: *Energy Technology Perspectives*, 2008, p. 400)

1 150/kW of installed capacity for large, high-quality resources, to USD 5 500/kW for small, low-quality resources.

The temperature of the geothermal fluids determines the electricity generation costs.

The operating costs are in the range of US Cents 2–5/kWh for flash and binary systems, excluding investment costs. In the case of the Geysers Field, California, the operating costs are US Cents 1.5–2.5/kWh. In Europe, generation costs range from US cents 6–11/kWh for traditional geothermal plants.

The costs of geothermal energy are given in Table 6.1 (source: *Energy Technology Perspectives*, 2008, p. 400).

6.5 RESEARCH & DEVELOPMENT

Enhanced Geothermal Systems (EGS) tap the heat from the hot, dry rock underground (vide further details under 11.4). Water becomes steam when it is pumped through boreholes and encounters the hot rock. When steam returns to the surface, it is used to generate electricity through a binary generator. The water is recirculated continuously. A number of countries are seeking EGS power – Australia (5.5 GW), USA (100 GW), China and India (100 GW). Switzerland is planning to build 50 EGS plants of 50 MW capacity (i.e, totaling 2.5 GW), to provide one-third of the electricity requirements of the country. EGS is not an unmixed blessing – an EGS plant near Basel, Switzerland, triggered a minor earthquake of magnitude 3.4 in Dec. 2006. Another problem with EGS is the large requirement of water – a small 5 MW plant requires 8500 t/d of water. A large scale plant may requires ten times more water.

Five km deep geothermal wells are highly productive, as the steam conditions are much more favourable (430–550°C; 230–260 bars), but drilling costs are prohibitively high (USD 5 million per well). Geothermal plants based on deep wells will become economical when the drilling costs come down (Bjarnason, 2007).

Chapter 7

Tidal power

U. Aswathanarayana

7.1 INTRODUCTION

Tidal barrages produce power for five to six hours during the spring tides, and three hours during the neap tides, within a tidal cycle lasting 12.4 hours. The problem with this kind of power generation is that power is produced in short bursts, depending upon the tidal ebb and flow timings. The power grid to which the tidal electricity is fed, should be capable of accommodating this burst.

The use of tidal energy to generate power is similar to that of hydroelectric power plants. A dam or barrage is built across a tidal bay or estuary where there is a difference of more than five metres between the high tide and low tide. Water flowing in and out of the dam runs the turbines installed along the dam or barrage, and generates electricity. Tidal plants have periods of maximum power generation every six hours. During periods of low electricity demand, extra water is pumped into the basin behind the barrage, on the analogy of pumped storage.

Apart from grid-connected electricity generation, ocean renewable energy could also be used for off-grid electricity generation in remote areas, aquaculture, desalination, production of compressed air for industrial applications, integration with other renewable energy resources, such as offshore wind power, solar PV, etc.

Tidal barrage projects are more environmentally intrusive than wave and marine current projects. The adverse environmental impact of tidal barrage projects is sought to be reduced by integrating oscillating water turbines with breakwater systems that convert water pressure into air pressure and use the compressed air to drive a Wells turbine. Such breakwaters linked projects (about 0.3 MW capacity) are being developed in Spain and Portugal. Portugal is also actively developing wave energy plants with the goal of achieving 23 MW by 2009.

Table 7.1 Some locations in the world for potential tidal power projects

Country	*Mean tidal Range (m)*	*Basin area (km²)*	*Installed Capacity (MW)*	*Approx. Annual output (TWh/yr)*	*Annual plant load factor (%)*
Argentina					
San Jose	5.8	778	5 040	9.4	21
Golfo Nuevo	3.7	2 376	6 570	16.8	29
Rio Deseado	3.6	73	180	0.45	28
Santa Cruz	7.5	222	2 420	6.1	29
Rio Gallegos	7.5	177	1 900	4.8	29
Australia					
Secure Bay	7.0	140	1 480	2.9	22
Walcott Inlet	7.0	260	2 800	5.4	22
Canada					
Cobequid	12.4	240	5 338	14.0	30
Cumberland	10.9	90	1 400	3.4	28
Shepody	10.0	115	1 800	4.8	30
India					
Gulf of Kutch	5.0	17.0	900	1.6	22
Gulf of Cambay	7.0	1 970	7 000	15.0	24
Korea (Rep)					
Garolim	4.7	100	400	0.836	24
Cheonsu	4.5	–	–	1.2	–
Mexico					
Rio Colorado	6–7	–	–	54	–
USA					
Passamaquoddy	5.5	–	–	–	–
Knik Arm	7.5	–	2 900	7.4	29
Turnagain Arm	7.5	–	6 500	16.6	29
Russian Feder.					
Mezeh	6.7	2 640	15 000	45	34
Tigur	6.8	1 080	7 800	16.2	24
Penzhinsk	11.4	20 530	87 400	190	25

(Source: Boyle, 2004, p. 226)

7.2 RESOURCE POSITION

The World Energy Council has estimated the world wave power at 2 TW. The realistically recoverable ocean energy resource is put at 100 GW. The estimated wave electricity potential is 300 TWh/yr. Table 7.1 (source: Boyle, 2004, p. 226) gives the locations of potential tidal power projects.

7.3 RANCE (FRANCE) AND SEVERN (UK) TIDAL BARRAGES

The 740 m-long Rance Barrage in France was built during 1961–67. It has 24 reversible turbines of 10 MW capacity, tidal range of up to 12 m, and typical head of approximately 5 m. Typically, the plant has been functional 90% of the time, and producing 480 GWh of electricity. Initially, there was adverse impact on fish and birds, but later the ecosystem got stabilized, and the impact got minimized. The 16 km-long barrage that is planned to be built across the Severn Estuary in U.K. would have a capacity of 8.6 GW, and would be capable of producing 17 TWh/yr, which would be roughly 5% of the electricity generated in U.K. in 2002. The *load factor*, which is the percentage of time a plant can deliver electricity, is about 23% for Severn Barrage, as against 77% for nuclear power stations, 84% for combined cycle gas turbines. The barrage would reduce the turbidity of water and thereby enhance the carrying capacity for migrating fish and migratory birds. The construction cost of the barrage will be huge (~USD 37 billion). The cost of electricity from the Severn Barrage has been estimated at US cents 8–11/kWh at 8% discount rate, and US cents 16–22/kWh at 15% discount rate (both at 1991 prices). Another view is that the economics of the project has to be computed on "total life cost" basis, as the barrage will have a life-time of more than 100 years, and as the turbines need to be replaced once in 30 years, and running costs are approximately 1%. Once the capital and interest costs have been paid off, the tidal barrage would be generating profits for the rest of the time.

Power plants based on tidal barrages have been in operation at La Rance in France (240 MW, built in 1960s), and Annapolis Royal in Canada (20 MW, built in 1980s).

Korea is constructing a 254 MW tidal energy plant, at the cost of USD 1 000/kW.

The potential for wave energy plants, typically 0.3 MW capacity, depends on wave heights. The wave potential increases towards the poles, but is site dependent. The European Atlantic coast, the North American Pacific Coast, and Australian south coast, hold promise.

Ocean Thermal Energy Conversion (OTEC) plants which are based on harnessing the temperature gradients in the ocean, are in operation in India. Heat pumps powered by oceanic thermal energy are being used for heating and cooling in a number of countries. OTEC plants are expected to become operational after 2030.

Norway is building a 10 MW demonstration plant to harness the energy based on salinity gradients.

7.4 RESEARCH & DEVELOPMENT AND COSTS

Considerable R&D effort is needed to ensure the commercial viability of ocean energy systems: Basic science research on wave behaviour and dynamics of wave absorption, applied science research on the design of supporting structures, turbines, foundations, engineering designs in regard to hull design, power takeoff systems, etc.

The design of tidal barrages has to take into account the possible adverse effects on mudflats and silt levels in the estuaries and wildlife living in and around the estuary.

The breakdown of the projected investment costs for shoreline and near shore ocean energy installations are as follows (in %): Civil works –55; Mechanical and electrical equipment –21%; Site preparation: 12%; Electrical transmission –5%; Miscellaneous –7%. Ocean energy projects are still in the development stage, and firm costs cannot be given. They are, however, in the range of USD 150/MWh to USD 300/MWh.

Investment and production costs of ocean energy are given in Table 7.2.

Table 7.2 Investment and production costs of ocean energy

	Investment cost (USD/kW)			*Production cost (USD/kW)*		
	2005	*2030*	*2050*	*2005*	*2030*	*2050*
Tidal barrage	2 000–4 000	1 700–3 500	1 500–3 000	60–100	50–80	45–70
Tidal current	7 000–10 000	5 000–8 000	3 500–6 000	150–200	80–100	45–80
Wave	6 000–15 000	2 500–5 000	2 000–4 000	200–300	45–90	40–80

(Source: *Energy Technology Perspectives*, 2008, p. 400).

Ocean energy technologies for the generation of electricity are in the early stages of development. Among ocean energy technologies, only wave energy and tidal energy have good potential, and are being actively developed in 25 countries. Technologies based on temperature and salinity gradients and marine biomass have little chance of becoming commercially viable in the near future.

Further details about Marine Energy can be had from chap. 11.3.

Chapter 8

Deployment of renewable energy technologies (RETs)

U. Aswathanarayana

8.1 CHARACTERISTICS AND COSTS OF COMMON RETs

Selected characteristics and costs of common renewable energy technologies (RETs) are given in Table 8.1. It may be noted that in general the costs of RETs are higher than conventional energy technologies which are typically around US cents 4 to 8/kWh. The position, however, is not static. The costs of many RETs are declining significantly due to technology improvements and market maturity. At the same time, the costs of some conventional energy technologies (for example, gas) are also declining. New kinds of gas deposits (such as, shale gas), new methods of mining (such as, horizontal drilling), and improvements in gas turbine technology, have brought down the costs of electricity production from gas.

8.2 POTENTIALS OF RETs

RETs are subject to constraints which determine what is achievable.

Theoretical potential: Natural energy flows which represent the theoretical upper limit of the amount of energy that can be generated from a specific source over a defined area.

For instance, solar insolation is high in low latitudes and low in high latitudes.

Technical potential: This is determined on the basis of technical boundary conditions, such as, conversion technologies or available land area for a particular installation.

The technical potential is dynamic – with improved R&D, conversion technologies and therefore the technical potential, may get enhanced.

Table 8.1 Key characteristics and costs of Reneweable Energy Technologies

Technology	*Typical characteristics*	*Typical current investment costs[1] (USD/kW)*	*Typical current Energy Production costs[2] (USD/MWh)*
POWER GENERATION			
Hydro			
Large hydro	Plant size: 10–18000 MW	1000–5500	30–120
Small hydro	Plant size: 1–10 MW	2500–7000	60–140
Wind			
Onshore wind	Turbine size: 1–3 MW Blade diameter: 60–100 meters	1200–1700	70–140
Offshore wind	Turbine size: 1.5–5 MW Blade diameter: 70–125 meters	2200–3000	80–120
Bioenergy[3]			
Biomass combustion for power (solid fuels)	Plant size: 10–100 MW	2000–3000	60–190
Municipal solid Waste (MSW) incineration	Plant size: 10–100 MW	6500–8500	n/a
Biomass CHP	Plant size: 0.1–1 MW (on-site) 1–50 MW (district)	3300–4300 (on-site) 3100–3700 (district)	n/a
Biogas (including landfill gas) digestion	Plant size: <200 kW–10 MW	2300–3900	n/a
Biomass co-firing	Plant size: 5–100 MW (existing); >100 MW (new plant)	120–1200 + power station costs	20–50
Biomass Integrated Gasifier Combined Cycle (BIGCC)	Plant size: 5–10 MW (demonstration); 30–200 MW (future)	4300–6200 (demonstration) 1200–2500 (future)	n/a
Geothermal Power			
Hydrothermal	Plant size: 1–100 MW; Types: Binary, single and double flash, Natural steam	1700–5700	30–100
Enhanced geothermal system	Plant size: 5–50 MW	5000–15,000	150–300 (projected)
Solar energy			
Solar PV	Power plants: 1–10 MW; Rooftop systems: 1–5 kWp	5000–6500	200–800[4]
Concentrating Solar power (CSP)	Plant size: 50–500 MW (trough), 10–20 MW (tower), 0.01–300 MW (future) (dish)	4000–9000 (trough)	130–230 (trough)[5]

Ocean energy			
Tidal and marine currents	Plant size: Several demonstration Projects up to 300 kW capacity; Some large scale projects under development	7000–10,000	150–200
Heating/Cooling			
Biomass heat (excluding CHP)	Size: 5–50 kW_{th} (residential)/ 1–5 MW_{th} (industrial)	120/kW_{th} (stoves); 380–1000/kW_{th} (furnaces)	10–60
Biomass heat from CHP	Plant size: 0.1–50 MW	1500–2000/kW_{th}	n/a
Solar hot water/heating	Size: 2–5^2 (household); 20–200 m^2 (medium/multifamily); 0.5–2 MW_{th} (large/district heating); Types: evacuated tube, Flat-plate	400–1250/m^2	20–200 (household); 10–150 (medium); 10–80 (large)
Geothermal heating/cooling	Plant capacity: 1–10 MW; types: Ground-source heat pumps, direct use, chillers	250–1450/kW_{th}	5–20
Biofuels (1st. Generation)			
Ethanol	Feedstocks: sugar cane, sugar beets, corn, cassava, sorghum, wheat (and cellulose in future)	0.3–0.3–0.6 billion per billion litres/ year of production capacity for ethanol	0.25–0.3/litre gasoline equivalent (sugar); 0.4–0.5/litre gasoline equivalent (corn)
Biodiesel	Feedstocks: soy, oilseed rape, mustard seed, palm, jatropha, tallow or waste vegetable oils	0.6–0.8 billion per billion litres/ year of production capacity	0.4–0.8/litre diesel equivalent
Rural (off-grid) Energy[6]			
Micro-hydro	Plant capacity: 1–100 kW	1000–2000	70–200
Pico-hydro	Plant capacity: 0.1–1 kW	n/a	200–400
Biomass gasifier	Size: 20–5000 kW	n/a	80–120
Small wind turbine	Turbine size: 3–100 kW	3000-5000	150–250
Household wind turbine	Turbine size: 0.1–3 kW	2000-3500	150–350
Village-scale Mini-grid	System size: 10–1000 kW	n/a	250–1000
Solar home system	System size: 20–100W	n/a	400–600

n/a – Not applicable

1. Using a 10% discount rate. The actual global range may be wider. Wind and solar include grid connection cost.
2. Costs in 2005 or 2006.
3. Wide range. Costs of delivered biomass feedstock vary by country and region due to factors such as variations in terrain, labour costs and crop yields.
4. Typical costs 20–40 US cents/kWh for low latitudes with high solar insolation of 2500 kWh/m^2/ year. 30–50 cents/kWh (typical of southern Europe) and 50–80 cents for higher latitudes.
5. Costs for parabolic trough plants. Costs decrease as plant size increases.
6. No infrastructure required which allows for lower costs per unit installed.

(Source: "*Deploying Renewables: Principles of effective Policies*", 2008, p. 80–83)

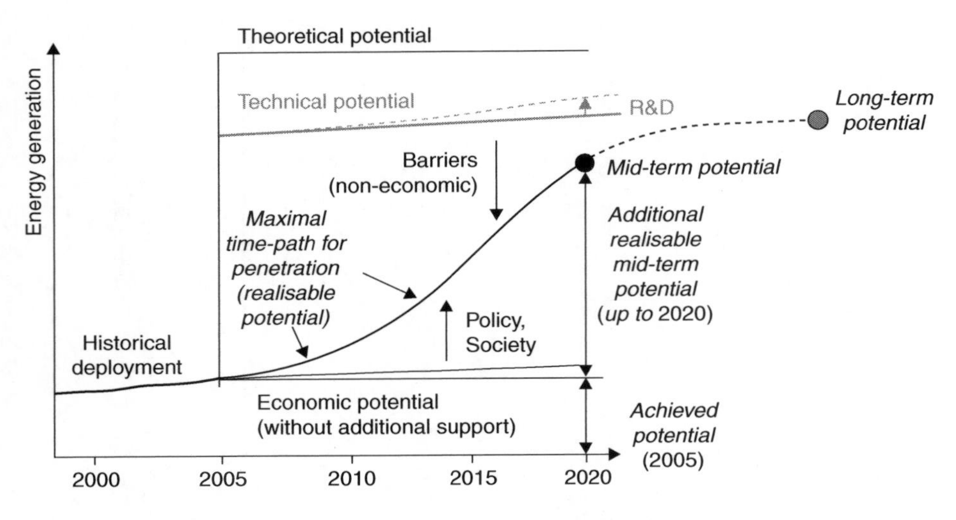

Figure 8.1 Metrics relating to RET potentials
(Source: "*Deploying Renewables: Principles for Effective Policies*", 2008, p. 62, © OECD – IEA)

Realisable potential: This corresponds to maximum achievable potential, assuming that the barriers can be overcome and development proceeds without hindrance. Realisable potential also depends of the market growth rates, and time of the year.

Economic potential: The potential that can be realized without the need for additional support, and which is economically competitive with conventional incumbent technologies.

The total realizable potential is the sum of the achieved potential (cumulative installed capacity) by 2005, plus the additional realizable potential in the period, 2005–2020.

This chapter discusses the realizable mid-term potential (to a time horizon of 2020) for RET options.

Fig. 8.1 shows the relationships among the different metrics of potential (source: "*Deploying Renewables: Principles for Effective Policies*", 2008, p. 62, ©OECD – IEA).

Models are developed for three kinds of situations:

- Country-specific cost resources curves for different RETs,
- Technology learning and associated experience curves
- Country- and technology-specific diffusion S curves

The following procedure is followed:

Static technical potential is calculated for a given RET on the basis of the current state-of-art and costs of that technology. Different categories of technical availabilities are examined in the context of the cost of exploitation which is a function of local geographical context. Where a technology has to take into account a limited resource, costs will rise with increasing utilization. For instance, in the case of wind energy, power

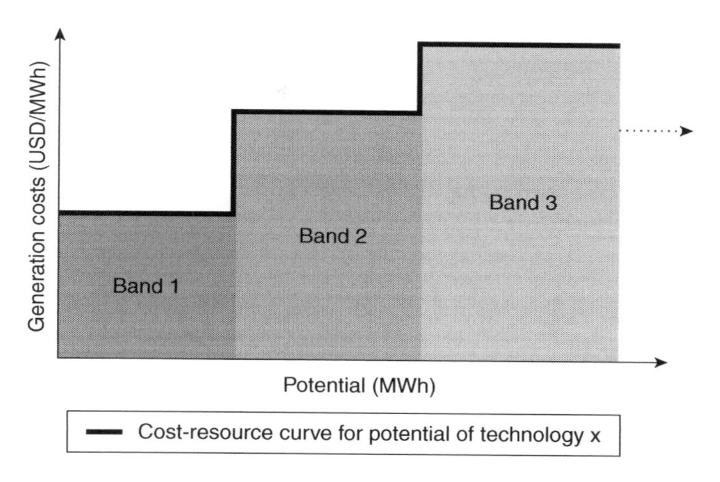

Figure 8.2 Cost-resource curve for the potential of a specific RET
(Source: "*Deploying Renewables: Principles for Effective Policies*", 2008, p. 63, © OECD – IEA)

plants are first located at places which have the highest wind density and largest number of yearly full-load hours, and therefore the lowest costs. After such sites are exhausted, plants have to be established in less optimal sites, which will be characterized by higher costs per kWh. In reality, the cost-resource curve is a continuous function of potential.

In order to simplify the picture, the model uses a stepped discrete function, whereby the technology potential is subdivided into different cost-resource bands.

Fig. 8.2 gives the cost-resource curve for potential of a specific RET (source: "*Deploying Renewables: Principles for Effective Policies*", 2008, p. 63, © OECD – IEA).

A static cost-resource curve does not take into consideration the benefits of technology learning/experience as we go along. Technology learning leads to reduction in costs, thereby making the costs of the later potential band of exploitation lower. An analysis of the economic consequences of technology innovation shows that costs decline by a constant percentage with each doubling of the produced/installed capacity. If the Learning Ratio (LR) is 10% for a given technology, it means that the costs per unit are reduced by 10% for each doubling of cumulative/ installed capacity. According to IEA, the learning rates for wind on-shore, wind offshore and solar photovoltaics have been found to be 7%, 9%, and 18% respectively. This explains as to why the solar PV costs are declining rapidly in China as the volume of their solar PV business increases exponentially.

Another aspect that has to be taken into consideration is the dynamics of a given RET. The market penetration of any technology typically follows the S-curve. Both technical and non-technical constraints have to be applied to the S-curve. It is possible that in some cases the technical constraint, such as, scaling up of component and technology manufacturing capacity, which takes time, may be in operation. Non-technical constraints include market and administrative barriers.

Suppose we wish to project the maximum possible potential of a technology for a particular country, using the S-curve. Let us say that the country has significant long-term

Table 8.2 Overview of alternate indicators of policy effectiveness

Indicator	*Formula*	*Advantage*	*Disadvantage*
Average annual growth rate	$g_n^i = \left(\frac{G_n^i}{G_{n-t}^i}\right)^{\frac{1}{t}} - 1$	Based on empirical values	No consideration of country-specific background
Absolute annual growth	$a_n^i = \frac{G_n^i - G_{n-1}^i}{n}$	Based on empirical values	No consideration of country-specific background
Effectiveness indicator	$E_n^i = \frac{G_n^i - G_{n-1}^i}{ADDPOT_n^i} = \frac{G_n^i - G_{n-1}^i}{POT_{2020}^i - G_{n-1}^i}$	Consideration of country specific background	Difficulties in the identification of additional mid-term potential

a_n^i: Absolute annual growth rate.
g_n^i: Average annual growth rate.
E_n^i: Effectiveness indicator for RES technology *i* for the year *n*.
G_n^i: Electricity generation by RES technology *i* in year *n*.
$ADDPOT_n^i$: Additional generation potential of RES technology *i* in year *n* until 2020.
POT_n^i: Total generation potential of RES technology *i* until 2020.
(Source: "*Deploying Renewables: Principles for Effective Policies*", 2008, p. 88, © OECD – IEA)

wind energy potential. Despite this, it has been found that its starting potential in 2005 has been low. This could mean that technical potential is a constraint, and the exploitation of whole technical potential is going to take time. Consequently, the realizable mid-term potential by 2020 may be lower than the long-term technical potential.

Ultimately, we will have to figure out the mid-term realizable potential for each country's resource.

8.3 MEASURING POLICY EFFECTIVENESS AND EFFICIENCY

The success of a policy of deployment of a given RET is quantified in terms of two parameters: impact on the market growth of the particular RET (*policy effectiveness*), impact on the associated cost of the policy support (*cost efficiency*).

Policy effectives Indicator "is calculated by dividing the additional renewable Energy deployment in a given year by the remaining mid-term assessed "realizable potential" to 2020 in the country concerned" (p. 88, "*Deploying Renewables: Principles of Effective Practice*", 2008). The merit of this indicator is that it allows unbiased comparisons across countries of different sizes, starting points in terms of renewable energy deployment, projected goals of renewable energy policies, and extent of availability of renewable energy resource. The characteristics of an incentive may vary with time, depending on whether they relate to upfront investment costs or operating returns. The remuneration for a given technology in a given country is expressed as a levelised return over a period of 20 years.

The various policy effectiveness indicators are shown in Table 8.2 (p. 88, "*Deploying Renewables: Principles of Effective Practice*", 2008. © OECD-IEA).

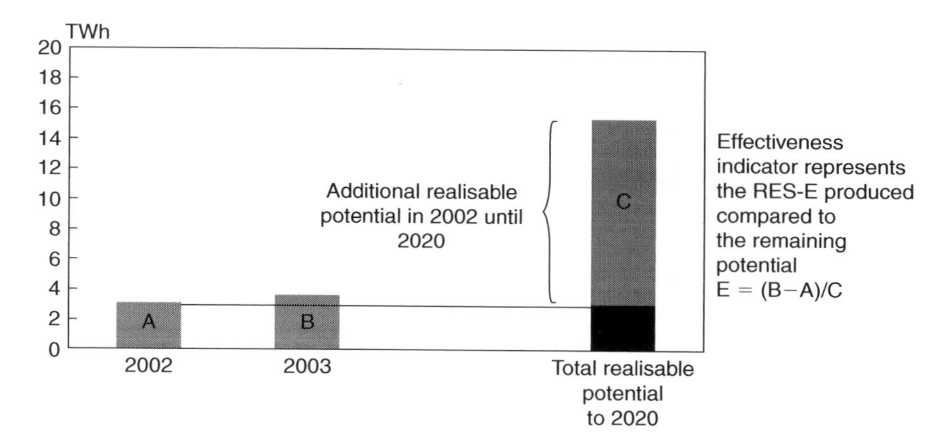

Figure 8.3 Example of the effectiveness indicators of policy effectiveness (Source: "*Deploying Renewables: Principles for Effective Policies*", 2008, p. 89, © OECD – IEA)

As the effectiveness indicator is a measure of the absolute market growth in relation to country and technology-specific opportunities, it permits comparison of support instruments.

Fig. 8.3 (source: "*Deploying Renewables: Principles of Effective Practice*", 2008, p. 89. © OECD-IEA) depicts an example of the effectiveness indicator for a specific RET in a specific country in a specific year. RESE stands for electricity generated from a Renewable Energy Source.

8.4 OVERVIEW OF SUPPORT SCHEMES

Two types of market instruments are used to subsidise electricity from renewable sources: (i) Investment support (capital grants, tax exemptions and reduction on the purchase of goods) and (ii) Operating support (price subsidies, green certificates, tender schemes and tax exemptions or reductions on the production of electricity. Experience shows that in the case of the renewable electricity, operating support, i.e. support per unit of electricity produced, is far more effective than investment support.

Operating support may take the form of fixing a quantity of renewable electricity to be produced or fixing a price to be paid for renewable electricity. Experience has shown that quantity-based instruments and price-based instruments have the same economic efficiency.

Quantity-based market instruments

Quota obligation is the mechanism used for the purpose. Governments set a particular target for renewables which obliges the producers, suppliers and consumers to source a certain percentage of their electricity from renewable energy. Tradable Green Certificates (TGCs) are used to facilitate this transaction. If an obligated party fails to meet its quota obligation, it is penalized. To avoid the penalty, an obligated party will have to make an investment in renewable electricity plants or buy green certificates from other producers or suppliers. The price of the TGC depends not only on the market, but also on the level of the quota target, the size of the penalty and the duration of the obligation. Quota obligation systems with TGCs are generally technology-neutral, in order to promote the most cost-efficient technology options. However, it may some times be necessary to provide technology-specific support, by providing separate quotas (bands) per technology which may be characterized by different duration of support or a different value per MWh.

In the case of **tendering systems**, tenders are called for the provision of certain amount of electricity from a certain technology source. Bidding for such tenders should ensure that the most economical option will be taken up.

Price-based market instruments

When operators of eligible domestic renewable electricity plants feed electricity to the grid, they are provided Feed-in Tariffs (FITs) and Feed-in Premiums (FIPs). FITs and FIPs are technology-specific and are regulated by the governments. FITs correspond to the total price per unit of electricity paid to the producers. FIPs are premiums (bonuses) paid to the producers over and above the electricity market price. Since FIPs have to be earned, they introduce an element of competition.

The tariff structure takes care of the cost to the grid operator. As FITs and FIPs are guaranteed for periods of 10–20 years, they constitute a long-term certainty and thus lower the market risk to investors. FITs and FIPS can be structured in such a way as to promote specific technologies and bring down costs to improve their market competitiveness.

Fiscal incentives

Producers of renewable electricity are provided some tax exemptions (e.g. carbon taxes) in order to enable them to compete in the market place as against conventional energy producers. The applicable tax rate would determine how effective such fiscal incentives would turn out to be. In the case of Nordic OECD countries where the taxes are high, tax exemptions are often adequate to stimulate the use of renewable electricity. In countries where taxes are low, tax exemptions need to be accompanied by other fiscal measures.

Another price-based mechanism is investment grants to reduce capital costs.

8.5 PUBLIC – PRIVATE PARTNERSHIP

Research, Development & Demonstration (RD&D) activities in the private sector tend to focus on near-term, and on applied RD&D. Their aim is to bring a particular product to market. As against this, public sector RD&D tend to be focused on long-term research, involving a large number of partners and often based in academia. This is possible because the public sector RD&D is less concerned with intellectual property rights – Fig. 8.4 Illustration of respective government and public sector RD&D roles in phases of research over time. Source: *Deploying Renewables: Principles of Effective policies, 2008, p. 170,* © OECD – IEA).

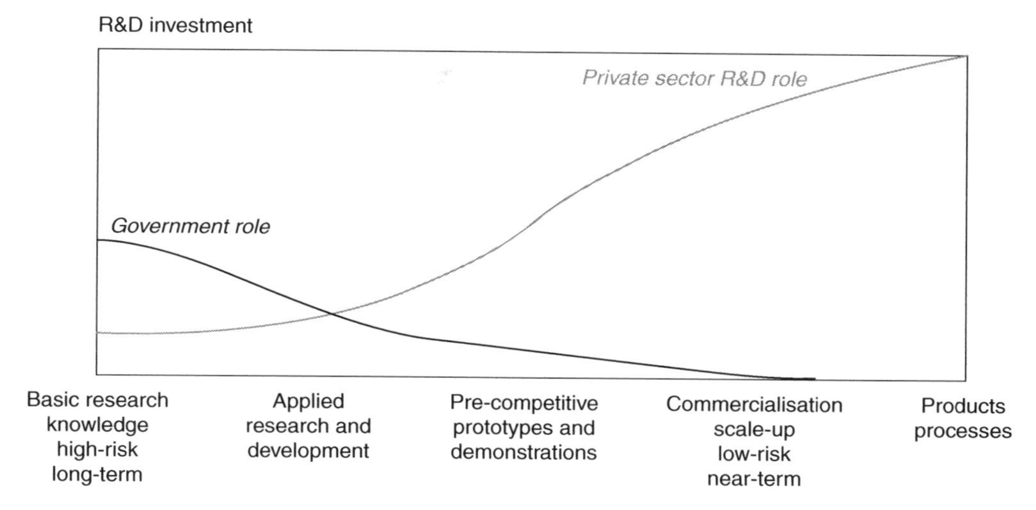

Figure 8.4 Illustration of respective government and public sector RD&D roles in phases of research over time

(Source: "*Deploying Renewables: Principles for Effective Policies*", 2008, p. 170, © OECD – IEA)

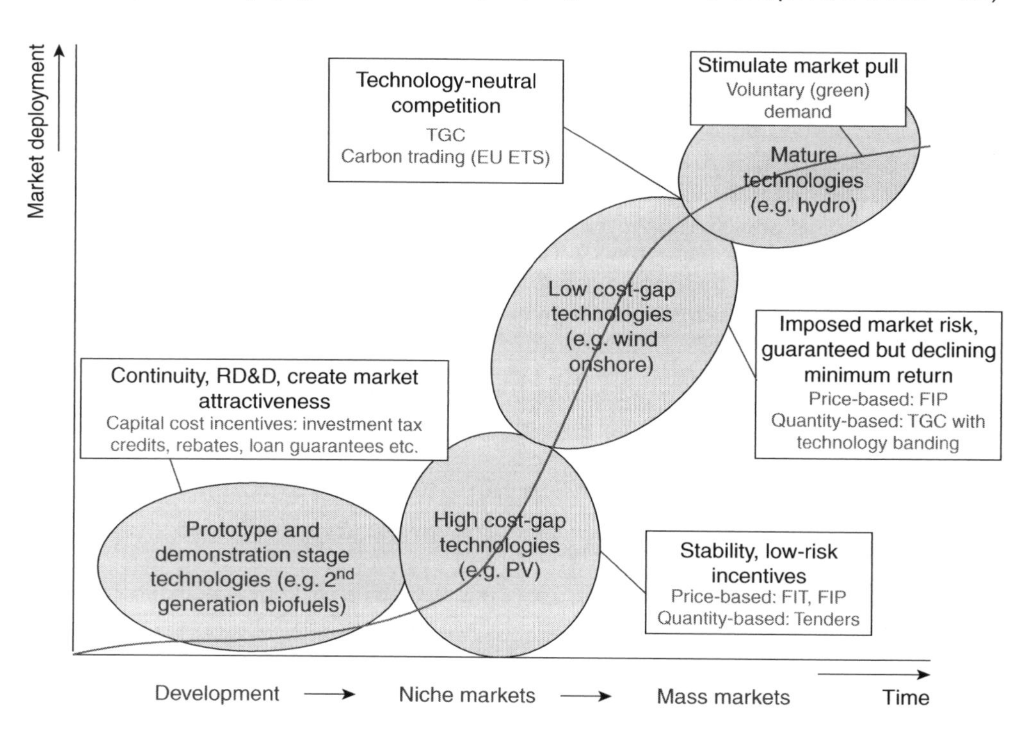

Figure 8.5 Combination of framework of policy incentives in function of technology maturity
(Source: "*Deploying Renewables: Principles for Effective Policies*", 2008, p. 25, © OECD – IEA)

The basic research phase is strongly public sector oriented, as it is unlikely to be interest to the private sector which is focused on the near-term. As technology develops and possibility of a product gets more defined, private sector research will increase and public sector role will decrease as intellectual property rights come to the fore.

8.6 AN INTEGRATED STRATEGY FOR THE DEPLOYMENT OF RETs

A strategy for the deployment of Renewable Energy Technologies (RETS) is schematically shown in Fig. 8.5 (Combination of framework of policy incentives in function of technology maturity, "*Deploying Renewables: Principles for Effective Policies*", 2008, p. 25, © OECD – IEA). The *S*-curve depicts the present position of technologies and incentive schemes. Countries have to decide upon the actual optimal mix of RETS, and timing of the policy incentives, depending upon their biophysical and socioeconomic situations. The level of competitiveness will depend upon the evolving prices of competitive technologies.

The deployment of Renewable Energy Technologies (RETs) has two concurrent goals : (i) exploit the "low-hanging" fruit of abundant RETs which are closest to market competitiveness, and (ii) developing cost-effective carbons for a low-carbon future in the long term. Instances are known whereby non-economic barriers, such as beauracratic red tape, complex administrative procedures, grid access, social acceptance of new technologies, lack of information or training, impeded the progress of RETs even when they are close to economic competitiveness with conventional technologies. High priority should be given for the removal of such impediments.

A policy is a market intervention intended to accomplish some goal that presumably would not be met if the policy did not exist (Paul Komor, 2004). The transition to mass market integration of renewables requires some policy corrections. For instance, the price placed on carbon and other externalities need to be enhanced. It should be realized that most renewables need economic subsidies, and the removal of non-economic barriers which are impeding the deployment of RETs. The policies should be able to lead to a future energy system in which RETs should be able to compete with other energy technologies on a level playing field. When once this is achieved, RETs would need no or few incentives for market penetration , and their deployment would be accelerated by consumer demand and general market forces.

Technology-specific support schemes need to be fashioned depending upon the level of maturity of a given RET at a given time, employing a range of policy instruments, including price-based, quantity-based, R&D support and regulatory mechanisms. Apart from continued R&D support, less mature technologies which have not yet achieved economic competitiveness generally need very stable low-risk incentives, such as capital cost incentives, feed-in-tariffs (FITs) or tenders. In the case of low-cost gap technologies, such as onshore wind and biomass combustion, more market-oriented instruments such as feedin-premiums may be used. Also, TGC (Tradable Green Certificates) systems may be used innovatively, by linking technology differentiation with quota obligation either by awarding technology multiples of TGCs or by introducing technology-specific obligation (known as technology banding).

Technology banding may sometimes be necessary as a transitional phase, or it may be bypassed by the adoption of a technology-neutral TGC system. When once a given RET is competitive with other CO_2-saving alternatives and is in a position to be deployed on a large scale, the support systems for the RET may no longer be necessary now that a level playing field with other energy technologies has been achieved. The position is not static. All RETs are evolving rapidly, in response to technology improvements and market penetration. Renewable Energy policy frameworks should be so structured as to facilitate technological RD&D and market development concurrently, within and across technology families.

8.7 RENEWABLE ENERGY DEVELOPMENT IN CHINA AND INDIA

China and to a lesser extent India, have shown how technology, economics and policy could be integrated to provide clean, reliable, secure and competitive energy supply. China has emerged as the largest maker of wind turbines and the largest manufacturer of solar panels in the world. It is building the most efficient types of coal power plants. China today has 9 GWe of nuclear power. It is preparing to build three times more nuclear power plants in the coming decade as the rest of the world put together. Renewable energy industries have created 1.12 million new jobs in 2008. China's top leadership is intensely focused on energy policy – it created a National Energy Commission which is a kind of "superministry" headed by Prime Minister Wen Jiabao himself. *New York Times* makes a prescient observation that just as the West has been dependent on Middle East for oil, it is destined to be dependent on China for renewable energy technologies.

India has established a new Ministry of New and Renewable Energy. The proposed outlay on Renewable Energy RD&D in India's Eleventh Five-year Plan (2007–2012) in terms of million INR, is: Bioenergy (1 500), Solar energy (3 600), Wind energy (2 000), Small hydropower (500), New Technologies (4 000), Solar Energy Centre (400), Centre for Wind Energy (400), National Institute of Renewable Technology (400), Others (1 820). Total, INR 14 620 million, eq. USD 353 million. A National Solar Mission has been established, with the target of 1 100 MW in the first phase. The activities under this mission include grid-connected solar power plants, high efficiency solar cells, solar PV and thermal power generation, etc. India plans to produce 20 GW of solar power by 2020, thereby creating 100 000 new jobs. All government buildings will be solar-powered by 2012. Microfinancing for solar power will be made available to 20 million households by 2020.

A National Mission on Enhanced Energy Efficiency has been established with an allocation of Rs. 75 000 crores. By 2015, this Mission will save 5% of energy, amounting to 100 Mt of CO_2.

REFERENCES

Aswathanarayana, U. (1985) *Principles of Nuclear Geology*. Rotterdam: A.A. Balkema,
Aswathanarayana, U. (2001) *Water Resources Management and the Environment*. The Netherlands: A.A. Balkema.

Aswathanarayana, U. (Ed.) (2009) *Energy Portfolios*. London: Taylor & Francis Group.
Bjarnason, B. (2007) IGA International Geothermal Association. Presentation in IHA World Congress, Antalya, Turkey, May, 2007.
Boyle, G. (2004) *Renewable Energy*, 2nd. Ed. Milton Keynes: The Open University.
BTM Consult ApS (2007) *International Wind Energy Development: World Market Update 2006*. BTM Consult ApS, Denmark.
DLR (German Aerospace Center) (2006) *Trans CSP – Trans-Mediterranean Interconnection for concentrating solar power*. Institute of Technical Thermodynamics, Stuttgart, Germany.
DLR (2007) Aqua-CSP – *Concentrating solar power for seawater desalination*. Institute of Chemical Thermodynamics, Stuttgart, Germany.
EUPVPLATF (European Photovoltaic Technology Platform) (2007) *Strategic Agenda*. EU PV Technology Platform, Brussels, 2007.
Fargione, J. et al. (2008) Land clearing and the Biofuel Carbon debt. *Science*, v. 319, no. 5867, p. 1235–1238.
Frankl, P., Menichetti, E., and Raugei, M. (2008) *Technical Data, Costs and Life Cycle Inventories of PV applications*. NEEDS (New Energy Technology Externalities Developments for Sustainability) Report prepared for the European Commission (under publication).
International Energy Agency (2008) *Energy Technology Perspectives*. Paris.
International Energy Agency (2008) *Deploying Renewables: Principles of Effective Policy*. Paris.
International Energy Agency (2009) *Technology Roadmap: Wind Energy*. Paris.
Lemming, J.K., Morthorst, P.E., and Clausen, N.E. (2007) *Offshore Wind Power Experiences, Potential and Key Issues of Development*. Report to the International Energy Agency, Ris? National Laboratory, Technical University of Denmark, Roskilde.
Paul Komor (2004) *Renewable Energy Policy*. Lincoln, NE, USA: iUniverse.
PV-TRAC (Photovoltaic Technology Research Advisory Council) (2005) *A Vision for Photovoltaic Technology*. PV-TRAC. European Commission.
Shiklomanov, I.A. (1998) *World Water Resources – A new appraisal and assessment for the 21st. century*. UNESCO, Paris.
Steinfeld, A. (2005) Thermochemical production of hydrogen: A Review. *Solar Energy*, **78**(5), 603–615.
Taylor, R. (2007) *Hydropower Potentials*. International Hydropower Association.
Vörösmarty , C. et al. (1997) The storage and aging of the continental runoff in large reservoir systems of the world, *Ambio*, 26, 210–219.

Section 3

Supply-side energy technologies

T. Harikrishnan (IAEA, Vienna), Editor

Chapter 9

Fossil fuels and CCS

Takashi Ohsumi, Toshima, Japan

9.1 INTRODUCTION

Potentially, there is a wide range of ways to reduce emissions of greenhouse gases. In the case of CO_2, reductions can be achieved by: reducing the demand for energy; altering the way in which it is used and changing the methods of production and delivering energy. Demand for energy can be influenced by a number of means including fiscal measures and changes in human behaviour. However, in the technical area, there are a number of distinct types of options for reducing emissions, as illustrated in Fig. 9.1 which are:

- Improving energy efficiency,
- Switching to low carbon fuels,
- Switching to no-carbon fuels, and
- Flue gas clean-up.

In most cases, the first two options are cost-effective and will deliver useful reductions, but on their own, are unlikely to be enough. Greater reductions could be attained by switching to no-carbon fuels such as renewable and nuclear power; however, the world is presently heavily dependent on the exploitation and use of fossil fuels. For this reason, it is important that there should also be technology options that will allow for the continued use of fossil fuels. However, continued use of fossil fuels needs to be undertaken without substantial emissions of CO_2. In this respect, one route forward would be the development and deployment of technologies for the capture and storage of CO_2 produced by the combustion of fossil fuels.

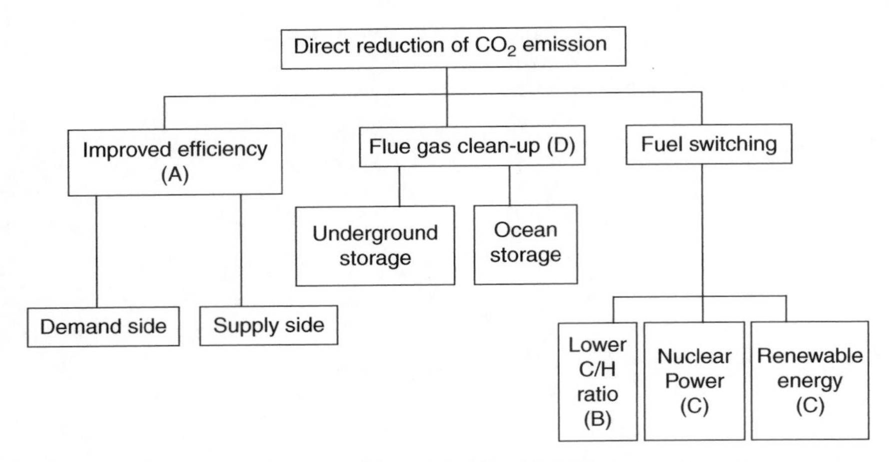

Figure 9.1 Options of measures of direct reduction of CO_2 emissions from the power generation systems

Table 9.1 Annual world consumption or production of fossil fuels in 2008 (BP, 2009)

Oil	*Natural Gas*	*Coal*
3.9 billion tons*	3 020 billion m^3*	6.8 billion tons**

*consumption; **production

It may be noted that an alternative way of reducing atmospheric levels of CO_2 is through the enhancement of natural CO_2 sinks. This could be achieved through enhancing the growth of terrestrial biomass (e. g. forests) or biomass in the oceans.

Power generation is the largest industry sector contributing to the emission of anthropogenic CO_2 in the world. Improving the efficiency of the fossil fuel power generation described in Sub-chapter 9.2 is regarded as a business-as-usual development of the technology. And therefore, its importance cannot be emphasized enough. Sub-chapter 9.3 will cover the fuel switching.

In the particular context of global warming, Marchetti (1977), showed that a straightforward measure in the continued use of fossil fuel is the CCS (Carbon dioxide Capture and Storage), which falls in a typical end-of pipe technology. We should note that the CCS is the ultimate technology of flue gas clean-up. At present, the world energy consumption is based approximately 80% on fossil fuels, and the large-scale infrastructure in the world for the mining and transporting of commercial fossil fuels has been installed to support the industrialized societies. The large-scale transport and storage systems will, therefore, be also necessary for CCS to handle a large amount of CO_2, comparable with the present annual consumption of fossil fuels shown in Table 9.1. CCS technology is treated in Subchapters 9.4 through 9.7.

Two thirds of the power generations are based on fossil fuels as shown in Figure 9.2. Examination on the reserve/production ratio of the fossil fuel resources shown in

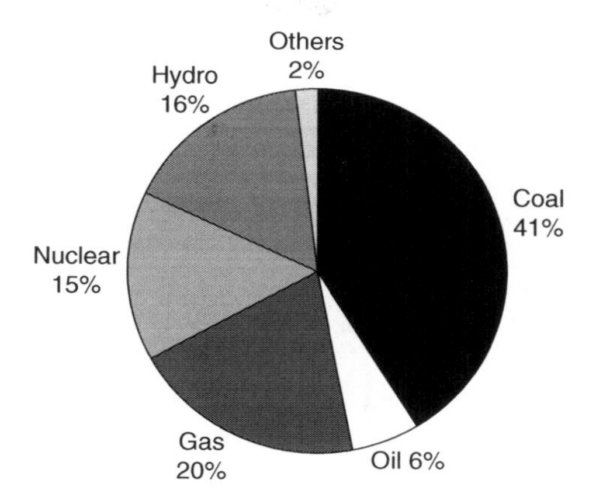

Figure 9.2 World power generation in 2006

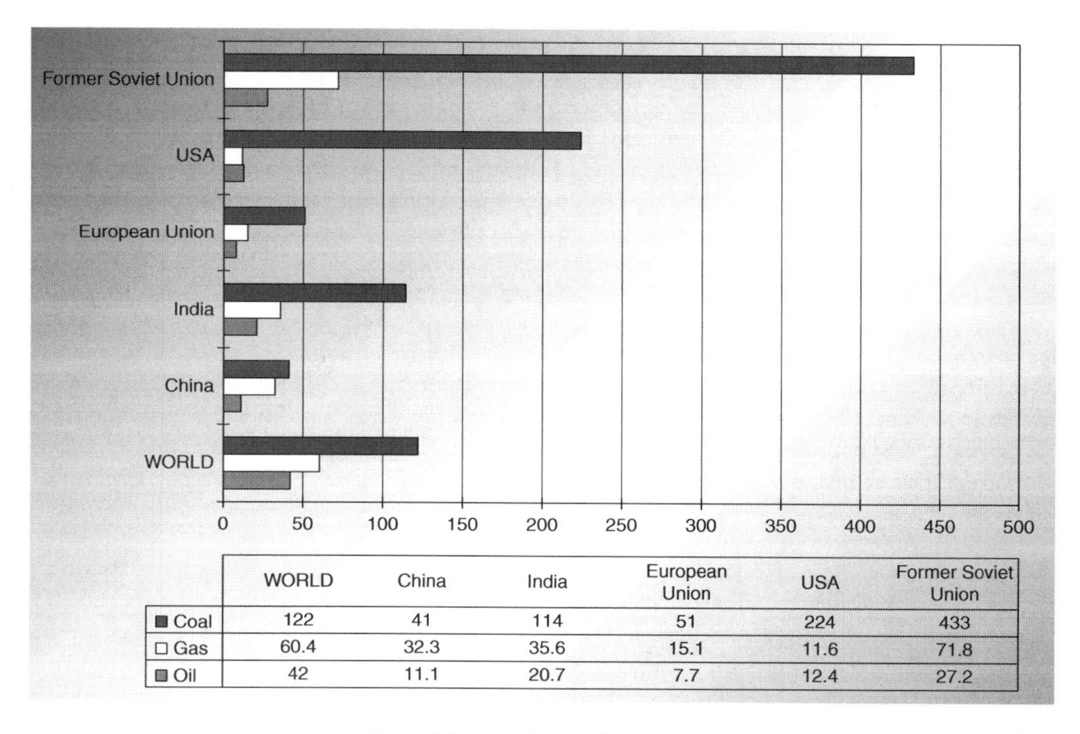

	WORLD	China	India	European Union	USA	Former Soviet Union
■ Coal	122	41	114	51	224	433
□ Gas	60.4	32.3	35.6	15.1	11.6	71.8
▩ Oil	42	11.1	20.7	7.7	12.4	27.2

Figure 9.3 R/P of fossil fuels (years)

Figure 9.3 tells us that the fuel switching and increased efficiency in the power generation are, over the long run, not sufficient and that the energy supply system in the world cannot reduce the CO_2 burden on the atmosphere without CCS technology.

9.2 EFFICIENCY IMPROVEMENT OF POWER GENERATION

The efficiency of hard coal-fired power plants averaged about 35% from 1992 to 2005 globally. The best available coal-fired plants can achieve 47%. The efficiency of brown coal-fired plants increased from 33% in 1992 to 35% in 2005.

The current efficiency of most coal-fired power plants is well below the levels that are already possible – and there is much potential for significant efficiency improvements in state-of-art technologies. Efficiency gains that can be realized by improving existing plant depends on the efficiency and age of the stock. The efficiency of power plants also depends on the quality of their fuel (especially in the case of coal), their environmental standards and their operation mode. All else being equal, power plants using high-ash, high-moisture coal (such as those used in India) have a lower efficiency than plants using low-ash, low-moisture coal. The cleaning of flue gases requires energy and, therefore, reduces power plant efficiency. Running plants below their rated output, a common practice in market-driven electricity supply systems, also substantially reduces plant efficiency.

Pulverized coal combustion (PCC) accounts for about 97% of the world's coal-fired capacity. Improving the efficiency of PCC plants has been the focus of considerable efforts by the industry as it seeks to stay competitive and to become more environmentally acceptable. PCC *subcritical* steam power plants, with steam pressure around 180 bar, temperature of 540°C and combustor-unit sizes up to 1 000 MW, are commercially available and in use worldwide. The average net efficiency (after in-plant power consumption) of larger subcritical plants burning higher quality coal is between 35% and 36%. New subcritical units with conventional environmental controls operate close 39% efficiency. The overall efficiency of older, smaller PCC plants that burn low quality coal, can be below 30%.

Supercritical steam-cycle plants with steam pressures of around 240 bar to 260 bar and temperatures of around 570°C have become the system of choice for new commercial coal-fired plants in many countries. Early *supercritical* units developed in Europe and United States in the 1970s lacked operational flexibility and reliability and experienced maintenance problems. These difficulties have been overcome. In Europe and Japan, plants with *supercritical* steam operate reliably and economically at net thermal efficiencies in the range of 42% to 45%, and even higher in some favourable locations. *Ultra-supercritical* plants are supercritical pressure units with steam temperatures of approximately 580°C and above.

Integrated coal gasification combined-cycle (IGCC) plants are a fundamentally different coal technology, and are now expected to become commercially available. A small number of plants that were initially built with public funding as demonstrators are currently operating, with the best one achieving 42% electric efficiency. Future coal-fired steam units and IGCC plants are expected to achieve efficiencies above 50% in demonstrator projects within ten years. The projection of the thermal efficiency of the new power plant in Japan is shown as an example in Figure 9.4.

Brown coal (lignite) is expected to increase its contribution to coal supply in some countries. It has a higher water content than hard coal, a lower heating value, and different boiler requirements. The optimal technology choice for hard coal and lignite may differ, as the availability and price of different coal types affects the power generation technology choice.

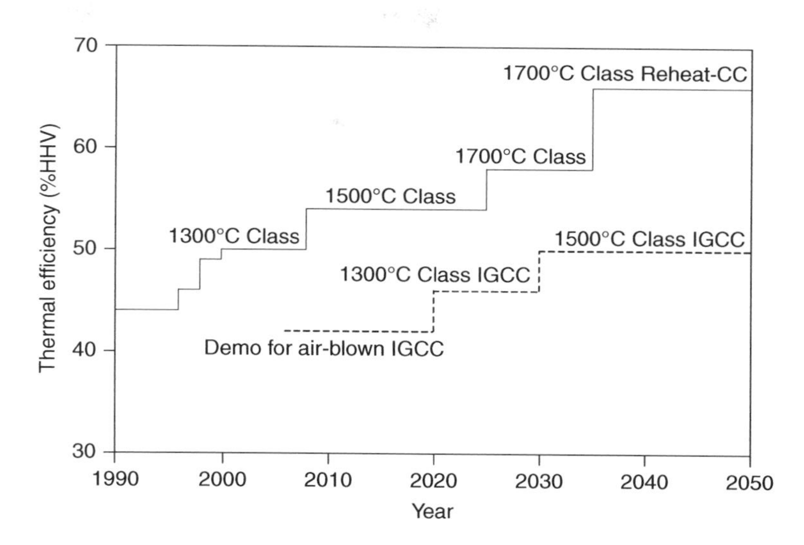

Figure 9.4 A future expectation of new power plants in Japan (modified from Moritsuka *et al.*, 2009); Solid line: natural gas combined-cycle, Dotted line: coal gasification combined-cycle

There is considerable scope to increase the efficiency of natural gas-fired generation, primarily by replacing gas-fired steam cycles with more efficient combined cycle plants, as depicted also in Figure 9.4. Because open-cycle plants are used as peaking plants, their annual use is low – which makes their low efficiency more acceptable from a cost perspective. A natural gas combined-cycle (NGCC) plant consists of a gas turbine and a steam cycle. A gas-fired steam cycle has efficiency similar to that of a coal-fired plant.

The average efficiency of natural gas-fired power plants increased 35% in 1992 to 42% in 2005. Most of the improvement in efficiency was a result of the introduction of large combined-cycle units, which now account for 38% of global gas-fired capacity.

In 2005, the average efficiency of natural gas-fired power plants ranged from about 33% in Russia to 49% in Western Europe. Average efficiencies in Europe have increased since 1990 with the introduction of natural gas, combined-cycle units. Since early 1990s, NGCC has been the preferred technology for new gas-fired generation plants. Efficiencies of the best available combined-cycle plants are 60%. Natural gas plant efficiency, however, falls considerably when plants are run at widely varying loads.

9.3 FUEL SWITCHING IN FOSSIL FUEL POWER PLANTS

Depending on the fuel type and application, the utilization of carbonaceous fuels causes direct and indirect emissions of one or more of the following: SO_x, NO_x, particulate matter, trace metal and elements, volatile organic carbon and greenhouse gases (e.g. CO_2, CH_4, N_2O). Direct emissions are usually confined to the point of combustion of the fuel. Indirect emissions include those that arise from upstream recovery, processing and distribution of the fuel. Life cycle analysis (LCA) can used to account for

Table 9.2 Direct CO_2 emission factors for some examples of carbonaceous fuels

Carbonaceous Fuel	*Heat Content (HHV)* *MJ kg^{-1}*	*Emission Factor* *gCO_2 MJ^{-1}*
Coal		
Anthracite	26.2	96.8
Bituminous	27.8	87.3
Sub-bituminous	19.9	90.3
Lignite	14.9	91.6
Biofuel		
Wood(dry)	20.0	78.4
Natural Gas	kJ m^{-3}	
	37.3	50
Petroleum Fuel	MJ m^{-3}	
Distillate Fuel Oil	38 650	68.6
Residual Fuel Oil	41 716	73.9
Kerosene	37 622	67.8
LPG(average for fuel use)	25 220	59.1
Motor Gasoline	–	69.3

From Table AI.13 in IPCC(2005)

all emissions (direct as well as indirect) arising from the recovery, processing, distribution and end-use of a fuel. Table 9.2 gives an idea of some direct emission of CO_2 anticipated, as an example.

Fossil fuels are made up of molecules of different sizes, with different proportions of a carbon and hydrogen. These differences determine the physical form of the fuel. At one extreme, natural gas is largely made up of methane, a small molecule consisting of one carbon and four hydrogen atoms. Methane is a gas at room temperature.

At the other extreme, coal consists of relatively large numbers of carbon atoms in the molecule, with roughly 11 hydrogen for every 10 carbon atoms. Coal is a complex solid which also contains water and other impurities, including sulphur and mineral matter, so it is not open to such simple description as with gaseous or liquid fuels. Typical US coals contain 70% and 80% carbon and 5% hydrogen by weight.

In the middle, crude oil contains a variety of compounds, some of which are gases but most are liquid, e.g. petrol or kerosene. The gases are small molecules like methane and pentane with a high proportion of hydrogen. The liquids are larger molecules with typically between seven and eleven carbon atoms per molecule and roughly two atoms of hydrogen for each atom of carbon. At room temperature some of these compounds (e.g. paraffin wax) are solid, a tendency which is emphasized by the increased number of carbon atoms in the molecule.

Any fuel has to be handled, moved and stored. Solid and liquid are easier to handle than gas but, for a given weight, the amount of energy available from a gas or liquid fuel is around twice as much as from coal; this factor influences how much work needs to be done to move the fuel. On the other hand, the amount of space occupied by a fuel is important with regard to storage, especially in a restricted space such as vehicle. In this respect, liquid fuels require one thousand times less space than gaseous fuels at room temperature and atmospheric pressure.

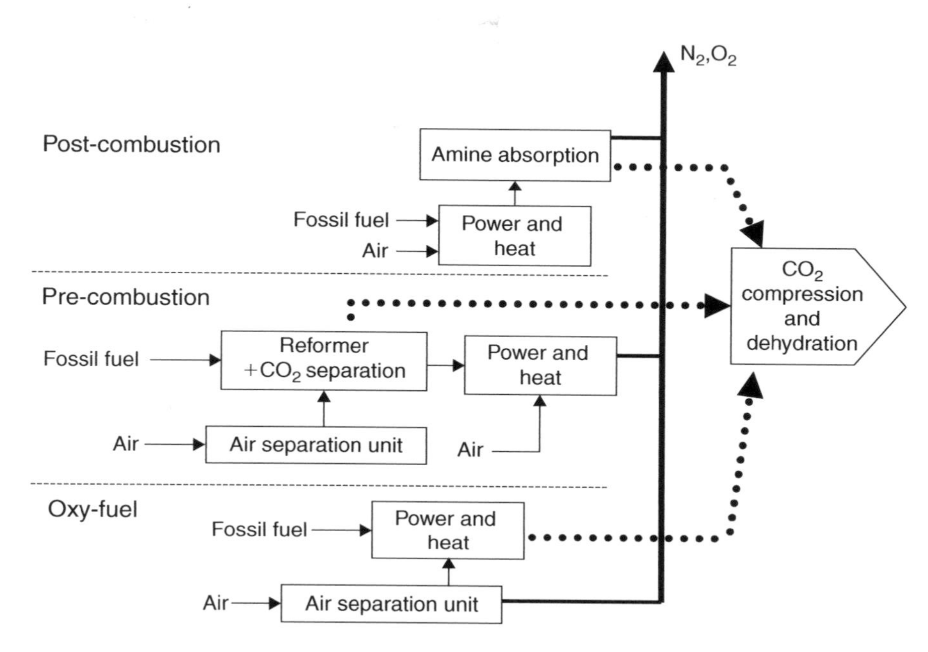

Figure 9.5 Diagram of the three methods of capturing CO_2 from power generation

In fuel switching among fossil fuels for the purpose pf making "green energy", e.g. from coal to natural gas, the above factors should be taken into consideration.

9.4 CAPTURE OF CO_2

There are three basic systems for capturing CO_2 from use of fossil fuels in power generation:

- Post-combustion capture;
- Pre-combustion capture;
- Oxy-fuel combustion capture.

These systems are shown in simplified form in Figure 9.5. In this chapter, only the post-combustion capture is treated, because it is very near to the commercialization. Chapter 3 of the IPCC Special Report on Carbon Dioxide Capture and Storage (2005) gives a good overview on all aspects of the capture technologies.

The use of post-combustion capture of CO_2 followed by geological storage holds the promise of significant CO_2-emission reductions from existing power plants. The main advantage of the technology is that it is the only practical technology able to reduce emissions from existing power plants and new ones. As it is an end-of-pipe technology fitted to a power plant, it is also the CO_2 capture technology, which is the easiest to implement. It produces a very pure CO_2 stream and is, in several ways, also a flexible technology with great adaptability. It allows for staged implementation, e.g. via the installation of separate modules, initially only treating a portion of the power plant flue gas. This would enable rapid uptake of technology improvement, such as new absorbents with better performance. During operation, it is also possible to vary the rate of CO_2-removal or to even switch off the capture process. Such flexibility is a desirable feature, which enables the power plant output to be varied in accordance with the fluctuations in demand and to match market conditions. This feature is technically and economically attractive, in liberalised electricity markets and this is not possible with other capture routes.

It is often said that the post-combustion capture has not yet demonstrated at full scale in a power plant. However, it is not the fact: in early 1980s a post-combustion amine scrubbing plant with a capacity of 3000 tons per day was installed in the United States to a thermal power plant for the purpose of selling its CO_2 to the enhanced oil recovery (EOR) business. The challenge exists that the capture process has not fully integrated and optimised into a power plant system. The application of commercially available post-combustion capture technologies, like all other capture routes, also leads to large increases in the power generation costs and decreases the power generation efficiency.

Although several different processes are currently under development for the separation of CO_2 from flue gases, absorption processes using aqueous solutions of chemical absorbents is the technology closest to commercialization at power plant scale, and likely to be technology used to make initial cut in global greenhouse gas emissions. Typical flow sheet of CO_2 recovery using aqueous solutions of chemical absorbent is shown in Figure 9.6. A blower is required to pump the gas through the absorber. At temperatures typically between 40 to 60°C, CO_2 is bound by the chemical absorbent in the absorber. After the absorption stage the flue gas passes through a water wash section to balance water in the system and to remove any solvent droplets or vapour carried over. It then leaves the absorber and is vented to the atmosphere. The "rich" absorbent solution, which contains the chemically bound CO_2 is then pumped to the top of a stripper, via a heat exchanger. The regeneration of the chemical absorbent is generally carried out in the stripper at elevated temperatures (100–140°C) and pressures between 1 and 4 bar. For generation to take place, heat is supplied to the reboiler, most likely in the form of steam from the low pressure steam turbine of the power plant. The thermal energy required for regeneration is significant and can be split into three main components: heating up the solution, providing the required energy for removing the chemically bound CO_2 from the solvent and for steam production with the stripper which act as a stripping gas. Steam is recovered in the condenser and the condensate fed back to the stripper, whereas the CO_2 product gas leaves the condenser. The CO_2 gas stream thus obtained is a relatively pure (>99%) product, with water vapour being the main other component. Due to the selective nature of the chemical absorption process, the concentration of inert gases is low. The "lean" absorbent solution, containing a low level of CO_2 is then pumped back to the absorber via the lean-rich heat exchanger and a cooler to bring it down to the absorber temperature level. CO_2 removal is typically between 85–95% of the flue gas composition.

Although there are already several commercial technologies for the separation of CO_2, flue gas scrubbing with amines projects itself as the most promising one when it comes to a large scale application, like a 500 MW power plant or bigger. A reason for this is the large experience behind this process thanks to its former, and current use in the oil industry for EOR. Nevertheless, it has long been since substantial progress was made with respect to the development of new solvents. The main problem is the

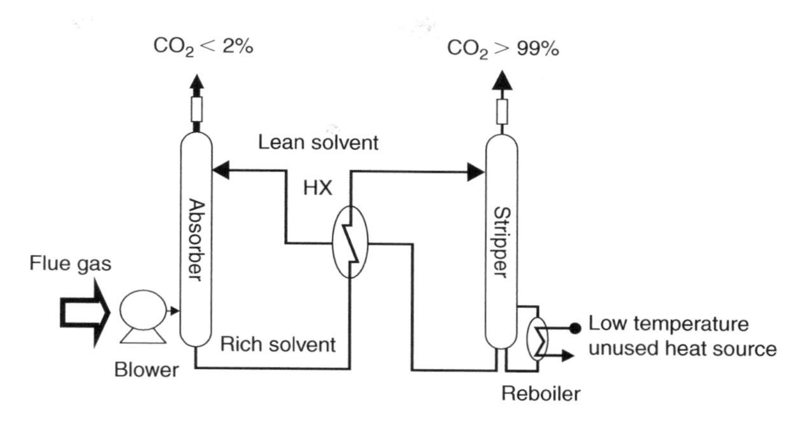

Figure 9.6 A scrubber system for capturing CO_2 in flue gas

amount of energy they require for their regeneration, which is the major power plant's efficiency drawback. Another hitch for this technology is the fact that the biggest existing commercial plants are designed to capture a maximum amount of 1000 ton/d in the case of EOR, 800 ton/day for chemical industry and 300 ton/day in the food industry. This number is too small for a power plant, in which case approximately 8 000 ton/day of CO_2 would have to be processed for a capture rate of 90%. While a lot of the research concerning the CO_2 capture has already been done, there are still some obstacles to overcome, before "near zero emissions' power plants" can actually be built. These include particularly integration of the capture with the power plant. Furthermore, once the CO_2 has been successfully separated from flue gases, it still has to be conditioned for its transport to the location of the storage field. This process also takes place at the power plant itself, resulting in another efficiency penalty.

9.5 COMPRESSION OF CO_2

In addition to scrubbing process and its integration with the power plant, the compression of the captured CO_2 should also a part of design of power plant. Not only does the power plant have to supply the energy for the capture process, but it also has to supply the energy for the compression of the CO_2 for transport purposes or liquefaction purpose. It is therefore important to have an idea of how this process is going to affect the power plant in terms of efficiency penalties due to the electricity needed by the compressor(s) and the amount of cooling water or coolant solvent required for this.

It is important to analyze how many stages would be required to reach a pressure of e.g. 100 bar and a temperature of e.g. 40°C. Under these conditions the CO_2 is at supercritical state. For transport purposes the pressure must be further increased until up to e.g. 200 bar depending on the pipeline length and the conditions of point that represents the CO_2 coming from the top of the stripper column of the capture plant. Several stages of compression would be necessary. But these are the specific design problems easily handled with a conventional engineering practice.

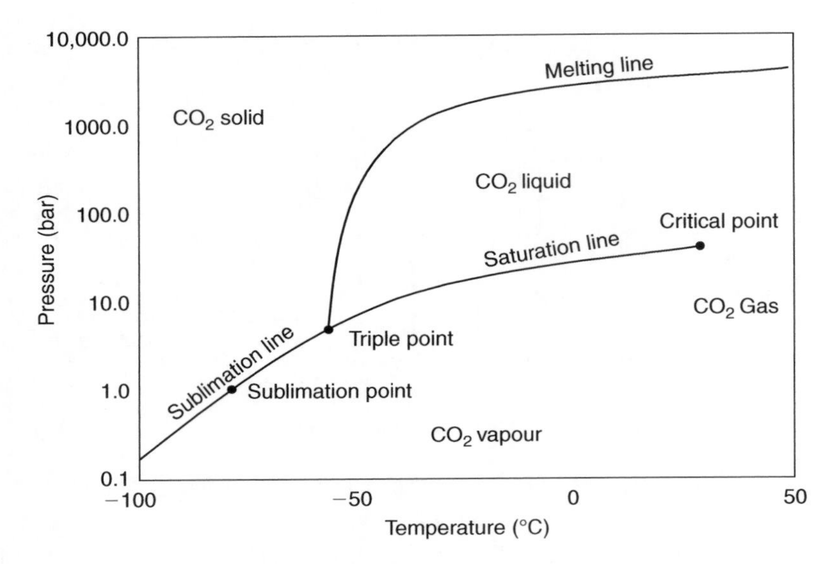

Figure 9.7 Phase diagram of CO_2

9.6 TRANSPORT OF CO_2 in CCS

As is the same in the case of the other waste disposal management, such as radioactive waste disposal, the engineering problems of transport to the disposal site, it is generally agreed, can be managed with presently available techniques. Still, the transport of CO_2 is the key for CCS deployment. Transport can be conducted in any form in the conditions of either near ambient temperature or ambient pressure. Phase diagram of CO_2 (Figure 9.7) tells us that under one atmospheric pressure we can choose the form of solid CO_2 (−78.5°C). If we use the pressure vessel with ambient temperature, liquid CO_2 phase is the choice (along the saturation line). The pipeline transport can be designed for supercritical state CO_2 above the critical point (31.1°C, 73.9 bar).

Taking these into account, Table 9.3 depicts the state of arts for various options of transport.

CO_2 transport by ship is an established technology. Shipping is less economical compared with properly-scaled pipeline, but is flexible to the change of plan (route, flow rate, growth of project, *etc*). It needs loading/unloading operations and is affected by sea state, but less sensitive in cost to water depth, sea bottom topography and transport distance.

Because Chapter 4 of the IPCC Special Report on Carbon Dioxide Capture and Storage (2005) has given the detailed analysis on the pipeline transport technology, this chapter puts an emphasis on the ship transport with a combined concept of the shuttle ship and socket buoy. Design parameters of ship such as cargo capacity, cargo conditions in pressure and temperature, speed, number of ships, *etc.* should be decided considering various trade-off relations anyway. However, we take note on that the state-of-art dimension of one capture unit is a million ton-CO_2 per year scale, i.e., the

Table 9.3 Summary of transport options of CO_2

State of CO_2	*Mean of transport*	*Remark*
Solid	Truck Train	Despite of its high energy penalty, a German scientist Seifritz discussed on the option in early days of CCS research, because he had concerns on the public perception issue of the pipeline laying in the densely populated area. For the purpose of experimental storage, solid CO_2 was once used by Nakashiki *et.al.* (1991)
Liquid	Tanker lorry Tanker ship	For the ocean storage options, ship transport was examined in detail. A large-scale pressure vessels containing the liquid CO_2 near the triple point can be installed on a carrier ship, which then transport the CO_2 up to 100 000 tons at a time.
Supercritical	Pipeline	There are over 5,800 km of pipelines dedicated to CO_2 transport in the USA, mainly for use in enhanced oil recovery (EOR) projects. For the fixed points of capture and storage sites, pipeline transport is considered to be the most efficient and economical option. Undersea pipelines are also being examined so far. Economies of scale and creation of a centralized pipeline network could lower costs marginally, but major cost reductions are unlikely.

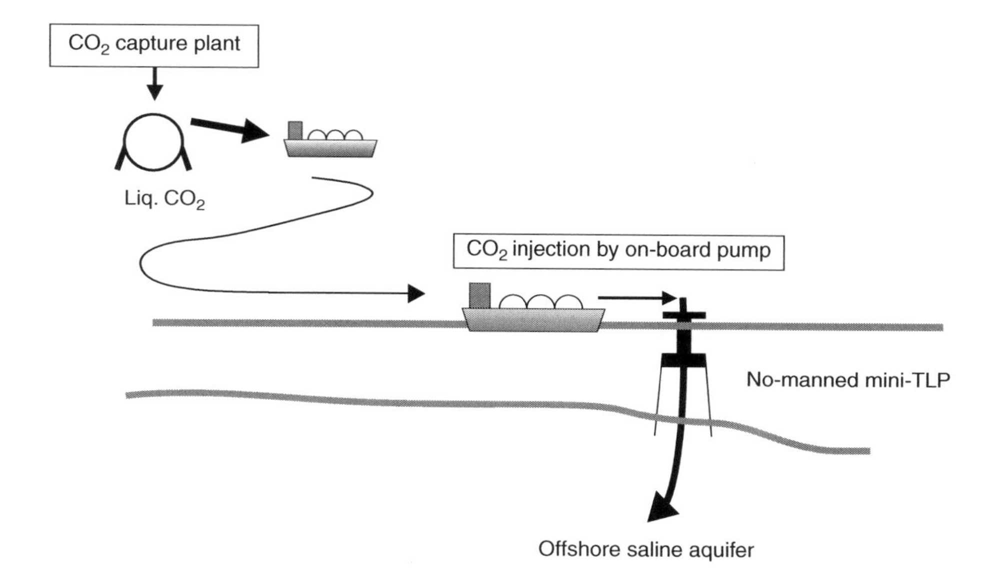

Figure 9.8 An example of shuttle tanker operation from a capture plant on-shore to an injection well offshore

average flow rate is about 3 000 ton per day. And the shipping will not take much time because the domestic transport distance is not too long in general. Hence we assumed that frequent shuttling with 3 000 ton sized ships is a practical option as schematically illustrated in Figure 9.8.

Table 9.4 An estimate of potential storage capacity in Japan

Categories	*Definitions*	*Capacity/Gton CO_2*
A1	Oil and Gas Reservoirs	3.5
A2	Drilled Aquifers with Structural Traps	5.2
A3	Undrilled Aquifers with Structural Traps	21.4
B1	Onshore dissolved-type Gas Fields	27.5
B2	Offshore Open Aquifers	88.5
Total		146.1

Modified from *Takahashi et al.* (2009)

The socket buoy proposed here could be 'mini-Tension Leg Platform' type, which is vertically taut moored with several number of 'tendons' (mooring elements) to be restricted its motion in winds and waves, and maintains the CO_2 pipe(s) steadily between the buoy at sea surface level and the sea bottom wellhead of the injection well.

9.7 STORAGE OF CO_2

The geological storage sites are identified in the sedimentary basins where suitable saline formations, oil or gas fields, or coal beds may be found (Bradshaw and Dance, 2004). In a macroscopic view of the plate tectonics theory, because Japan Archipelago, as an example, is located on a complex boundary area overlying the major four plates (i.e., Philippine Plate, Pacific Plate, North American Plate and Eurasian Plate), one might expect that it provides a poor opportunity for identification of storage sites. However, we were able to find out the suitable storage sites, if we seek regionally into offshore marine areas as identified and compiled by Takahashi *et al.* (2009).

CO_2 storage in offshore geologic formations is not ocean storage proposed by Marchetti (1977). The CO_2 injected into the geological formation under the sub-seabed is stored deep beneath the marine floor, avoiding the hazard of direct ocean injection, including effects on ocean ecology. Furthermore, the marine sedimentary basins offer enormous storage potential. For example, in the potential estimate in Japan shown in Table 9.4, the capacity potential in the marine sedimentary basins are in offshore and occupies the 60% of the estimated storage potential.

In many marine settings, the upper parts of geological layers, if it is dominated by clay, is unconsolidated, which means that faults and fractures do not persist as high-permeability pathways for CO_2 escape.

The advantage of offshore carbon dioxide storage includes the potential to manage pressure within the geologic formation. Carbon dioxide storage is different form EOR in that EOR involves both the injection of CO_2 and the extraction of fluid – usually a mixture of water, CO_2 and oil (the CO_2 is usually reinjected). Injection of CO_2 into saline aquifers on land, however, without extraction of the saline water, increases pore pressures and changes the way CO_2 migrates in the subsurface. If the permeability of the reservoir is extremely high, then management of pressure is not a problem, because pressure transients near the injection well are rapidly dispersed. As the scale of CCS increases, with as much as 100 million tons per year injected in a single formation, management of subsurface pressure will become a greater and greater challenge. Indeed, displacement of saline water and pressure management may prove the greatest overall challenge for CO_2 storage.

An approach of large scale storage of CO_2 in the Utsira formation into which CO_2 from Sleipner project is now being injected is proposed (Lindeberg et al., 2009): injecting CO_2 in several wells evenly distributed in the whole formation. This was tested with a set of scenarios with 210 injection wells and another set with 70 injection wells distributed in a rectangular pattern with a second grid of water producers placed in between, corresponding to the 5-spot pattern known from onshore oil production. The distance between injection wells and producers was 7 km in the first and 13 km in the second scenario. The location of the wells was strictly geometrical and no measures were taken to optimise their location with respect to the topography of the cap rock. In the simulation tests CO_2 was injected in the bottom of the formation (similar to the actual Sleipner case) and also the water producers were perforated the bottom to reduce the risk of CO_2 break through. The injection period was selected to 300 years corresponding to a time horizon well within an expected era for fossil fuels. The total injection rate was 0.15 Gt/year and the resulted typical distribution of CO_2 might be acceptable. It was shown that an emission permit for 160 Mt per year Utsira formation water (3% brine) would be required. Millions tonnes of water containing up to 40 ppm oil is today being disposed into the North Sea from oil production. A permit for emission for the relatively pure brine could therefore be expected. Thus, the key challenge in this proposal is to control the pressure and at the same time minimising co-production of CO_2.

Beyond the technical advantages, there are numerous social, political, and economic reasons why offshore storage of CO_2 is likely to be important during early deployment of CCS, at least in densely populated areas. In some developed countries, locating storage sites near populated areas where most CO_2 is created may be practically impossible.

In general, working in the offshore environment is more expensive; drilling rigs, seismic surveys, and well manifolds are all much more expensive than for a comparable situation on land. The overall economics of CCS, however, make offshore storage very attractive. CCS costs are dominated by the cost of capture and compression. If it is easier to get permits and finance for offshore sites, thanks to 13 years of demonstration at Sleipner, then the extra cost for characterization is more than justified.

With all the advantages, including enormous capacity and the actively managed pressure, CO_2 injected deep beneath the sub-seabed is probably the best option for large population centers near the coast. It may take a long time before people are comfortable storing vast quantities of CO_2 near to where they live, even if the best science suggests that it is perfectly safe.

Chapter 10

Nuclear power

T. Harikrishnan (IAEA, Vienna)

10.1 INTRODUCTION

As nuclear power and green energies, such as wind, solar and tidal emit zero or very negligible carbon, they can play potentially important role in curbing emissions, if they can replace fossil fuels in future generations. In fact nuclear power has been included with the renewable energy power sources for some time now.

This also has been the argument put forward to support the recent nuclear renaissance seen in China and India as well as over 50 nuclear "new comer" states (states actively thinking to introduce nuclear power for the first time). Many established nuclear states where nuclear power has been put on hold or being rolled back are also rethinking to increase its share in electricity generation.

A spin-of of nuclear power generation or otherwise the nuclear technologies provide for vital benefits towards the improvement of human well-being. Nuclear applications help support and improve human health, food and agriculture, and development and management of natural resources.

Nuclear power presently contributes 15% of world electricity requirements. At present 436 nuclear reactors provide 370 GWe to thirty counties (Table 10.1). Further 56 reactors are under construction with 5 GWe installed capacity (IAEA, 2010). This displaces about 3 giga tonnes of CO_2/year, compared to similar scale of generation by burning coal (OECD/NEA, 2009).

10.1.1 Future projections

Criticisms and public antipathy in many developed countries had prevented accelerated growth of nuclear power after 1980s. While public opinion was shaped by two

Table 10.1 World nuclear power production

	Operational		*Under Construction*	
Country	*No. of Units*	*Total MW(e)*	*No. of Units*	*Total MW(e)*
Argentina	2	935	1	692
Armenia	1	376		
Belgium	7	5 863		
Brazil	2	1 884		
Bulgaria	2	1 906	2	1 906
Canada	18	12 573		
China	11	8 438	21	20 920
Czech Republic	6	3 678		
Finland	4	2 696	1	1 600
France	58	63 130	1	1 600
Germany	17	20 480		
Hungary	4	1 889		
Iran			1	915
India	18	3 984	5	2 708
Japan	54	46 823	1	1 325
Korea, Republic Of	20	17 705	6	6 520
Mexico	2	1 300		
Netherlands	1	482		
Pakistan	2	425	1	300
Romania	2	1 300		
Russian Federation	31	21 743	9	6 894
Slovak Republic	4	1 762	2	810
Slovenia	1	666		
South Africa	2	1 800		
Spain	8	7 450		
Sweden	10	8 992		
Switzerland	5	3 238		
Ukraine	15	13 107	2	1 900
United Kingdom	19	10 137		
United States Of America	104	100 683	1	1 165
Total:	436	370 394	56	51 855

(Includes long term shutdown in Canada (4 units; 2 530 Mwe) and Japan (1 unit; 246 MWe; Total includes Taiwan, China which has 6 (4 949 MW(e)) operating reactors and 2 (2 600 MW(e)) reactors under construction.) (based on IAEA – Power Reactor Information System)

major accidents involving nuclear reactors, Three Mile Island in USA and Chernobyl in erstwhile USSR (presently Ukraine), economic pragmatism due to very low oil prices prevailed over energy planners. Many countries took decisions to roll back nuclear power and replace it other forms of energy in the late 80s and 90s (Cohen, 1990).

That the nuclear aversion was really short-sighted dawned up on the energy planners during the last decade when three factors became apparent. The foremost was the fact that fossil fuel such as oil and coal are being exhausted faster than it was ever imagined. Their prices are no more very low nor their supply assured.

Secondly the reality of global warming and the fact that the planet has only very limited capacity of accommodate more carbon was established by scientific studies.

Thirdly emerging economies of China and India amongst few others are breaking out into a phase high economic growth, which needs vast amounts of added energy supply.

The International Atomic Energy Agency (IAEA) projects that the global nuclear power capacity will reach between 473 GWe (low projection) to 748 GWe (high projection) in 2030. The International Energy Agency (IEA) has a reference projection of 433 GWe in 2030 (IAEA, 2009a).

The IEA has published two climate-policy scenarios. The '550 policy scenario', which corresponds to long-term stabilization of the atmospheric greenhouse gas concentration at 550 parts per million of CO_2, equates to an increase in global temperature of approximately 3°C. The '450' policy scenario equates to a rise of around 2°C. In the 550 policy scenario, installed nuclear capacity in 2030 is 533 GWe. In the 450 policy scenario the nuclear share is 680 GWe.

The OECD Nuclear Energy Agency has projected 404–625 GWe in 2030 and 580–1400 GWe in 2050. The US Energy Information Administration has a reference projection of 498 GWe of nuclear power in 2030.

All the above projections tend to be generally revised upward in the present scenario of accelerated nuclear growth and heavy energy demand anticipated in some of the emerging economies such as China and India.

10.1.2 Nuclear power and green energies

Considering the vast resources of uranium and thorium, the two fissionable materials widely available on the surface of earth, and its energy content, nuclear energy could be considered as a renewable source of energy. This could be multiplied many times if extraction of uranium from sea water is also taken in to account.

Fast breeder reactors effectively utilize all the fissionable content of in uranium and thorium fuel and therefore generate very little waste. It is 100 times more efficient that current generation of light water and heavy water reactor technologies. This fact, combined with negligible emission of carbon, makes nuclear power a renewable and sustainable source of energy.

Apart from being a source of power, nuclear energy could also contribute to production of hydrogen, desalination of seawater, thus compliment green energies. Small nuclear reactor designs such as Pebble Bed Modular Reactors (PBMR) and Compact High Temperature Reactors (CHTR) could support a decentralized model of power generation and provide process heat for hydrogen production or desalination of water (IAEA, 2008a).

Nuclear power was recognized as a reliable, safe, clean and cheap source of energy since the mid 20th century when the first successful generation of electricity was demonstrated on December 20, 1951 at Experimental Breeder Reactor (EBR-1), Arco, Idaho (Michal, 2001). Before this reactor was shut down in 1964, it sufficiently laid the sustainable roadmap for nuclear power to utilize not only the uranium resources of the plant, but also the vast thorium resources, as well as the possibility of extracting power out of the used fuel by burning most of the long-lived isotopes.

But the developments that dominated the first and second generation nuclear reactors thereafter was only based on use of uranium and utilization of only about 1% of the fissile and fissionable content of the fuel and discard the rest as waste to be stored and ultimately disposed of in deep geological repositories.

Third generation reactors today recycle part of the fissile content as Mixed Oxide Fuel (MOX) and the Fourth Generation reactors to a large extent will follow up with breeder design of EBR-1 to utilize thorium also in a fuel cycle, which will create or breed more fuel than it actually burns and thus elevating nuclear power to the status of renewable energy or green energy.

10.2 NUCLEAR FISSION

Radioactivity was discovered in 1896 by Henri Becquerel. Additional work by Marie Curie, Pierre Curie, Ernest Rutherford and others proved that unstable atomic nucleus spontaneously loses energy by emitting ionizing particles and radiation. It was E. Rutherford who in 1917 demonstrated the possibility of splitting atom and emission of particles with high energies.

Nuclear fission got its break-through when Otto Hahn and Fritz Strassmann in 1938 split the uranium atom by bombarding it with neutrons and proved that the elements barium and krypton were formed. Importance of nuclear fission started gaining attention when it became apparent that fission of heavy elements is an exothermic (heat emitting) reaction which can release large amounts of energy, both as electromagnetic radiation and as kinetic energy of the fragments (DOE, 1993).

The amount of energy released by nuclear fission was found to be several orders of magnitude higher than exothermic chemical reactions such as burning of wood, coal, oil or gas. Typically a fission event releases about ~200 MeV (million electron volt) of energy. On the other hand, most chemical oxidation reactions such as burning coal or wood, release a few eV per event. Fission of a kilogram of ^{235}U can produce 7.2×10^{13} Joules of energy, whereas only 2.4×10^{7} Joules is obtained by burning one kilogram of coal.

Therefore nuclear fuel contains more than twenty million times energy, than does a chemical fuel. The energy of nuclear fission is released as kinetic energy of the fission products and fragments and as electromagnetic radiation in the form of gamma rays. In a nuclear reactor this energy is converted to heat as the particles and gamma rays collide with the atoms that make up the reactor and its coolant, such as light water, heavy water or liquid metal.

When the isotope ^{235}U fissions into two nuclei fragments a total mean fission energy 202.5 MeV is released. Typically ~169 MeV appears as the kinetic energy of the daughter nuclei. Additionally an average of 2.5 neutrons are emitted with a kinetic energy of ~2 MeV each (total of 4.8 MeV).

Many heavy isotopes are fissionable in the sense that they can undergo fission when struck by free neutrons. But isotopes that sustain a fission chain reaction when struck by low energy neutrons are also called fissile. A few particularly fissile and readily obtainable isotopes, such as ^{235}U and ^{239}Pu, are called nuclear fuels (Bodansky, 2003).

10.2.1 Fission chain reaction

A nuclear chain reaction can occur when one nuclear reaction causes an average of one or more nuclear reactions, thus leading to a self-propagating number of these reactions (Fig 10.1). All fissionable and fissile isotopes undergo a small amount of spontaneous fission (a form of radioactive decay) which releases a few free neutrons.

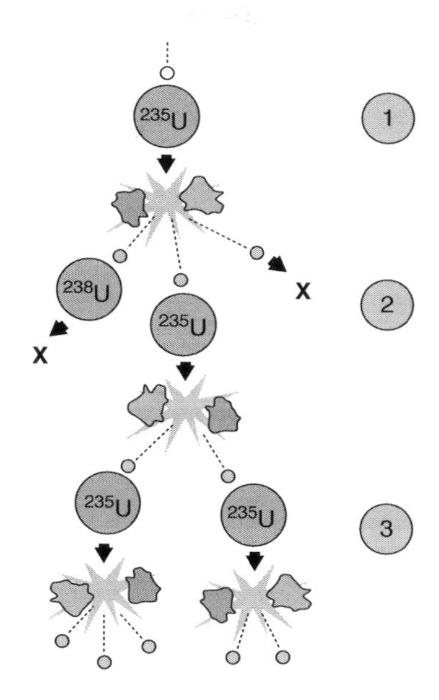

Figure 10.1 A nuclear fission chain reaction

The neutrons interact with other nuclei in the vicinity causing further fission and thus releasing more neutrons. If these freshly released neutrons outnumber the neutrons that escape from the sample or assembly, and a sustained nuclear chain reaction will take place.

Not all fissionable isotopes can sustain a chain reaction. For example, ^{238}U, the most abundant form of uranium, is fissionable but not fissile. It undergoes fission when hit by a fast neutron with over 1 MeV of kinetic energy. But too few of the ~1.7 neutrons produced by ^{238}U fission are energetic enough to induce further fissions.

Instead ^{238}U will absorb the slow neutrons to produce ^{239}Pu, which can be used as a nuclear fuel. In fact up to half of the power produced by a present uranium fuelled light and heavy reactor is by the fission of ^{239}Pu produced or "bred" in place.

10.2.2 Natural fission reactors

The technology required to drive a sustained chain reaction in a reactor is not complicated or difficult to develop and deploy. If conditions were appropriate, it could have taken place naturally on the surface of earth. It has been discovered that that nuclear fission chain reactions was sustained on earth in the geological past (Meshik, 2005).

Evidences suggest that in three uranium ore deposits in Oklo, Gabon self-sustaining nuclear fission was sustained by natural elements about 2 billion years ago. This natural light water moderated process was possible because 2 billion years ago natural uranium was richer in the shorter-lived fissile isotope ^{235}U (about 3%), than natural uranium available today (0.7%). Natural uranium has to be enriched today to 3% to be usable

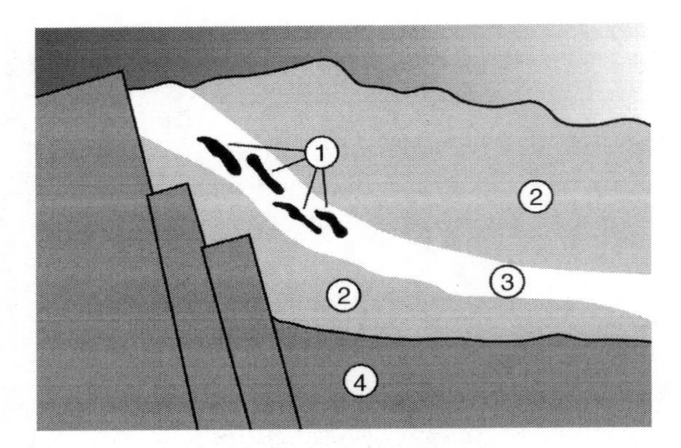

Figure 10.2 Oklo natural fission reactors (1. Nuclear reactor zone; 2. Sandstone; 3. Uranium ore zones; 4. Granite)

in light-water reactors. For that reason natural uranium fission chain reactions would not be possible at present.

The ^{235}U abundance in Oklo uranium ore was found to be only 0.44%. This low ^{235}U abundance and presence of neodymium and other elements suggest that a natural nuclear reactor existed in the past. It was apparent that considerable amount of ^{239}PU was also produced. The approximated shape of the reactor zone and hydraulic gradient allowed moderation and reflection of neutrons produced by spontaneous fission or cosmic ray induced fission. These conditions allowed the reactor to achieve criticality (Fig 10.2).

As the reactor power increased, the water moderator would heat, reducing its density and its effectiveness as a moderator and reflector. The reactors thus could have operated cyclically, operating for half hour until accumulated heat boiled away the water, then shutting down for up to 2.5 hours until the rocks cooled sufficiently to allow water saturation. Based on the amount of fission products generated, the Oklo reactors are estimated to have operated for more than 150 000 years.

It is estimated that the average operating power was about 100 KW, similar to that of some modern research reactors. The reactors produced a total of 15 GW yr of thermal energy and consumed an estimated 5–6 tonnes of ^{235}U and produced an equal mass of fission products. Majority of the fission products have remained in place for nearly 2 billion years, in spite of their location in fractured, porous, and water-saturated sandstone for most of the time.

10.2.3 Nuclear reactors

A nuclear reactor is a device or system in which nuclear chain reactions are initiated, controlled, and sustained. Nuclear reactors are usually used for many purposes, but production of electrical power is the most dominant commercial application.

It can be also used for production radio-isotopes for medical use, to power ships, submarines and ice-breakers, and for nuclear research. The production of electricity by a nuclear reactor is accomplished by utilizing the heat from the fission reaction to drive steam turbines.

Current nuclear reactors technology is based on a sustained nuclear fission chain reaction to induce in a fissile material fuel, releasing both energy and free neutrons. A reactor encloses nuclear fuel or reactor core surrounded by a neutron moderator such as light water, heavy water or graphite and control rods that control the rate of the reaction (DOE, 1993).

In a nuclear reactor, the neutron flux at a given time is a function of the rate of fission neutron production and the rate of neutron losses due to non-fission absorption and leakage from the system. When a reactor's neutron population remains steady, so that as many new neutrons are produced as lost, the fission chain reaction will be self-sustaining and the reactor is referred as "critical". When the reactor's neutron production exceeds the loss is called "supercritical", and when losses dominate, it is considered "subcritical".

For the sustained chain reaction to be possible the uranium-fueled reactors must include a neutron moderator that interacts with newly produced fast neutrons from fission events to reduce their kinetic energy from several MeV to several eV, making them more likely to induce fission. This is because ^{235}U is much more likely to undergo fission when struck by one of these thermal neutrons than by a freshly-produced neutron from fission.

Any element that strongly absorbs neutrons is called a reactor poison, because it tends to shut down an ongoing fission chain reaction. Some reactor poisons are deliberately inserted into fission reactor cores to control the reaction. Boron or cadmium control rods are usually used for this purpose. Many reactor poisons are produced by the fission process itself, and buildup of neutron-absorbing fission products affects both the fuel economics and the controllability of nuclear reactors.

While many fissionable isotopes exist in nature, the useful fissile isotope found in any sufficient quantity is ^{235}U. It is about 0.7% of the naturally occurring uranium ore. The rest about 99.3% is the fissionable ^{238}U isotope. Therefore in most of the light-water reactors uses ^{235}U must be enriched artificially up to 3–5%. Chemical properties of ^{235}U and ^{238}U are identical, so physical processes such as gaseous diffusion, gas centrifuge or mass spectrometry must be used for isotopic separation based on small differences in mass.

Nuclear reactors with heavy water moderation can operate with natural uranium, eliminating altogether the need for enrichment. The Pressurized Heavy Water Reactors (PHWR) are an example of this type. Some graphite moderated reactor designs can also use natural uranium as fuel (Table 10.2).

In the reactor core major part of the heat is generated due to conversion of the kinetic energy of fission products to thermal energy, when the nuclei collide with nearby atoms. Some of the gamma rays produced during fission are absorbed by the reactor and their energy converted to heat. Heat is also produced by the radioactive decay of fission products and materials that have been activated by neutron absorption. This decay heat source will remain for some time even after the reactor is shutdown.

A nuclear reactor coolant is circulated through the reactor core to absorb the heat that it generates. Coolant is usually water but sometimes a gas or a liquid metal or molten salt is also used. The heat is carried away from the reactor and is then used to generate steam, which drives a turbine coupled with an electrical generator to produce electricity.

Table 10.2 Current nuclear power reactor types

Reactor Type	*Moderator*	*Coolant*	*Fuel*	*Comments*
Pressurized Water Reactor (PWR)	Light water	Light water	Enriched uranium	Pressure vessel contains the fuel, control rods, moderator, and coolant. The hot water transfers heat to primary and then to a secondary steam generator that drives the turbines.
Boiling Water Reactor (BWR)	Light water	Light water	Enriched uranium	Pressure vessel contains the fuel, control rods, moderator, and coolant. Water is allowed to boil inside the pressure vessel producing the steam that runs the turbines.
Pressurized Heavy Water Reactor (PHWR)	Heavy water	Heavy water	Natural uranium/ Slightly enriched uranium	Fuel is contained in pressure tubes and hot heavy water transfers heat to steam generator that drives the turbines.
Gas Cooled Reactor (GCR) and Advanced Gas Cooled Reactors (AGR)	Carbon dioxide	Graphite	Natural uranium (GCR)/Enriched uranium (AGR)	Pressure vessel contains the fuel, control rods, moderator, and coolant. Heat exchanger that produces steam is housed out of the pressure vessel in GCR and inside the pressure vessel in AGR.
Light Water cooled Graphite moderated Reactor (LWGR)	Light Water	Graphite	Natural Uranium/ Low Enriched Uranium	Pressure tubes running through graphite moderator allows water to boil producing steam that drives the turbines.
Fast Breeder Reactor (FBR)	No moderator	Liquid metal (sodium, lead)	Highly Enriched Uranium/MOX (U + Pu). ^{238}U/ ^{232}Th blanket	Heat exchangers immersed in a liquid sodium pool (pool type design) or outside the reactor vessel (loop type design) transfer heat to a secondary coolant circuit having a steam generator that drives the turbines. "Breeds" more fuel (^{239}Pu or ^{233}U) than it consumes.

Table 10.3 Current world nuclear reactors

Reactor Type	Operating Reactors		Reactors Under Construction	
	Units	*Installed Capacity (MWe)*	*Units*	*Installed Capacity (MWe)*
Pressurized Water Reactor	265	244 337	47	44 689
Boiling Water Reactor (BWR)	92	83 690	3	3 925
Pressurized Heavy Water Reactor (PHWR)	45	22 639	3	1 096
Gas Cooled Reactor (GCR)	18	8 949	–	–
Light water cooled Graphite Moderated Reactor (LWGR)	45	22 639	1	925
Fast Breeder Reactor (FBR)	1	560	2	1 220
Total	436	370 394	56	51 855

(Based on IAEA Power Reactor Information System)

In some reactors the coolant acts as a neutron moderator too. A moderator increases the power of the reactor by causing the fast neutrons that are released from fission to lose energy and become thermal neutrons. Thermal neutrons are more likely than fast neutrons to cause fission, so more neutron moderation means more power output from the reactors.

The power output of the reactor is controlled by controlling how many free neutrons are able to create more fission. Control rods that are made of a nuclear poison are used to absorb neutrons, so that there are fewer neutrons available to cause fission. Inserting the control rod deeper into the reactor will reduce its power output, and extracting the control rod will increase it.

Depending on the type of nuclear reaction, reactors are classified as thermal reactors and fast reactors. Thermal reactors use slow or thermal neutrons. Almost all current reactors are of this type. These contain neutron moderator materials that slow neutrons until their neutron temperature is thermalized, that is, until their kinetic energy approaches the average kinetic energy of the surrounding particles.

Thermal neutrons have a far higher cross section or probability of fissioning the fissile nuclei ^{235}U, ^{239}Pu and ^{241}Pu and relatively lower probability of capture by ^{238}U, compared to the faster neutrons that originally result from fission. This allows the use of low-enriched uranium or even natural uranium fuel in thermal reactors. The moderator is often also the coolant, such as water under high pressure to increase the boiling point.

Fast reactors use fast neutrons to cause fission in the fuel. Fast reactors do not require a neutron moderator, and use less-moderating coolants. But maintaining a chain reaction in a fast reactor requires the fuel to be enriched to about 20% or more in fissile material. This is due to the relatively lower probability of fission versus capture by ^{238}U. Fast reactors have the potential to produce less transuranic waste because all actinides are fissionable with fast neutrons.

Pressurized Water Reactors, Boling Water Reactors and Pressurized Heavy Water Reactors are the mainstay of world nuclear power programme as can be seen from Table 10.3 (IAEA, 2010).

Table 10.4 Nuclear fuel cycle stages and activities (Adapted from IAEA, 2009b)

Sub-cycle	*Stage*	*Activity*
FRONT END	Uranium Mining and Milling	Uranium Mining Uranium Ore Processing U Recovery from Phosphates
	Conversion	Conversion to UO_2 Conversion to UO_3 Conversion to UF_4 Conversion to UF_6 Re-Conversion to U_3O_8 (Depleted U) Conversion to U Metal
	Enrichment	Uranium Enrichment
	Uranium Fuel Fabrication	Re-conversion to UO_2 Powder Fuel Fabrication (U Pellet-Pin) Fuel Fabrication (U Assembly) Fuel Fabrication (Burnable Poison Pellet-Pin) Fuel Fabrication (Research Reactors) Fuel Fabrication (Pebble)
	IRRADIATION IN REACTORS	
BACK END	Spent Fuel Reprocessing and Recycling	Spent Fuel Reprocessing
		Re-Conversion to U_3O_8 (Rep U) Co-conversion to MOX Powder Fuel Fabrication (MOX Pellet-Pin) Fuel Fabrication (MOX Assembly) Fuel Fabrication (RepU-ERU(Enriched Recycled uranium Pellet-Pin) Fuel Fabrication (RepU-ERU Assembly)
	Spent Fuel Storage	AR Spent Fuel Storage
		AFR Wet Spent Fuel Storage AFR Dry Spent Fuel Storage
	Spent Fuel Conditioning	Spent Fuel Conditioning
	Spent Fuel Disposal	Spent Fuel Disposal

10.3 SUSTAINABLE NUCLEAR FUEL CYCLE OPTIONS

The nuclear fuel cycle may be broadly defined as the set of processes and operations needed to manufacture nuclear fuel, its irradiation in nuclear power reactors and storage, reprocessing, recycling or disposal (Table 10.4). The nuclear fuel cycle starts with uranium exploration and ends with disposal of the materials used and generated during the cycle. Several nuclear fuel cycles can be considered depending on the type of reactor and the type of fuel used and whether or not the irradiated fuel is reprocessed and recycled.

The Nuclear fuel cycle has been further subdivided into the front-end and the back-end sub-cycles. The front-end of the fuel cycle occurs before irradiation and the back-end begins with the discharge of spent fuel from the reactor (IAEA, 2009b).

If spent fuel is not reprocessed, the fuel cycle is referred to as an open fuel cycle (or a once-through fuel cycle); if the spent fuel is reprocessed, it is referred to as a closed fuel cycle. Choosing the 'closed' or 'open' fuel cycle is a matter of national policy. Some countries have adopted the 'closed' fuel cycle solution, and some others have chosen the 'open' fuel cycle. Combination of solutions or on hold (wait and see) is a position of other nuclear power countries. (IAEA, 2005a, b)

In the open fuel cycle nuclear material passes through the reactor just once. After irradiation, the fuel is kept in at-reactor pools until it is sent to away from reactor storage. It is planned that the fuel will be conditioned and put into a final repository in this mode of operation. No final repositories for spent fuel have yet been established anywhere in the world.

In the closed fuel cycle, the spent fuel is reprocessed to extract the remaining uranium and plutonium from the fission products and other actinides. The reprocessed uranium and plutonium is then reused in the reactors. This strategy has been adopted by some countries mainly in light water reactors in the form of mixed oxide (MOX) fuel.

Another closed fuel cycle practice is the recycle of nuclear materials in fast reactors in which, reprocessed uranium and plutonium are used for production of fast reactor fuel. Such a reactor can produce more fissile plutonium than it consumes.

In reprocessing stage, the fission products, minor actinides, activation products, and reprocessed uranium are separated from the reactor-grade plutonium, which can then be fabricated into MOX fuel. The proportion of the non-fissile even-mass isotopes of plutonium rises with recycle. So reuse plutonium from used MOX fuel beyond three recycles is not usually done in thermal reactors. This is not a limitation in fast reactors.

10.3.1 Thorium fuel cycle

The most potential sustainable fuel cycle option for the future is that of thorium. Abundance of uranium and its relative ease of handling was the reason much attention was not paid in past in developing thorium fuel cycle. But the recent concerns about constraints in uranium supply well into future have promoted renewed attention to thorium. The historical thorium utilization details are given in Table 10.5.

In thorium fuel cycle, the naturally abundant isotope of thorium, ^{232}Th, is fertile material which is transmuted into the fissile artificial uranium isotope ^{233}U which is the nuclear fuel. The sustained fission chain reaction could be started with existing ^{233}U or some other fissile material such as ^{235}U or ^{239}Pu. Subsequently a breeding cycle similar to but more efficient than that with ^{238}U – ^{239}Pu can be created (IAEA, 2005b).

Thorium is at least 3–4 times more abundant in nature than all uranium isotopes and is fairly evenly spread on the surface of Earth. Unlike uranium, naturally occurring thorium consists of only a single isotope (^{232}Th) in significant quantities. Consequently, all mined thorium is useful in thermal reactors without the need for an enrichment process.

Thorium based fuels exhibit several attractive nuclear properties relative to uranium-based fuels such as:

- fertile conversion of thorium is more efficient in a thermal reactor.
- fewer non-fissile neutron absorptions and improved neutron economy.
- can be the basis for a thermal breeder reactor.

Thorium-based fuels also display favorable physical and chemical properties which improve reactor and repository performance. Compared to the predominant reactor fuel, uranium dioxide (UO_2), thorium dioxide (ThO_2) has a higher melting point, higher thermal conductivity, and lower coefficient of thermal expansion. Thorium dioxide also exhibits greater chemical stability.

Because the ^{233}U produced in thorium fuels is inevitably contaminated with ^{232}U, thorium-based used nuclear fuel possesses inherent proliferation resistance. Elimination of at least the transuranic portion of the nuclear waste problem is possible in thorium fuel cycle. But there are some long-lived actinides that constitute a long term radiological impact, especially ^{231}Pa.

If thorium is used in an open fuel cycle (i.e. utilizing ^{233}U in-situ), higher burnup is necessary to achieve a favorable neutron economy. Although thorium dioxide has performed well at burnups of 170 000 MWd/t and 150 000 MWd/t, there are challenges associated with achieving this burnup in light water reactors.

The challenge associated with a once-through thorium fuel cycle is the comparatively long time scale over which ^{232}Th breeds to ^{233}U. The half-life of ^{233}Pa is about 27 days, which is an order of magnitude longer than the half-life of ^{239}Np in the uranium fuel cycle. As a result substantial ^{233}Pa builds into thorium-based fuels. ^{233}Pa is a significant neutron absorber, and although it eventually breeds into fissile ^{235}U, this requires two more neutron absorptions, which degrades neutron economy and increases the likelihood of transuranic production.

If thorium is used in a closed fuel cycle in which ^{233}U is recycled, remote handling is necessary because of the high radiation dose resulting from the decay products of ^{232}U. This is also true of recycled thorium because of the presence of ^{228}Th, which is part of the ^{232}U decay sequence. Although there is substantial worldwide experience recycling uranium fuels (e.g. PUREX), similar technology for thorium (e.g. THOREX) is still under development.

Historical thorium utilization in various reactors is given in Table 10.5.

10.3.2 Uranium resources and production

Uranium is an element that is widely distributed within the earth's crust. Its principal use is as the primary fuel for nuclear power reactors. Naturally occurring uranium is composed of about 99.3% ^{238}U, 0.7% ^{235}U and traces of ^{234}U. In order to use uranium in the ground, it has to be extracted from the ore and converted into a form which can be used in the nuclear fuel cycle.

A deposit of uranium discovered by various exploration techniques is evaluated to determine the amounts of uranium materials that are extractable at specified costs. Uranium resources are the amounts of ore that are estimated to be recoverable at stated costs.

IAEA Uranium 2007 Resources, Production and Demand (Red Book) reports that the total Identified Resources in 2007 is about 5 469 000 tonnes U in the <USD 130/kgU category (Table 10.6). Total Additionally Undiscovered Resources (Prognosticated Resources and Speculative Resources) amounts to another 10 500 000 tU (OECD/NEA-IAEA, 2008).

The reported Identified Resources (~5.5 million tonnes natural uranium) can last 83 years at the current rate of consumption of about 70 000 tonnes per year. Moreover,

Table 10.5 Thorium utilization in different experimental and power reactors (Source IAEA, 2005b)

Name and Country	*Type*	*Power*	*Fuel*	*Operation period*
AVR, Germany	HTGR, Experimental (Pebble bed reactor)	15 MW(e)	Th+^{235}U Driver Fuel, Coated fuel particles, Oxide & dicarbides	1967–1988
THTR-300, Germany	HTGR, Power (Pebble Type)	300 MW(e)	Th+^{235}U, Driver Fuel, Coated fuel particles, Oxide & dicarbides	1985–1989
Lingen, Germany	BWR Irradiation-testing	60 MW(e)	Test Fuel (Th,Pu)O2 pellets	Terminated in 1973
Dragon, UK OECD-Euratom also Sweden, Norway & Switzerland	HTGR, Experimental (Pin-in-Block Design)	20 MWt	Th+^{235}U Driver Fuel, Coated fuel particles, Oxide & Dicarbides	1966–1973
Peach Bottom, USA	HTGR, Experimental (Prismatic Block)	40 MW(e)	Th+^{235}U Driver Fuel, Coated fuel particles, Oxide & dicarbides	1966–1972
Fort St Vrain, USA	HTGR, Power (Prismatic Block)	330 MW(e)	Th+^{235}U Driver Fuel, Coated fuel particles, Dicarbide	1976–1989
MSRE ORNL, USA	MSBR	7.5 MWt	^{233}U Molten Fluorides	1964–1969
Shippingport & Indian Point 1, USA	LWBR PWR, (Pin Assemblies)	100 MW(e), 285 MW(e)	Th+^{233}U Driver Fuel, Oxide Pellets	1977–1982, 1962–1980
SUSPOP/KSTR KEMA, Netherlands	Aqueous Homogenous Suspension (Pin Assemblies)	1 MWt	Th+HEU, Oxide Pellets	1974–1977
NRU & NRX, Canada	MTR (Pin Assemblies)		Th+^{235}U, Test Fuel	Irradiation–testing of few fuel elements
KAMINI; CIRUS; & DHRUVA, India	MTR Thermal	30 kWt; 40 MWt; 100 MWt	Al+^{233}U Driver Fuel, 'J' rod of Th & ThO_2, 'J' rod of ThO_2	All three research reactors in operation
KAPS 1&2; KGS 1&2; RAPS 2, 3&4, India	PHWR, (Pin Assemblies)	220 MW(e)	ThO_2 Pellets (For neutron flux flattening of initial core after start-up)	Continuing in all new PHWRs
FBTR, India	LMFBR, (Pin Assemblies)	40 MWt	ThO_2 blanket	In operation

Table 10.6 Total identified uranium resources

Country	*Identified Resources (tonnes U)*	%
Australia	1 243 000	22.73
Canada	423 200	7.74
Kazakhstan	817 300	14.94
Russia	545 600	9.98
Brazil	278 400	5.09
Jordan	111 800	2.04
Namibia	275 000	5.03
Niger	274 000	5.01
South Africa	435 100	7.96
Ukraine	199 500	3.65
USA	339 000	6.20
Uzbekistan	111 000	2.03
Others	415 900	7.60
Total	5 468 800	

the reported uranium resource figures presented in the Red Book are only a part of the already known resources and are not an inventory of the total amount of recoverable uranium. Examples where uranium resources are known, but not reported, are the Russian Federation, the USA and Australia.

Conventional resources are defined as resources from which uranium is recoverable as a primary product, a co-product or an important by-product, while unconventional resources are resources from which uranium is only recoverable as a minor by-product, such as uranium associated with phosphate rocks, non-ferrous ores, carbonatite, black schists, and lignite.

Most of the unconventional uranium resources reported to date are associated with uranium in phosphate rocks, but other potential sources exist (e.g., seawater and black shale). Historically phosphate deposits are the only unconventional resources from which a significant amount of uranium has been recovered. Estimates of potentially recoverable uranium associated with phosphates, non-ferrous ores, carbonatite, black schist and lignite are of the order of 10 million tonnes U.

Seawater contains an estimated 4 500 Mt U, but at a very low concentration of 3.3 parts per billion (ppb). Thus, 330 000 t of water would have to be processed to produce one kg of uranium. Currently, such production is too expensive. Research was carried out in Germany, Italy, Japan, UK and USA in the 1970s and 1980s. Current bench scale marine experiments in Japan indicate that uranium might possibly be extracted with braid type adsorbents moored to the sea floor, with a production capacity of 1 t U per year at an estimated cost of about $300/kg U. Laboratory scale research is also being carried out in France and India.

Uranium ore can be extracted through conventional mining by open cut and underground methods. In some cases uranium is recovered as a by-product, for example of copper mining. In-situ leach (ISL) mining methods also are used to recover uranium. In this technology, uranium is leached from the in-place ore through an array of regularly spaced wells and is then recovered from the leach solution at a surface plant.

Table 10.7 World uranium production

Country	*U Production in 2008 (tonnes U)*
Canada	9 000
Kazakhstan	8 521
Australia	8 430
Namibia	4 366
Russia*	3 521
Niger	3 032
Uzbekistan	2 338
USA	1 430
Ukraine*	800
China*	769
South Africa	655
Brazil	330
India*	271
Czech Repub.	263
Romania (est)	77
Pakistan*	45
France	5
Total	43 853

*estimated

Mined uranium ores normally are processed by grinding the ore materials to a uniform particle size and then treating the ore to extract the uranium by chemical leaching. The milling process commonly yields dry powder-form material consisting of natural uranium, "yellowcake," which is sold on the uranium market as U_3O_8.

In 2008, uranium production worldwide was 43 853 tonnes U. Canada, Kazakhstan and Australia accounted for almost 60% of world production in 2008. These three together with Namibia, Niger, the Russian Federation, Uzbekistan and the USA accounted for 93% of production (Table 10.7) (WNA, 2009)

Uranium production in 2008 covered only about 75% of the world's reactor requirements of 58 685 tonnes U. The remainder was covered by secondary sources such as stockpiles of natural uranium, stockpiles of enriched uranium, reprocessed uranium from spent fuel, mixed oxide (MOX) fuel with ^{235}U partially replaced by ^{239}Pu from reprocessed spent fuel, and re-enrichment of depleted uranium tails.

10.3.3 Thorium resources

Thorium, abundant and widely dispersed, could also be used as a nuclear fuel resource. Most of the largest identified thorium resources were discovered during the exploration of carbonatites and alkaline igneous bodies for uranium, rare earth elements, niobium, phosphate, and titanium.

Today, thorium is recovered mainly from the mineral monazite as a by-product of processing heavy-mineral sand deposits for titanium-, zirconium-, or tin-bearing minerals. Worldwide thorium resources, which are listed by major deposit types in

Table 10.8 World resources of thorium

Country	*Identified Resources Th (tonnes)*	*Prognosticated Resources Th (tonnes)*
Brazil	302 000	330 000
Turkey	344 000	400 000–500 000
India	319 000	–
United States	400 000	274 000
Norway	132 000	132 000
Greenland	54 000	32 000
Canada	44 000	128 000
Australia	452 000	–
South Africa	18 000	130 000
Egypt	100 000	280 000
Other Countries	33 000	81 000
Russia	75 000	–
Venezuela	300 000	–
World Total	**2 573 000**	**1 787–1 887**

Table 10.8, are estimated to total about 4.4 million tonnes Th (OECD/NEA-IAEA, 2008).

The primary source of the world's thorium is the rare-earth and thorium phosphate mineral, monazite. Monazite itself is recovered as a by-product of processing heavy-mineral sands for titanium and zirconium minerals.

10.3.4 Uranium conversion, enrichment and fuel fabrication

Milled uranium oxide, U_3O_8, must be converted to uranium hexafluoride, UF_6, which is the form required by most commercial uranium enrichment facilities currently in use. A solid at room temperature, uranium hexafluoride can be changed to a gaseous form at moderately higher temperature of 57°C. The uranium hexafluoride conversion product contains only natural, not enriched, uranium.

Triuranium octaoxide (U_3O_8) is also converted directly to ceramic grade uranium dioxide (UO_2) for use in reactors not requiring enriched fuel, such as PHWR. The volumes of material converted directly to UO_2 are typically quite small compared to the amounts converted to UF_6.

Total global conversion capacity is about 75 000 tonnes of natural uranium per year (tU/yr) for uranium hexafluoride (UF_6) and 4 500 tU/yr for uranium dioxide (UO_2). Current demand is about 70 000 tU/yr (IAEA, 2009a).

Natural UF_6 thus must be enriched in the fissionable isotope for it to be used as nuclear fuel in most of the light water reactors. The different levels of enrichment required for a particular nuclear fuel application are specified. Light water reactor fuel normally is enriched to 3.5% ^{235}U, but uranium enriched to lower concentrations also is required.

Enrichment is accomplished using some one or more methods of isotope separation. Gaseous diffusion and gas centrifuge are the commonly used uranium enrichment technologies. About 96% of the byproduct from enrichment is depleted uranium (DU), which can be used for armor, kinetic energy penetrators, radiation shielding and ballast.

Enrichment requirements are expressed in Separative Work Units (SWU). It is a function of the concentrations of the feedstock, the enriched output, and the depleted tailings; and is expressed in units which are so calculated as to be proportional to the total input and to the mass processed. The same amount of separative work will require different amounts of energy depending on the efficiency of the separation technology.Total global enrichment capacity is currently about 50 million separative work units per year (SWU/yr) compared to a total demand of approximately 45 million SWU/yr.

For use as nuclear fuel, enriched uranium hexafluoride is converted into uranium dioxide (UO_2) powder that is then processed into pellet form. The pellets are then fired in a high temperature sintering furnace to create hard, ceramic pellets of enriched uranium. The cylindrical pellets then undergo a grinding process to achieve a uniform pellet size.

The pellets are stacked, according to each nuclear reactor core's design specifications, into tubes of corrosion-resistant metal alloy. The tubes are sealed to contain the fuel pellets and these tubes are called fuel rods. The finished fuel rods are grouped in special fuel assemblies that are then used to build up the nuclear fuel core of a power reactor.

The metal used for the tubes depends on the design of the reactor. Stainless steel was used in the past, but most reactors now use zirconium. For the most common types of reactors, boiling water reactors (BWR) and pressurized water reactors (PWR), the tubes are assembled into bundles with the tubes spaced precise distances apart. These bundles are then given a unique identification number, which enables them to be tracked from manufacture through use and into disposal.

Total global fuel fabrication capacity is currently about 11 500 tU/yr (enriched uranium) for light water reactor (LWR) fuel and about 4 000 tU/yr (natural uranium) for pressurized heavy water reactor (PHWR) fuel. Total demand is about 12 000 tU/yr.

10.3.5 Spent fuel management and reprocessing

After its operating cycle, the reactor is shut down for refuelling. The spent fuel or used fuel discharged is stored either at the reactor site, commonly in a spent fuel pool or, in a common facility away from reactor sites. If on-site pool storage capacity is exceeded, it may be desirable to store the now cooled aged fuel in modular dry storage facilities known as Independent Spent Fuel Storage Installations (ISFSI) at the reactor site or at a facility away from the site.

The spent fuel rods are usually stored in water or boric acid, which provides both cooling, the spent fuel continues to generate decay heat as a result of residual radioactive decay, and shielding to protect the environment from residual ionizing radiation, although after several years of cooling they may be moved to dry cask storage (IAEA, 2008b).

The total amount of spent fuel discharged globally was projected to reach 324 000 tonnes heavy metal (tHM) by the end of 2008. Of this amount, about 95 000 tHM have already been reprocessed, 16 000 tHM are currently stored to be reprocessed and 213 000 tHM are stored in spent fuel storage pools at reactors or in away-from-reactor (AFR) storage facilities. AFR storage facilities are being regularly expanded both by adding modules to existing dry storage facilities and by building new facilities.

Spent fuel discharged from reactors contains appreciable quantities of fissile (^{235}U and ^{239}Pu), fertile (^{238}U), and other radioactive materials, including reaction poisons, which is why the fuel had to be removed. These fissile and fertile materials can be chemically separated and recovered from the spent fuel. The recovered uranium and plutonium can, if economic and institutional conditions permit, be recycled for use as nuclear fuel. This is currently not done for civilian spent nuclear fuel in the US.

Mixed oxide, or MOX fuel, is a blend of reprocessed uranium and plutonium and depleted uranium which behaves similarly, although not identically, to the enriched uranium feed for which most nuclear reactors were designed. MOX fuel is an alternative to low-enriched uranium (LEU) fuel used in the light water reactors which predominate nuclear power generation. Total global reprocessing capacity is about 6 000 tHM/yr.

10.4 ADVANCED AND NEXT GENERATION REACTORS

About a dozen advanced reactors are in various stages of development. Some are evolutionary from the PWR, BWR and PHWR designs and others some are more radical departures. The former include the Advanced Boiling Water Reactor (ABWR), two of which are now operating with others under construction, and the planned passively safe ESBWR and AP1000 units.

Advanced Heavy Water Reactor is proposed with heavy water moderator, that will be the next generation design of the PHWR type in India. Thorium utilization and breeding is planned in this reactor. The design includes a number of passive safety systems. India is also planning to build fast breeder reactors using the ^{232}Th – ^{233}U fuel cycle.

10.4.1 Generation IV reactors

Generation IV reactors are a set of theoretical nuclear reactor designs currently being researched. These designs are generally not expected to be available for commercial construction before 2030. Current reactors in operation around the world are generally considered second- or third-generation systems, with the first-generation systems having been retired some time ago.

Research into these reactor types was officially started by the Generation IV International Forum (GIF) based on eight technology goals. The primary goals are to improve nuclear safety, improve proliferation resistance, minimize waste and natural resource utilization, and to decrease the cost to build and run such plants (DOE, 2002).

The designs being researched are:

1. *Very-high-temperature reactor (VHTR):* The reactor concept utilizes a graphite-moderated core with a once-through uranium fuel cycle. This reactor design envisions an outlet temperature of 1 000°C. The reactor core can be either a prismatic-block or a pebble bed reactor design. The high temperatures enable applications such as process heat or hydrogen production via the thermochemical iodine-sulfur process. It would also be passively safe.
2. *Supercritical-water-cooled reactor (SCWR):* A concept that uses supercritical water as the working fluid. SCWRs are basically light water reactors (LWR)

operating at higher pressure and temperatures with a direct, once-through cycle. It could operate at much higher temperatures than both current PWRs and BWRs.

3. *Molten-salt reactor (MSR):* A reactor design where the coolant is a molten salt. The nuclear fuel dissolved in the molten fluoride salt as uranium tetrafluoride (UF_4), the fluid would reach criticality by flowing into a graphite core which would also serve as the moderator. Many current concepts rely on fuel that is dispersed in a graphite matrix with the molten salt providing low pressure, high temperature cooling.
4. *Gas-cooled fast reactor (GFR):* This system features a fast-neutron spectrum and closed fuel cycle for efficient conversion of fertile uranium and management of actinides. The reactor is helium-cooled, with an outlet temperature of 850°C and using a direct Brayton cycle gas turbine for high thermal efficiency.
5. *Sodium-cooled fast reactor (SFR):* This design builds on two closely related existing projects, the liquid metal fast breeder reactor and the Integral Fast Reactor. The goals are to increase the efficiency of uranium usage by breeding plutonium and eliminating the need for transuranic isotopes ever to leave the site.
6. *Lead-cooled fast reactor (LFR):* This design features a fast-neutron-spectrum lead or lead/bismuth eutectic (LBE) liquid-metal-cooled reactor with a closed fuel cycle. Options include a range of plant ratings, including a "battery" of 50 to 150 MW of electricity that features a very long refueling interval, a modular system rated at 300 to 400 MW, and a large monolithic plant option at 1 200 MW.

10.4.2 Generation V reactors

Generation V and V+ reactors are defined as designs which are theoretically possible, but which are not being actively considered or researched at present. Though such reactors could be built with current or near term technology, they trigger little interest for reasons of economics, practicality, or safety:

- Liquid Core reactor where the fissile material is molten uranium cooled by a working gas pumped in through holes in the base of the containment vessel.
- Gas core reactor where the fissile material is gaseous uranium-hexafluoride contained in a fused silica vessel. A working gas such as hydrogen would flow around this vessel and absorb the UV light produced by the reaction.
- Gas core EM reactor with photovoltaic arrays converting the UV light directly to electricity.
- Fission fragment reactor that generates electricity by decelerating an ion beam of fission byproducts instead of using nuclear reactions to generate heat.

10.4.3 Fusion reactors

Fusion power is the power generated by nuclear fusion reactions. Two light atomic nuclei fuse together to form a heavier nucleus and in doing so, release a large amount of energy. Most design studies for fusion power plants involve using the fusion reactions to create heat, which is then used to operate a steam turbine, which drives generators to produce electricity (Atzeni and Meyer-ter-Vehn, 2004).

Several fusion reactors have been built, but as yet none has 'produced' more thermal energy than electrical energy consumed. Despite research having started in the 1950s, no commercial fusion reactor is expected before 2050. The ITER project is currently leading the effort to commercialize fusion power (Nuttall, 2008).

In 1997 Joint European Torus (JET) produced a peak of 16.1 MW of fusion power (65% of input power), with fusion power of over 10 MW sustained for over 0.5 sec. The High Power laser Energy Research facility (HiPER) is undergoing preliminary design for possible construction in the European Union starting around 2010.

10.4.4 Accelerator Driven System

A subcritical reactor is a nuclear fission reactor that produces fission without achieving criticality. Instead of a sustaining chain reaction, a subcritical reactor uses additional neutrons from an outside source. The neutron source can be a nuclear fusion machine or a particle accelerator producing neutrons by spallation. Such a device with a reactor coupled to an accelerator is called an Accelerator Driven System (ADS).

The long-lived transuranic elements in nuclear waste can in principle be fissioned, releasing energy in the process and leaving behind the fission products which are shorter-lived. This would shorten considerably the time for disposal of radioactive waste. The three most important long-term radioactive isotopes that could advantageously be handled that way are ^{237}Np, ^{241}Am and ^{243}Am (IAEA, 2003).

ADS design propose a high-intensity proton accelerator with an energy of about 1 GeV, directed towards a spallation target made of thorium that is cooled by liquid lead-bismuth in the core of the reactor. In that way, for each proton interacting in the target, an average 20 neutrons are created to irradiate the surrounding fuel. Thus, the neutron balance can be regulated such as the reactor would be below criticality if the additional neutrons by the accelerator were not provided. Whenever the neutron source is turned off, the reaction ceases.

There are technical difficulties to overcome before ADS can become economical and eventually be integrated into future nuclear waste management. The accelerator must provide a high intensity and be highly reliable. There are concerns about the window separating the protons from the spallation target, which is expected to be exposed to stress under extreme conditions.

The chemical separation of the transuranic elements and the fuel manufacturing, as well as the structure materials, are important issues. Finally, the lack of nuclear data at high neutron energies limits the efficiency of the design.

Some laboratory experiments and many theoretical studies have demonstrated the theoretical possibility of such a plant. CERN, was one of the first to conceive a design of a subcritical reactor, the so-called "energy amplifier". In 2005, several large-scale projects are going on in Europe and Japan to further develop subcritical reactor technology.

10.5 NUCLEAR ECONOMICS

Nuclear power plants have a 'front-loaded' cost structure, i.e. they are relatively expensive to build but relatively inexpensive to operate. Thus existing well-run operating nuclear power plants continue to be a competitive and profitable source of electricity.

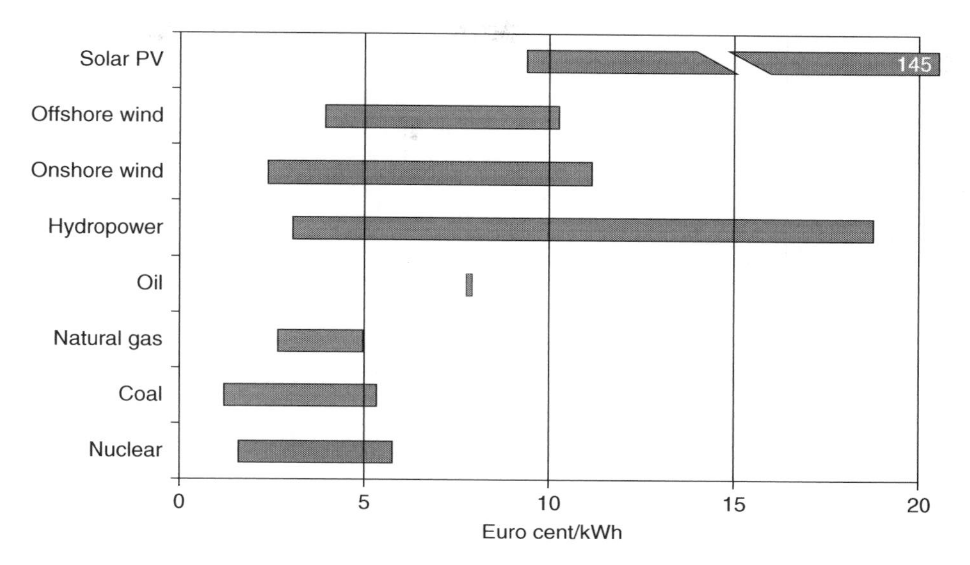

Figure 10.3 Ranges of levelized costs associated with new nuclear reactor construction

For new construction, however, the economic competitiveness of nuclear power depends on the alternatives available, on the overall electricity demand in a country and how fast it is growing, on the market structure and investment environment, on environmental constraints, and on investment risks due to possible political and regulatory delays or changes. Thus economic competitiveness is different in different countries and situations.

Figure 10.3 summarize estimates from seven recent studies of electricity costs for new power plants with different fuels. The ranges incorporate only internalized costs. If high enough priority is given to improving national energy self-sufficiency, for example, the preferred choice in a specific situation might not be the least expensive (IAEA, 2006).

Different technologies have different costs. Proven designs may cost less than first-of-a kind reactors, and building a first-of-a-kind reactor will likely cost more than building subsequent reactors of the same design. Different estimates also incorporate different learning rates in anticipating how costs will decrease with experience.

Different perspectives can also lead to different estimates. A 2006 report by the UK Sustainable Development Commission stated that vendors of reactor systems had a clear market incentive, especially ahead of contractual commitments, to underestimate costs. Utilities may have a tendency to be more conservative.

Another contributor to higher overall cost estimates may be the fact that the greater share of those estimates come from Europe and especially North America, where the lack of recent construction experience relative to Asia and new reactor designs likely contribute to the higher estimates.

Construction delays can add significantly to the cost of a plant. Because a power plant does not yield profits during construction, longer construction times translate directly into higher finance charges.

In some countries in the past unexpected changes in licensing, inspection and certification of nuclear power plants added delays and increased construction costs. However, the regulatory processes for siting, licensing, and constructing have been standardized, streamlining the construction of newer and safer designs.

At the end of a nuclear plant's lifetime (estimated at between 40 and 60 years), the plant must be decommissioned. Operators are usually required to build up a fund to cover the decommissioning costs while the plant is operating, to limit the financial risk from operator bankruptcy.

The cost per unit of electricity produced (kWh) will vary according to country, depending on costs in the area, the regulatory regime and consequent financial and other risks, and the availability and cost of finance. Costs will also depend on geographic factors such as availability of cooling water, earthquake likelihood, and availability of suitable power grid connections. So it is not possible to accurately estimate costs on a global basis (NEI, 2007).

In 2003, the Massachusetts Institute of Technology (MIT) issued a report entitled, "The Future of Nuclear Power". They estimated that new nuclear power in the US would cost 6.7 cents per kWh. However, the Energy Policy Act of 2005 includes a tax credit that should reduce that cost slightly. In 2009, MIT updated its 2003 study, concluding that inflation and rising construction costs had increased the overnight cost of nuclear power plants to about $4 000/kWe, and thus increased the power cost to 8.4 cents/kWh.

The lifetime cost of new generating capacity in the United States was estimated in 2006 by the U.S. government at 5.93 cents per kWh. A 2008 study based on historical outcomes in the U.S. said costs for nuclear power can be expected to run $0.25–0.30 per kWh. A 2008 study concluded that if carbon capture and storage was required then nuclear power would be the cheapest source of electricity even at $4 038/kW in overnight capital cost (WNA, 2010).

10.6 NUCLEAR SAFETY

Nuclear safety covers the actions taken to prevent nuclear and radiation accidents or to limit their consequences. This covers nuclear power plants as well as all other nuclear facilities, the transportation of nuclear materials, the use and storage of nuclear materials for medical, power, industry, and military uses.

Modern nuclear power plants have a defense-in-depth plan for safety. First layer of defense is the inert, ceramic quality of the uranium oxide itself. Second layer is the airtight zirconium alloy of the fuel rod. Third layer is the reactor pressure vessel made of steel more than a dozen centimeters thick. Fourth layer is the pressure resistant, airtight containment building. The fifth layer is the reactor building or in newer power plants a second outer containment building (IAEA, 2009c).

Two major accidents involving nuclear reactors are the Three Mile Island accident in USA and the Chernobyl accident in the then USSR (now in Ukraine). These two accidents contributed to slowing the growth of nuclear energy since 1990s.

The Three Mile Island accident was a partial core meltdown in Unit 2, a pressurized water reactor, in Dauphin County, Pennsylvania near Harrisburg. It was the most significant accident in USA, resulting in the release of up to 481 PBq (13 million curies) of radioactive gases, but less than 740 GBq (20 curies) of the particularly dangerous ^{131}I.

The accident began at 4:00 a.m. on Wednesday, March 28, 1979, with failures in the non-nuclear secondary system, followed by a stuck-open pilot-operated relief valve (PORV) in the primary system, which allowed large amounts of reactor coolant to escape. The mechanical failures were compounded by the initial failure of plant operators to recognize the situation as a loss of coolant accident due to inadequate training and human factors industrial design errors relating to ambiguous control room indicators in the power plant's user interface (Walker, 2004).

There were no human fatalities in this accident.

Chernobyl nuclear accident occurred on 26 April 1986 at the Chernobyl Nuclear Power Plant in the Ukrainian Soviet Socialist Republic (then part of the Soviet Union), now in Ukraine. It is considered to be the worst nuclear power plant disaster in history and resulted in a severe release of radioactivity following a massive power excursion that destroyed the reactor. The accident raised concerns about the safety of the Soviet nuclear power industry as well as nuclear power in general, slowing its expansion for a number of years while forcing the Soviet government to become less secretive.

On 26 April 1986 at 1:23 a.m., reactor 4 suffered a massive, catastrophic power excursion. This caused a steam explosion, followed by a second (chemical, not nuclear) explosion from the ignition of generated hydrogen mixed with air, which tore the top from the reactor and its building, and exposed the reactor core. This dispersed large amounts of radioactive particulate and gaseous debris containing fission products including ^{137}Cs and ^{90}Sr and other highly radioactive reactor waste products (IAEA, 1992).

The open core also allowed atmospheric oxygen to contact the super-hot core containing 1 700 tonnes of combustible graphite moderator. The burning graphite moderator increased the emission of radioactive particles, carried by the smoke. The reactor was not contained by any kind of hard containment vessel.

The radioactive plume drifted over large parts of the western Soviet Union, Eastern Europe, Western Europe, and Northern Europe, with some nuclear rain falling as far away as Ireland. Large areas in Ukraine, Belarus, and Russia were badly contaminated, resulting in the evacuation and resettlement of over 336 000 people.

Most fatalities from the accident were caused by radiation poisoning. The 2005 report prepared by the Chernobyl Forum, led by the IAEA and World Health Organization (WHO), attributed 56 direct deaths (47 accident workers, and nine children with thyroid cancer), and estimated that there may be 4 000 extra cancer deaths among the approximately 600 000 most highly exposed people. Although the Chernobyl Exclusion Zone and certain limited areas remain off limits, the majority of affected areas are now considered safe for settlement and economic activity.

Safety indicators, such as those published by World Association of Nuclear Operators (WANO) improved dramatically in the 1990s (Figs. 10.4 and 10.5) (IAEA, 2009a). However, in some areas improvement has stalled in recent years, as in the case of unplanned scrams shown in Fig. 10.4.

The gap between the best and worst performers is still large, providing substantial room for continuing improvement. Since the 1986 accident at Chernobyl, enormous efforts have been made in upgrading reactor safety features, but facilities still exist at which nuclear safety assistance should be made a priority.

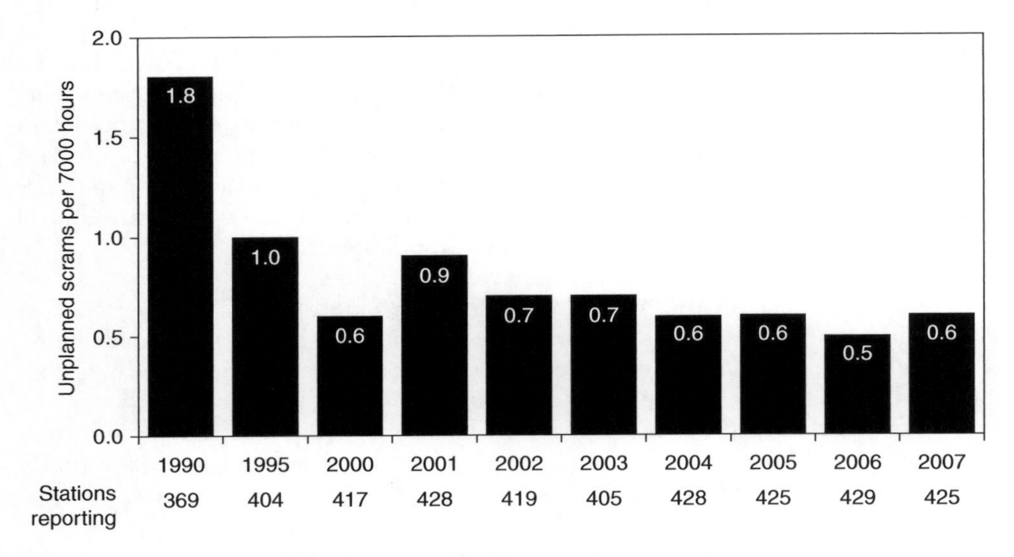

Figure 10.4 Unplanned scrams per 7 000 hours critical
(Source: WANO 2007 Performance Indicators.)

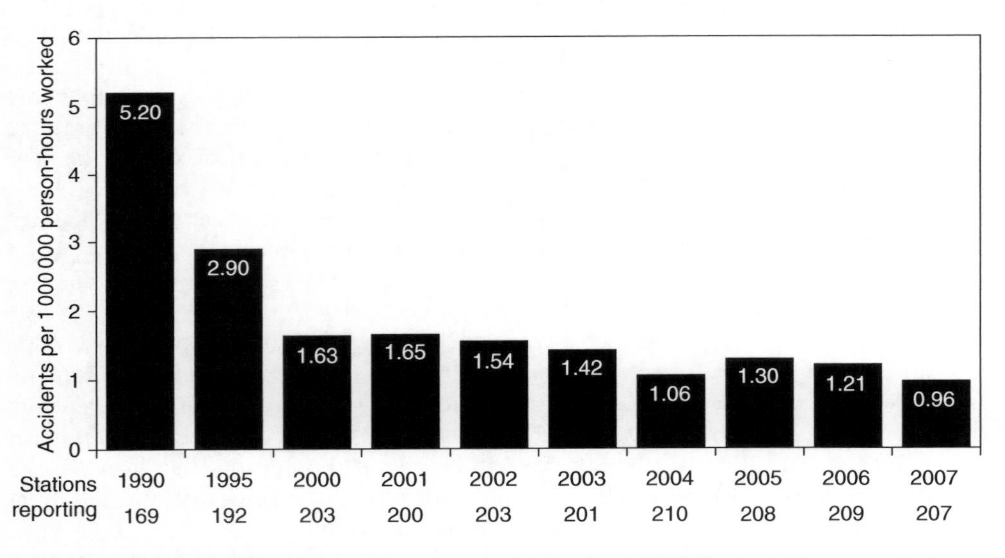

Figure 10.5 Industrial accidents at nuclear power plants per 1 000 000 person-hours worked
(Source: WANO 2007 Performance Indicators.)

In 2008, the International Nuclear Safety Group (INSAG) published Improving the International System for Operating Experience Feedback (IAEA, 2008d). INSAG noted that in all fields of human activity, serious accidents are nearly always preceded by less serious precursor events. If lessons can be learned from the precursors and these lessons

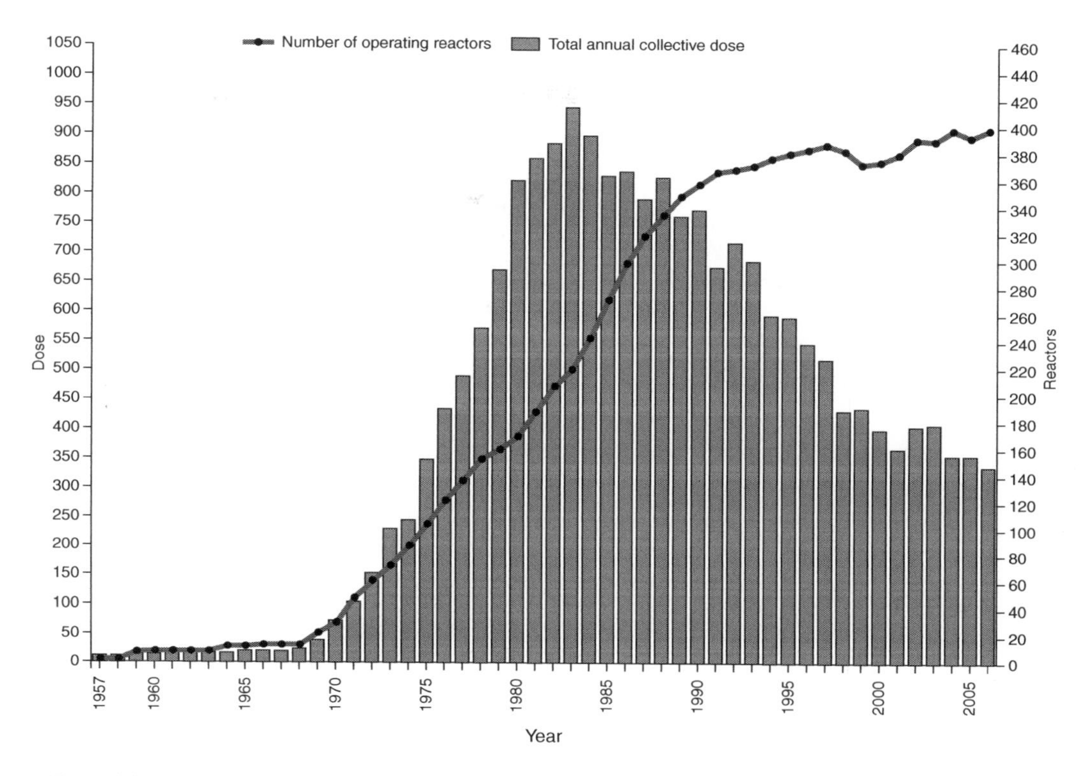

Figure 10.6 Evolution of the total annual collective dose (man Sv) and number of operating reactors

put into practice, the probability of a serious accident occurring can be significantly reduced.

In general, occupational radiation protection in nuclear installations around the world is well managed and few workers in these installations receive significant radiation doses. Figure 10.6 shows the trend for total annual collective dose received by NPP workers. It should be noted that the recent levelling off of the collective dose over the past three years is mainly the result of the completion of earlier successful and significant efforts at optimization of radiation protection over the past ten years (IAEA, 2009c).

Contrary to other exposures to ionizing radiation, which have remained constant or decreased over the past decade, medical exposures have increased at a remarkable rate. After natural background radiation, medical uses constitute the next largest source of ionizing radiation to the world's population (Fig 10.7).

10.7 DISPOSAL OF NUCLEAR WASTES

Storing high level nuclear waste above ground for a century or so is considered appropriate. This allows the material to be more easily observed and any problems detected

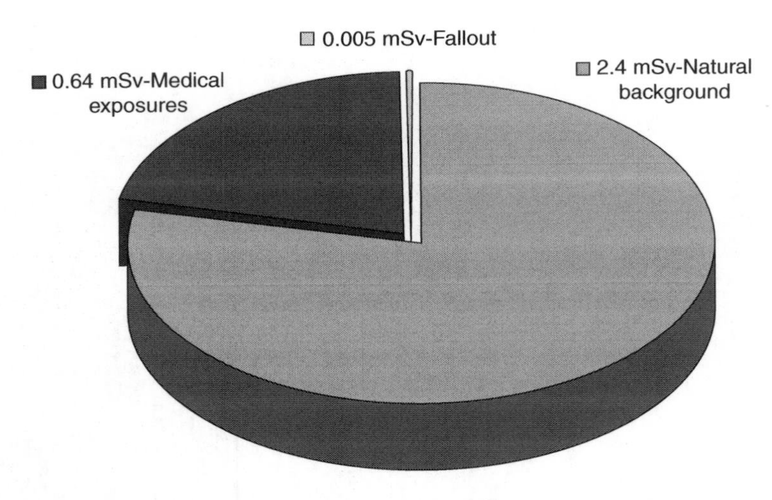

Figure 10.7 World average radiation exposure for a person

and managed, while decay of radionuclides over this time period significantly reduces the level of radioactivity and associated harmful effects to the container material.

It is also considered likely that over the next century newer materials will be developed which will not break down as quickly when exposed to a high neutron flux, thus increasing the longevity of the container once it is permanently buried (Saling and Fentiman, 2002).

The process of selecting appropriate deep final repositories for high level waste and spent fuel is now under way in several countries. The basic concept is to locate a large, stable geologic formation and use mining technology to excavate a tunnel, or large-bore tunnel boring machines to drill a shaft 500–1 000 meters below the surface where rooms or vaults can be excavated for disposal of high-level radioactive waste. The goal is to permanently isolate nuclear waste from the human environment.

Because some radioactive species have half-lives longer than one million years, even very low container leakage and radionuclide migration rates must be taken into account. It may require more than one half-life until some nuclear materials lose enough radioactivity to no longer be lethal to living things.

Sea-based options for disposal of radioactive waste include burial beneath a stable abyssal plain, burial in a subduction zone that would slowly carry the waste downward into the Earth's mantle, and burial beneath a remote natural or human-made island. They are currently not being seriously considered because of the legal barrier of the Law of the Sea and because in North America and Europe sea-based burial has become taboo from fear that such a repository could leak and cause widespread damage.

The proposed land-based subductive waste disposal method disposes of nuclear waste in a subduction zone accessed from land and therefore is not prohibited by international agreement. This method has been described as the most viable means of disposing of radioactive waste and as the state-of-the-art in nuclear waste disposal technology.

Another approach termed "Remix and Return" would blend high-level waste with uranium mine and mill tailings down to the level of the original radioactivity of the uranium ore, then replace it in inactive uranium mines. This approach has the merits of providing jobs for miners who would double as disposal staff, and of facilitating a cradle-to-grave cycle for radioactive materials.

There have been proposals for reactors that consume nuclear waste and transmute it to other, less-harmful nuclear waste. In particular, the Integral Fast Reactor was a proposed nuclear reactor with a nuclear fuel cycle that produced no transuranic waste and in fact, could consume transuranic waste.

Another option is to find applications of the isotopes in nuclear waste so as to re-use them. Already, ^{137}Cs, ^{90}Sr and a few other isotopes are extracted for certain industrial applications such as food irradiation and radioisotope thermoelectric generators. While re-use does not eliminate the need to manage radioisotopes, it may reduce the quantity of waste produced.

Space disposal is an attractive notion because it permanently removes nuclear waste from the environment. However, it has significant disadvantages, not least of which is the potential for catastrophic failure of a launch vehicle. Furthermore, the high number of launches that would be required, due to the fact that no individual rocket would be able to carry very much of the material relative to the material needed to be disposed of.

In the future, alternative, non-rocket space launch technologies may provide a solution. It has been suggested that through the use of a stationary launch system many of the risks of catastrophic launch failure could be avoided. A promising concept is the use of high power lasers to launch "indestructible" containers from the ground into space.

Chapter 11

Next generation green technologies

T. Harikrishnan (IAEA)

11.1 INTRODUCTION

Renewable energy technologies are essential contributors to sustainable energy as they contribute to world energy security by reducing dependence on fossil fuel resources, and providing opportunities for mitigating greenhouse gases. The three generations of renewable technologies, reaching back more than 100 years are:

- *First-generation technologies* include hydropower, biomass combustion, and geothermal power and heat.
- *Second-generation technologies* include solar heating and cooling, wind power, modern forms of bioenergy, and solar photovoltaics.
- *Third-generation technologies* are still under development and include advanced biomass gasification, enhanced geothermal system, and marine energy.

First- and second-generation technologies have entered the markets. Third-generation technologies are not yet widely demonstrated or commercialized. They may have potential comparable to other renewable energy technologies (IEA, 2007).

Bioenergy or biofuel technologies being developed today, notably cellulosic ethanol biorefineries, could allow biofuels to play a much bigger role in the future. Crop residues, such as corn stalks, wheat straw and rice straw and wood waste and municipal solid waste are potential sources of cellulosic biomass. Dedicated energy crops, such as switchgrass, are also promising cellulose sources that can be sustainably produced in many regions of the United States.

Biomass gasification is potentially more efficient than direct combustion of the original fuel. Syngas can be burned directly in internal combustion engines, used to produce methanol and hydrogen. It can be converted by the Fischer-Tropsch process into synthetic fuel.

There is an increased interest in alga-culture or farming algae for making vegetable oil, biodiesel, bioethanol, biogasoline, biomethanol, biobutanol and other biofuels, using land that is not suitable for agriculture. Algal fuels do not affect fresh water resources. They can be produced using ocean and wastewater, and are biodegradable and relatively harmless to the environment.

Department of Energy, USA estimates that if algae fuel replaced all the petroleum fuel in the United States, it would require an area of 40 000 km^2. This is less than 15% the area of corn harvested.

Marine energy encompasses wave energy and tidal energy obtained from oceans, seas, and other large bodies of water. Portugal has the world's first commercial wave farm, the Aguçadora Wave Park. The farm will initially generate 2.25 MW of power. Funding for a wave farm in Scotland was announced in 2007 at a cost of over 4 million pounds. The farm will be the world's largest with a capacity of 3 MW.

In 2007, the world's first turbine to create commercial amounts of energy using tidal power was installed in the narrows of Strangford Lough in Ireland. The 1.2 MW underwater tidal electricity generators take advantage of the fast tidal flow, which can be up to 4 m/s.

Enhanced Geothermal Systems (EGS) systems are currently being developed and tested in France, Australia, Japan, Germany, the U.S. and Switzerland. EGS do not require natural convective hydrothermal resources. The largest EGS project in the world is a 25 megawatt demonstration plant currently being developed in the Cooper Basin, Australia. The Cooper Basin has the potential to generate 5 000–10 000 MW.

11.2 BIOMASS GASIFICATION

Biomass has been a major energy source, prior to the discovery of fossil fuels like coal and petroleum. Even though its role is presently diminished in developed countries, it is still widely used in rural communities of the developing countries. There are five accepted technologies for converting biomass fuels into electrical energy:

- *Conventional steam cycle* – biomass is burned to produce steam which is then used to drive a turbine
- *Gasification* – biomass is converted to a gas using a high temperature oxygen starved environment
- *Pyrolysis* – biomass is converted to a liquid rather than a gas
- *Anaerobic digestion* – typically sewage sludge is digested to produce methane
- *Landfill gas* – collection of gas from landfill sites

Gasification is the most attractive of the technologies, but also one of the least developed. Gasification is the process of converting solid fuels to gaseous fuel. There are a number of practical and engineering issues with gasification which, until now, have been a barrier to full commercial roll out of this technology.

The biomass integrated gasifier/gas turbine combined cycle (BIG/GTCC) is not yet commercially employed. Substantial demonstration and commercialisation efforts are ongoing worldwide. Overall economics of biomass-based power generation should improve considerably with BIG/GTCC systems.

Biomass gasification is a process that converts biomass into carbon monoxide and hydrogen at high temperatures with a controlled amount of oxygen. The resulting gas mixture is called synthesis gas or syngas and is itself a fuel. Gasification relies on chemical processes at elevated temperatures >700°C, which distinguishes it from biological processes such as anaerobic digestion that produce biogas.

Pyrolysis is only one of the steps in the conversion process. The other steps are combustion with air and reduction of the product of combustion, water vapour and carbon dioxide into combustible gases, carbon monoxide, hydrogen, methane, some higher hydrocarbons and inert gases, carbon dioxide and nitrogen.

Using the syngas is potentially more efficient than direct combustion of the original fuel. Syngas can be burned directly in internal combustion engines, used to produce methanol and hydrogen. It can be converted by the Fischer-Tropsch process into synthetic fuel.

Any type of organic material can be used as the raw material for gasification, such as wood, biomass, or even plastic waste. High-temperature combustion refines out corrosive ash elements such as chloride and potassium, allowing clean gas production from otherwise problematic fuels. Gasification of fossil fuels is currently widely used on industrial scales to generate electricity.

11.2.1 Biomass

Biomass, a renewable energy source, is biological material derived from living, or recently living organisms, such as wood, waste, and alcohol fuels. Biomass is commonly plant matter grown to generate electricity or produce heat. Forest residues such as dead trees, branches and tree stumps, as well as yard clippings, wood chips and garbage may be used as biomass.

Biomass also includes plant or animal matter used for production of fibers or chemicals. Biomass may also include biodegradable wastes that can be burnt as fuel. It excludes organic materials such as fossil fuels which have been transformed by geological processes into substances such as coal or petroleum.

Although fossil fuels have their origin in ancient biomass, they are not considered biomass because they contain carbon that has been excluded of the carbon cycle for a very long time. Their combustion therefore disturbs the carbon dioxide content in the atmosphere.

Biomass is a natural substance available, which stores solar energy by the process of photosynthesis in the presence of sunlight. It chiefly contains cellulose, hemicellulose and lignin, with an average composition of $CH_6H_{10}O_5$.

Industrial biomass can be grown from numerous types of plants, including miscanthus, switchgrass, hemp, corn, poplar, willow, sorghum and sugarcane. It can come from a variety of tree species, ranging from eucalyptus to oil palm. The particular plant used is usually not important to the end products, but it does affect the processing of the raw material (Volk et al, 2000).

There are five basic categories of material:

- *Virgin wood:* from forestry, arboriculture activities or from wood processing;
- *Energy crops:* high yield crops grown specifically for energy applications;

Table 11.2.1 Gasification processes

Process	*Products*	*Remarks*
Pyrolysis	H_2, CH_4, char	Dependent on the properties of the carbonaceous material and determines the structure and composition of the char (solid carbon residue), which will then undergo gasification reactions.
Combustion	CO_2, CO	Char reacts with oxygen to form to form the products $C + \frac{1}{2} O_2 \rightarrow CO$
Gasification	CO, H_2	Char reacts with carbon dioxide and steam
Water gas shift reaction	CO_2, H_2	Reaches equilibrium very fast at the temperatures in a gasifier and balances the concentrations of carbon monoxide, steam, carbon dioxide and hydrogen. $CO + H_2O \leftrightarrow CO_2 + H_2$

- *Agricultural residues:* residues from agriculture harvesting or processing
- *Food waste:* from food and drink manufacture, preparation and processing, and post-consumer waste
- *Industrial waste and co-products:* from manufacturing and industrial processes.

The largest source of energy from wood is pulping liquor or "black liquor", a waste product from processes of the pulp, paper and paperboard industry. Waste energy is the second largest source of biomass energy. The main contributors of waste energy are municipal solid waste (MSW), manufacturing waste, and landfill gas.

11.2.2 Gasification

Gasification process was originally developed in the 1800s in Europe and elsewhere, to produce town gas for lighting and cooking. Electricity and natural gas later replaced town gas since the 1920s. The gasification process has been utilized for the production of synthetic chemicals and fuels since then.

In a gasifier, the carbonaceous material undergoes several different processes (Table 11.2.1). Limited amount of oxygen or air is introduced into the reactor to allow some of the organic material to be "burned" to produce carbon monoxide and energy, which drives a second reaction that converts further organic material to hydrogen and additional carbon dioxide.

Gasification is a two-stage reaction consisting of oxidation and reduction processes. These processes occur under sub-stoichiometric conditions of air with biomass. The first part of sub-stoichiometric oxidation leads to the loss of volatiles from biomass and is exothermic. It results in peak temperatures of 1 400 to 1 500 K and generation of gaseous products like carbon monoxide, hydrogen in some proportions and carbon dioxide and water vapour.

The products are in turn are reduced in part to carbon monoxide and hydrogen by the hot bed of charcoal generated during the process of gasification. Reduction reaction is an endothermic reaction to generate combustible products like CO, H_2 and CH_4. Since char is generated during the gasification process the entire operation is self-sustaining.

Four types of gasifiers are currently available:

1. *Counter-current fixed bed*: This consists of a fixed bed of carbonaceous fuel through which the "gasification agent" (steam, oxygen and/or air) flows in counter-current configuration. The ash is either removed dry or as a slag. The slagging gasifiers have a lower ratio of steam to carbon, achieving temperatures higher than the ash fusion temperature.
2. *Co-current fixed bed*: This is similar to the counter-current type, but the gasification agent gas flows in co-current configuration with the fuel (downwards). Heat needs to be added to the upper part of the bed, either by combusting small amounts of the fuel or from external heat sources. The produced gas leaves the gasifier at a high temperature, and most of this heat is often transferred to the gasification agent added in the top of the bed.
3. *Fluidized bed*: Here the fuel is fluidized in oxygen and steam or air. The ash is removed dry or as heavy agglomerates that defluidize. The temperatures are relatively low in dry ash gasifiers, so the fuel must be highly reactive; low-grade coals are particularly suitable. The agglomerating gasifiers have slightly higher temperatures, and are suitable for higher rank coals.
4. *Entrained flow*: Here a dry pulverized solid, an atomized liquid fuel or fuel slurry is gasified with oxygen or air in co-current flow. The gasification reactions take place in a dense cloud of very fine particles. Most coals are suitable for this type of gasifier because of the high operating temperatures and because the coal particles are well separated from one another.

Advantages and disadvantages of the above gasifiers are shown in Table 11.2.2.

Gasification can utilize any organic material, including biomass and plastic waste. The resulting syngas can be combusted. Alternatively, if the syngas is clean enough, it may be used for power production in gas engines, gas turbines or even fuel cells, or converted efficiently to dimethyl ether (DME) by methanol dehydration, methane via the Sabatier reaction, or diesel-like synthetic fuel via the Fischer-Tropsch process.

Demonstration projects of biomass gasification include those of the Renewable Energy Network Austria, including a plant using dual fluidized bed gasification that has supplied the town of Güssing with 2 MW of electricity and 4 MW of heat, generated from wood chips, since 2003. Where the wood source is sustainable, 250–1 000 kWe and new zero carbon biomass gasification plants have been installed in Europe that produce tar free syngas.

Gasifiers offer a flexible option for thermal applications, as they can be retrofitted into existing gas fuelled devices such as ovens, furnaces, boilers, etc., where syngas may replace fossil fuels. Heating values of syngas are generally around 4–10 MJ/m^3.

Diesel engines can be operated on dual fuel mode using producer gas. Diesel substitution of over 80% at high loads and 70–80% under normal load variations can easily be achieved. Spark ignition engines and SOFC fuel cells can operate on 100% gasification gas. Mechanical energy from the engines may be used for e.g. driving water pumps for irrigation or for coupling with an alternator for electrical power generation.

Small-scale rural biomass gasifiers have been applied in India to a large extent. Most of the applications are 9 kWe systems used for water pumping and street lighting. The open top, twin air entry, re-burn gasifier developed at Combustion, Gasification and

Table 11.2.2 Advantages and disadvantages of different types of gasifiers

Gasifier	*Advantages*	*Disadvantages*
Counter-current fixed bed	High thermal efficiency.	Relatively low throughput. Relatively low gas exit temperatures. Fuel must have high mechanical strength and must be non-caking. Tar and methane production significant (so product gas must be cleaned before use).
Co-current fixed bed	High gas exit temperatures. High energy efficiency. Tar levels low.	
Fluidized bed	Low-grade coal can be used in dry ash gasifiers and higher rank coal in agglomerating gasifiers Useful for fuels that form highly corrosive ash such as biomass	Low conversion efficiency.
Entrained flow	Most coals are suitable. High throughput. Tar and methane are not present.	Low thermal efficiency as the gas must be cooled before it can be cleaned with existing technology High oxygen requirement Certain types of biomasses form corrosive slag Some fuels must be pulverized and mixed with limestone

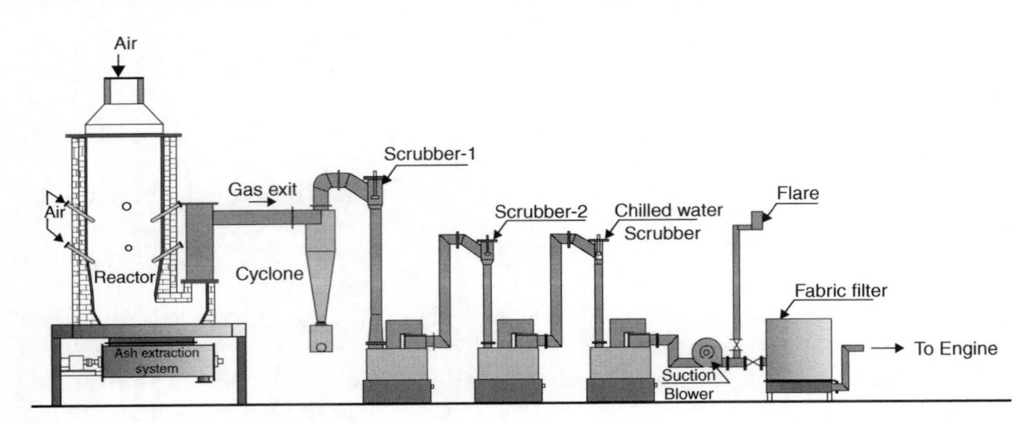

Figure 11.2.1 Open top twin air entry re-burn gasifier

Propulsion Laboratory (CGPL) of Indian Institute of Science generates superior quality producer gas (Fig 3.3.2.1). More than 40 plants are successfully operating in India and elsewhere for heat and power applications.

11.2.3 Syngas

Syngas is the gas mixture that contains varying amounts of carbon monoxide and hydrogen. Production methods include steam reforming of natural gas or liquid hydrocarbons to produce hydrogen, the gasification of coal, biomass, and in some types of waste-to-energy gasification facilities. Syngas consists primarily of hydrogen, carbon monoxide, and very often some carbon dioxide, and has less than half the energy density of natural gas. Syngas is combustible and often used as a fuel source or as an intermediate for the production of other chemicals.

Syngas is also used as intermediate in producing synthetic petroleum for use as a fuel or lubricant via Fischer-Tropsch synthesis and previously the Mobil methanol to gasoline process (Beychok, 1974). Syngas for use as a fuel is most often produced from coal, first by pyrolysis to coke (impure carbon), aka destructive distillation, followed by alternating blasts of steam and air, or from biomass or municipal waste mainly by the following paths:

$$C + H_2O \rightarrow CO + H_2$$

$$C + O_2 \rightarrow CO_2$$

$$CO_2 + C \rightarrow 2CO$$

The syngas produced in large waste-to-energy gasification facilities can be used to generate electricity. Coal gasification processes were used for many years to manufacture illuminating gas (coal gas) for gas lighting and to some extent, heating, before electric lighting and the natural gas infrastructure became widely available.

Syngas can be used in the Fischer-Tropsch process to produce diesel, or converted into methane and dimethyl ether in catalytic processes.

11.2.4 Fisher–Tropsch process

The Fischer–Tropsch process is a catalyzed chemical reaction in which syngas is converted into liquid hydrocarbons of various forms. The most common catalysts are based on iron and cobalt, although nickel and ruthenium have also been used. The principal purpose of this process is to produce a synthetic petroleum substitute, typically from coal, natural gas or biomass, for use as synthetic lubrication oil or as synthetic fuel.

This synthetic fuel can run truck, car, and some aircraft engines. The use of diesel is increasing in recent years. Combination of biomass gasification (BG) and Fischer-Tropsch (FT) synthesis is a possible route to produce renewable transportation fuels (biofuels) (Inderwildi et al, 2008).

The Fischer–Tropsch process involves a variety of competing chemical reactions, which lead to a series of desirable products and undesirable byproducts. The most important reactions are those resulting in the formation of alkanes. These can be described by chemical equations of the form:

$$(2n + 1)H_2 + nCO \rightarrow CnH(2n + 2) + nH_2O$$

where ‘n’ is a positive integer.

The simplest of these $(n = 1)$, results in formation of methane, which is generally considered an unwanted byproduct (particularly when methane is the primary feedstock used to produce the synthesis gas). Process conditions and catalyst composition are usually chosen to favor higher order reactions (n>1) and thus minimize methane formation.

The process was invented in petroleum-poor but coal-rich Germany in the 1920s, to produce liquid fuels. It was used in Germany and Japan during World War II to produce substitute fuels. By early 1944 production reached more than 124 000 barrels per day (19 700 m^3/d) from 25 plants ~6.5 million tons per year.

Currently a handful of companies have commercialized FT technology. Shell in Bintulu, Malaysia, uses natural gas as a feedstock, and produces primarily low-sulfur diesel fuels and food-grade wax. Choren Industries has built an FT plant in Germany.

The process is used in South Africa, a country with large coal reserves but lacking in oil. Sasol in South Africa uses coal and natural gas as a feedstock, and produces a variety of synthetic petroleum products. Sasol production meets most of the country's diesel fuel requirements.

This process has received renewed attention in the quest to produce low-sulfur diesel fuel in order to minimize environmental degradation from the use of diesel engines. US-based company, Rentech, is currently focusing on converting nitrogen-fertiliser plants from using a natural gas feedstock to using coal or coke, and producing liquid hydrocarbons as a co-product.

11.2.5 Biomass Integrated-gasifier/gas turbine combined cycle

The biomass integrated-gasifier/gas turbine combined cycle (BIG/GTCC) technology was first identified in 1990s as an advanced technology. It has the potential to be cost-competitive with conventional condensing-extraction steam-turbine (CEST) technology using biomass as fuel, while dramatically increasing the electricity generated (Larson et al., 2001).

The basic elements of a BIG/GTCC power plant include a biomass dryer, a gasifier for converting the biomass into a combustible fuel gas, a gas cleanup system, a gas turbine-generator fuelled by combustion of the biomass-derived gas, a heat recovery steam generator (HRSG) to raise steam from the hot exhaust of the gas turbine, and a steam turbine-generator to produce additional electricity.

Three variations of this basic configuration are under commercial development. The principal differences among the variants arise from the design of the gasifier. Variant 1 involves a fluidized-bed reactor operating at atmospheric pressure using air for partial oxidation of the biomass.

TPS, Sweden, uses a second gasification stage with dolomite catalyst to reduce the content of tars. The elimination of tars is required to prevent downstream operating difficulties that can arise from tar condensation. The product gas from the tar "cracker" is cooled from about 900°C to near-ambient temperature, cleaned, and compressed to the pressure needed for injection into the gas turbine combustor.

Variant 2 involves operating the gasifier near atmospheric pressure and using some form of indirect heating of the biomass, rather than partial combustion. In the design of Battelle Columbus Laboratory, USA, hot sand carries heat to gasify most of the biomass, leaving behind some char.

The char and sand are circulated to a second reactor, where air is introduced to burn the char and thereby reheat the sand. The product gas passes through a tar cracking unit and is then cooled, cleaned and compressed to fuel the gas turbine. The heat in the combustion products leaving the char-burning reactor is recovered to raise steam, to dry biomass fuel, or for other useful purposes.

Variant 3 involves operating a fluidized-bed gasifier under elevated pressure using air for the partial oxidation. The gasifier product gas is cooled only modestly, cleaned at elevated temperature using a ceramic or sintered metal filter, and then passed to the gas turbine combustor. Leading developers of the pressurized gasifier concept include Foster-Wheeler, USA and Carbona, Finland.

11.2.6 Environmental benefits of gasification

Regardless of the final fuel form, gasification itself and subsequent processing neither directly emits nor traps greenhouse gasses such as carbon dioxide. Combustion of syngas or derived fuels emits the exact same amount of carbon dioxide as would have been emitted from direct combustion of the initial fuel.

Biomass gasification and combustion could play a significant role in a renewable energy economy, because biomass production removes the same amount of CO_2 from the atmosphere as is emitted from gasification and combustion. While other biofuel technologies such as biogas and biodiesel are carbon neutral, gasification in principle may run on a wider variety of input materials and can be used to produce a wider variety of output fuels.

Gasification plants produce significantly lower quantities of air pollutants. Gasification can reduce the environmental impact of waste disposal because it can use waste products as feedstocks—generating valuable products from materials that would otherwise be disposed as wastes.

Gasification's byproducts are non-hazardous and are readily marketable. Gasification plants use significantly less water than traditional coal-based power generation, and can be designed so they recycle their process water, discharging none into the surrounding environment.

CO_2 can be captured from an industrial gasification plant using commercially proven technologies. Great Plains Substitute Natural Gas plant in North Dakota has been capturing the same amount of CO_2 as a 400 MW coal power plant would produce and sending that CO_2through pipeline to Canada for enhanced oil recovery.

11.3 MARINE ENERGY

Marine energy or ocean energy refers to the energy carried by ocean waves and tides. The oceans represent a vast and largely untapped source of energy in the form of fluid flow (currents, waves, and tides) and thermal and salinity gradients. There are a number of approaches to extracting energy from the ocean, though most remain in the investigation or demonstration phase (Twidell et al, 2006).

The term marine energy encompasses both wave energy and tidal energy obtained from oceans, seas, and other large bodies of water. The oceans have a tremendous amount of energy and are close to many concentrated population centers. Many researches show that marine energy has the potentiality of providing for a substantial amount of new renewable energy around the world (Table 11.3.1) (Wick and Schmitt, 1977).

Table 11.3.1 Marine renewable resources

Resource	*Power (TW)*	*Energy Density (m)*
Ocean Currents	0.05	0.05
Ocean Waves	2.7	1.5
Tides	0.03	10
Thermal Gradient	2	210
Salinity Gradient	2.6	240

In the past marine energy technologies received relatively little research funding. However, there is renewed interest in the technology, and several concepts now envisage full-scale demonstration prototypes. But marine energy technologies must still solve some major problems such as,

(i) proving the energy conversion potential;
(ii) overcoming a very high technical risk from a dynamic environment;
(iii) resource assessment;
(iv) energy production forecasting and design tools;
(v) test and measurement standards;
(vi) environmental impacts;
(vii) arrays of farms of ocean energy systems; and
(viii) dual-purpose plants that combine energy and other structures.

11.3.1 Marine current power

Marine current power is a form of power based on the harnessing of the kinetic energy of ocean currents. US Department of the Interior estimates that capturing just 1/1 000 of the available energy from the Gulf Stream, would supply Florida with 35% of its electrical needs (DOI, 2006).

Although not widely used at present, marine current power has an important potential for future electricity generation. Marine currents are more predictable than wind energy and solar power. Ocean currents are driven by wind and solar heating of the waters near the equator, although some ocean currents result instead from variations in water density and salinity.

These currents are relatively constant and flow in one direction only, in contrast to the tidal currents closer to shore. Some examples of ocean currents are the Gulf Stream, Florida Straits Current, and California Current. Ocean currents tend to be concentrated at the surface, although significant current continues at depths.

Ocean current speeds are generally lower than wind speeds. This is important because the kinetic energy contained in flowing bodies is proportional to the cube of their velocity. However, another more important factor in the power available for extraction from a flowing body is the density of the material. Water is about 835 times denser than wind, so for the same area of flow being intercepted, the energy contained in a 20 Km/h water flow is equivalent to that contained in an air mass moving at about 177 Km/h.

Ocean currents represent a significant, currently untapped, reservoir of energy. The total worldwide power in ocean currents has been estimated to be about 5,000 GW, with power densities of up to 15 kW/m^2.

In large areas with powerful currents, it would be possible to install water turbines in groups or clusters to create a marine current facility. Turbine spacing would be determined based on wake interactions and maintenance needs. A 30 MW demonstration array of vertical turbines in a tidal fence is being investigated in the Philippines (WEC, 2001).

However a number of potential problems need to be addressed, including avoidance of drag from cavitations (air bubble formation that creates turbulence and decreases the efficiency of current-energy harvest), prevention of marine growth build up, corrosion control, and overall system reliability. Because the logistics of maintenance are likely to be complex and the costs potentially high, system reliability is of high importance.

Ocean currents flow relatively steadily throughout the year and in some cases the flow is considerable. An example is the Straits of Florida where the Gulf Stream flows out of the Caribbean Sea and into the North Atlantic on its way to northern Europe. The speed of the current is around 7.4 km/h at the surface, but it decreases with depth. There is a potential extractable power of 1 kW/m^2 near the surface.

A 300 kW full-scale plant installed by Marine Current Turbines (MCT) has been operating at Lynmouth, Devon (UK) since May 2003. MCT has also been planning deep sea marine current systems, which could be constructed in large farms and thus use economies of scale both in construction and maintenance and in the infrastructure for bringing the electricity to shore. Another approach which has identified the potential of the Gulf Stream is the Gorlov helical turbine, a vertical-axis turbine which is being currently prototyped in South Korea.

No currently operating commercial turbines are connected to an electric-power transmission or distribution grid; however, a number of configurations are being tested on a small scale. Because no commercial turbines are currently in operation, it is difficult to assess the costs of current-generated energy and its competitiveness with other energy sources. Initial studies suggest that for economic exploitation, velocities of at least 2 m/s would be required, although it is possible to generate energy from velocities as low as 1 m/s.

Major costs of these systems would be the cables to transport the electricity to the onshore grid. There are many similarities and common problems with tidal-current energy extraction.

Potential environmental impacts of ocean current energy extraction include:

- Impacts on marine ecology and conflicts with other potential uses of the same area of the ocean;
- Resource requirements associated with the construction and operation; and
- Protection of species, particularly fish and marine mammals.

The slow blade velocities should allow water and fish to flow freely and safely through the structure. Protective fences and sonar-activated brakes could prevent larger marine mammals from harm. In the siting of the turbines, consideration of impacts on shipping routes, and present as well as anticipated uses such as commercial and recreational fishing and recreational diving, would be required.

The need to introduce possible mitigating factors, such as the establishment of fishery exclusion zones has to be considered. Concerns have been raised about risks from slowing the current flow by extracting energy. Local effects, such as temperature and salinity changes in estuaries caused by changes in the mixing of salt and fresh waters, would need to be considered for their potential impact on estuary ecosystems (Charlier and Justus 1993).

Damage to seabed flora is also potentially dangerous and designs are being explored which are anchored to the seabed but operate at a distance, rather than having towers built on the bed. Since there are at present no firm plans for deployment of these devices, it is difficult to evaluate whether this will be a serious problem.

11.3.2 Ocean thermal energy

Ocean thermal energy conversion (OTEC) uses the temperature difference that exists between deep and shallow waters to run a heat engine. The greatest efficiency and power is produced with the largest temperature difference. This temperature difference generally increases near the equator. The ocean surface contains a vast amount of solar energy, which can potentially be harnessed for human use. If this extraction could be made cost effective on a large scale, it could be a source of renewable energy (Avery and Wu, 1994).

The technical challenge of OTEC is to generate significant amounts of power efficiently from this very small temperature ratio. Changes in efficiency of heat exchange in modern designs allow performance approaching the theoretical maximum efficiency.

The earth's oceans are continually heated by the sun and cover nearly 70% of the surface. This makes them the world's largest solar energy collector and energy storage system. On an average day, 60 million km^2 of tropical seas absorb an amount of solar radiation equal in heat content to about 250 billion barrels of oil.

The total energy available is one or two orders of magnitude higher than other ocean energy options such as wave power. But the small magnitude of the temperature difference makes energy extraction comparatively difficult and expensive, due to low thermal efficiency. Earlier OTEC systems had an overall efficiency of 1 to 3%. The theoretical maximum efficiency lies between 6 and 7%.

Current designs under review will operate closer to the theoretical maximum efficiency. The energy carrier, seawater, is free, though it has an access cost associated with the pumping materials and pump energy costs. An OTEC plant can be configured to operate continuously to supply base load power.

As long as the temperature between the warm surface water and the cold deep water differs by about 20°C, an OTEC system can produce a significant amount of power. The oceans are thus a vast renewable resource, with the potential to help us produce billions of watts of electric power. The cold, deep seawater used in the OTEC process is also rich in nutrients, and it can be used to culture both marine organisms and plant life near the shore or on land.

This cold seawater is an integral part of the three types of OTEC systems: closed-cycle, open-cycle, and hybrid. To operate, the cold seawater must be brought to the surface. This can be accomplished through direct pumping. A second method is to desalinate the seawater near the sea floor; this lowers its density, which will cause it to rise up through a pipe to the surface.

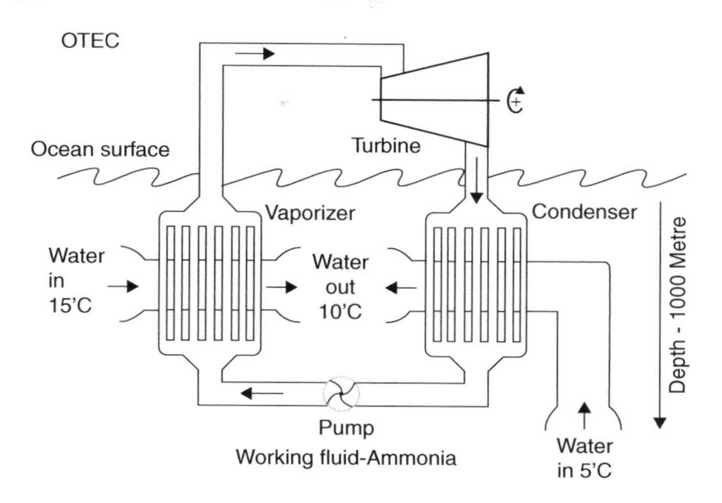

Figure 11.3.1 Scheme of closed cycle OTEC plant

Closed-cycle systems use fluid with a low boiling point, such as ammonia, to rotate a turbine to generate electricity. Warm surface seawater is pumped through a heat exchanger where the low-boiling-point fluid is vaporized. The expanding vapor turns the turbo-generator. Then, cold, deep seawater—pumped through a second heat exchanger—condenses the vapor back into a liquid, which is then recycled through the system (Fig 11.3.1).

In 1979 the Natural Energy Laboratory (NEL) and several private-sector partners developed the mini OTEC experiment, which achieved the first successful at-sea production of net electrical power (Trimble and Owens, 1980). The mini OTEC vessel was moored 2.4 km off the Hawaiian coast and produced enough net electricity to illuminate the ship's light bulbs, and run its computers and televisions. NEL in 1999 tested a 250 kW pilot closed-cycle plant.

Open-cycle OTEC uses the tropical oceans' warm surface water to make electricity. When warm seawater is placed in a low-pressure container, it boils. The expanding steam drives a low-pressure turbine attached to an electrical generator. The steam, which has left its salt and contaminants behind in the low-pressure container, is pure fresh water. It is condensed back into a liquid by exposure to cold temperatures from deep-ocean water. This method has the advantage of producing desalinized fresh water, suitable for drinking water or irrigation.

In 1984 National Renewable Energy Laboratory developed a vertical-spout evaporator to convert warm seawater into low-pressure steam for open-cycle plants. Energy conversion efficiency of 97% was achieved for the seawater-to-steam conversion process. The overall efficiency of an OTEC system was few per cent. In 1993, an open-cycle OTEC plant at Keahole Point, Hawaii, produced 50 000 watts of electricity during a net power-producing experiment.

Hybrid cycle combines the features of both the closed-cycle and open-cycle systems. In a hybrid OTEC system, warm seawater enters a vacuum chamber where it is flash-evaporated into steam, similar to the open-cycle evaporation process. The steam vaporizes the ammonia working fluid of a closed-cycle loop on the other side of an ammonia vaporizer. The vaporized fluid then drives a turbine to produce electricity. The steam condenses within the heat exchanger and provides desalinated water. The electricity produced by the system can be delivered to a utility grid or used to manufacture methanol, hydrogen, refined metals, ammonia, and similar products.

OCEES International, Inc. is working with the U.S. Navy on a design for a proposed 13 MW OTEC plant in Diego Garcia, which would replace the current power plant running diesel generators. The OTEC plant would also provide 1.25 MGD of potable water to the base. Another U.S. company has proposed building a 10 MW OTEC plant in Guam.

Lockheed Martin's Alternative Energy Development team is currently in the final design phase of a 10 MW closed cycle OTEC pilot system which will become operational in Hawaii during 2012–2013. This system is being designed to expand to 100 MW commercial systems in the near future.

OTEC has important benefits other than power production. The 5°C cold seawater made available by an OTEC system creates an opportunity to provide large amounts of cooling to operations that are related to or close to the plant. The cold seawater from an OTEC plant can be used in chilled-water coils to provide air-conditioning for buildings.

OTEC technology also supports chilled-soil agriculture. When cold seawater flows through underground pipes, it chills the surrounding soil. The temperature difference between plant roots in the cool soil and plant leaves in the warm air allows many plants that evolved in temperate climates to be grown in the subtropics.

Aquaculture can be a byproduct of OTEC. Deep ocean water contains high concentrations of essential nutrients that are depleted in surface waters due to biological consumption. This "artificial upwelling" mimics the natural upwelling that is responsible for fertilizing and supporting marine ecosystems.

Desalinated water can be produced in open- or hybrid-cycle plants using surface condensers. In a surface condenser, the spent steam is condensed by indirect contact with the cold seawater. Studies indicate that a 2 MWe net plant could produce about 4 300 m^3 of desalinated water each day.

Hydrogen can be produced via electrolysis using electricity generated by the OTEC process. The steam generated can be used as a relatively pure medium for electrolysis with electrolyte compounds added to improve the overall efficiency.

It will be possible to extract many elements contained in salts and other forms and dissolved in sea water. In the past, most economic analyses concluded that mining the ocean for trace elements dissolved in solution would be unprofitable, in part because much energy is required to pump the large volume of water needed. The Japanese recently began investigating the concept of combining the extraction of uranium dissolved in seawater with wave-energy technology.

The economics of energy production today have delayed the financing of a permanent, continuously operating OTEC plant. OTEC is very promising as an alternative energy resource for tropical island communities that rely heavily on imported fuel. OTEC could provide the islands with much-needed power, as well as desalinated water and a variety of aquaculture products.

Because OTEC systems have not yet been widely deployed, estimates of their costs are uncertain. One study estimates power generation costs as low as US $0.07 per kilowatt-hour, compared with $0.05–$0.07 for subsidized wind systems.

Future research needed to accelerate the development of OTEC systems include:

- Characterization of cold-water pipe technology;
- Advanced heat exchanger systems to improve heat transfer performance and decrease costs; and
- Innovative turbine concepts for the large machines required for open-cycle systems

11.3.3 Salinity gradient power

Salinity gradient power is the energy retrieved from the difference in the salt concentration between seawater and river water. Two practical methods for this are reverse electro-dialysis (RED) and pressure retarded osmosis (PRO). Both processes rely on osmosis with ion specific membranes. Osmotic pressure is the chemical potential of concentrated and dilute solutions of salt.

All energy that is proposed to use salinity gradient technology relies on the evaporation to separate water from salt. Solutions with higher concentrations of salt have higher osmotic pressure.

The technologies have been tested in laboratory conditions. They are being developed on commercial scales in the Netherlands (RED) and Norway (PRO). Though the cost of the membrane is quite high, a new cheap membrane, based on an electrically modified polyethylene plastic, has been proposed. The world's first osmotic plant with capacity of 4 kW was established in 2009 in Tofte, Norway.

Other methods have been proposed and are currently under development include that based on electric double layer capacitor and vapor pressure difference technologies. (Olsson et al, 1979; Brogioli, 2009).

The osmotic pressure difference between fresh water and seawater is equivalent to 240 m of hydraulic head. Theoretically a stream flowing at 1 m^3/s could produce 1 MW of electricity. The worldwide fresh to seawater salinity resource is estimated at 2.6 TW. This is comparable to the ocean thermal gradient estimated at 2.7 TW. Inland highly saline lakes have higher potential. The Dead Sea osmotic pressure differential corresponds to a head of 5 000 m, which is almost twenty times greater than seawater.

Salinity gradient power is a specific renewable energy alternative that creates renewable and sustainable power by using naturally occurring processes. This practice does not contaminate or release CO_2 emissions. Vapor pressure methods will release dissolved air containing CO_2 at low pressure, but these non-condensable gases can be re-dissolved.

In PRO, a membrane separates two solutions, salt water and fresh water. Only water molecules can pass the semi-permeable membrane. As a result of the osmotic pressure difference between both solutions, fresh water will diffuse through the membrane in order to dilute the solution. The pressure drives the turbines and powers the generator that produces the electrical energy (Brauns, 2007).

RED is the salinity gradient energy retrieved from the difference in the salt concentration between seawater and river water. A salt solution and fresh water are let through a stack of alternating cathode and anode exchange membranes. The chemical potential difference between salt and fresh water generates a voltage over each membrane and the total potential of the system is the sum of the potential differences over all membranes.

RED process works through difference in ion concentration instead of an electric field, which has implications for the type of membrane needed. As in a fuel cell, the cells are stacked. A module with a capacity of 250 kW has the size of a shipping container.

In the Netherlands more than 3 300 m^3 fresh water runs into the sea per second on average. The membrane halves the pressure differences which results in a water column of approximately 135 meters. The energy potential is 4.5 GW.

There has generally been a lack of systematic research and development activity in this area. Early technical advances were not considered promising, mainly because they relied on expensive membranes. Membrane technologies have advanced, but to date, they remain the technical barrier to economical energy production. Efforts are underway to address those issues and alternatively develop designs that eliminate membrane. Additional challenges include high capital costs and low efficiency (Jones and Rowley, 2003).

Principal advantages are no fuel cost, no CO_2 emissions or other significant effluents that may interfere with global climate. Inefficient extraction would be acceptable as long as there is an adequate return on investment. Salts are not consumed in the process. Systems could be non-periodic, unlike wind or wave power. Systems can be designed for large or small-scale plants and could be modular in layout.

11.3.4 Tidal power

Tidal power is a form of hydropower that converts the energy of tides into electricity or other useful forms of power. Tidal power has potential for future electricity generation. Tides are more predictable than wind energy and solar power (Baker, 1991).

Tidal power is the only form of energy which derives directly from the relative motions of the earth–moon system, and to a lesser extent from the earth–sun system. The tidal forces produced by the moon and sun, in combination with earth's rotation, are responsible for the generation of the tides.

For producing significant amount of energy out of tidal water turbines, range of tides should be high. Substantial amount of water should be there for pushing water through the turbine. Approximately 4 to 5 r meters range of tides is require producing significant amount electricity.

It is significantly important to spot the appropriate place which provides suitable and sustainable conditions to produce tidal energy. There are plenty of places around the globe which provide good conditions for installing water turbines. The Bay of Fundy in Canada and the Bristol Channel between England and Wales are two particularly noteworthy examples.

The magnitude of the tide at a location is the result of the changing positions of the moon and sun relative to the earth, the effects of earth rotation, and the local shape of the sea floor and coastlines. The stronger the tide, either in water level height or tidal current velocities, the greater the potential for tidal electricity generation (Hammons, 1993).

Tidal power can be classified into three main types:

- *Tidal stream systems* make use of the kinetic energy of moving water to power turbines.

Table 11.3.2 Operating and proposed tidal power facilities

Country	*Facility*	*Type*	*Capacity (MW)*	*Start year*
France	La Rance	Barrage	240	1966
Canada	Annapolis Royal Generating Station, Nova Scotia	Barrage	18	1984
Canada	Race Rocks Tidal Power Demonstration Project, Vancouver Island	Tidal stream	–	2006
Russia	Kislaya Guba on the Barents Sea	Barrage	0.5	2006
Russia	Penzhinskaya Bay	Tidal stream	–	Proposed
Russia	Kislaya Guba	Tidal stream	12	Under construction
Republic of Korea	Jindo Uldolmok Tidal Power Plant	Tidal stream	90	2009
Republic of Korea	Sihwa Lake Tidal Power Plant	Tidal stream	254	Under construction
Republic of Korea	Islands west of Incheo	Tidal stream	1 320	Proposed
United Kingdom	Strangford Lough in Northern Ireland.	Tidal stream	1.2	2008
United Kingdom	River Severn	Barrage	8 000 (max) 2 000 (av)	Proposed
China	Jiangxaia	Tidal lagoon	3.2	1980
China	Yalu river	Tidal lagoon	300	Proposed
Philippines	San Bernardino Strait	Tidal stream	2200	Proposed

- *Barrages* make use of the potential energy in the difference in height or head between high and low tides.
- *Tidal lagoons* can be constructed as self contained structures, not fully across an estuary.

Tidal stream generators draw energy from currents in much the same way as wind turbines. Tidal stream turbines may be arrayed in high-velocity areas where natural tidal current flows are concentrated such as the west and east coasts of Canada, the Strait of Gibraltar, the Bosporus, and numerous sites in Southeast Asia and Australia. Some of the operating and proposed facilities are shown in Table 11.3.2.

The higher density of water means that a single generator can provide significant power at low tidal flow velocities. Water velocities at about one-tenth of the speed of wind provide the same power for the same size of turbine system. However this limits the application in practice to places where the tide moves at speeds of at least 1m/s even at neap tides (Lecomber, 1979).

Tidal stream generators are an immature technology. Only a few commercial scale production facilities are yet routinely supplying power. No standard technology has yet emerged as the clear winner. But large varieties of designs are being experimented with, some very close to large scale deployment.

Several prototypes have shown promise, but they have not operated commercially for extended periods to establish performances and rates of return on investments. The

Table 11.3.3 Prototype tidal stream generators

Device	*Principle/Description*	*Examples*
Axial Turbines	Similar to the concept of traditional windmills; operating under the sea	1. Kvalsund, south of Hammerfest, Norway with 300 kW capacity. 2. Seaflow, off the coast of Lynmouth, Devon, England with 300 kW capacity. 3. Verdant Power, in the East River between Queens and Roosevelt Island, New York City. 4. SeaGen, in Strangford Lough in Northern Ireland has connected 150 kW into the grid. 5. OpenHydro, being tested at the European Marine Energy Centre (EMEC), in Orkney, Scotland.
Vertical and horizontal axis cross-flow turbines	Deployed either vertically or horizontally.	1. Gorlov turbine being commercially piloted on a large scale in S. Korea; starting with a 1 MW plant that started in May 2009 and expanding to 90 MW by 2013. 2. Proteus, which uses a barrage of vertical axis cross flow turbines for use mainly in estuaries. 3. Turbine-Generator Unit (TGU) prototype at Cobscook Bay and Western Passage tidal sites near Eastport, Maine. 4. Trials in the Strait of Messina, Italy, started in 2001 of the Kobold concept.
Oscillating devices	No rotating component. Aerofoil sections which are pushed sideways by the flow.	1. Stingray, tested off the Scottish coast with 150 kW capacity. 2. Pulse Tidal, in the Humber estuary.
Venturi effect	Uses a shroud to increase the flow rate through the turbine. Mounted horizontally or vertically.	1. Tidal Energy, commercial trials in the Gold Coast, Queensland (2002). 2. Hydro Venturi, is to be tested in San Francisco Bay.

devices could be classified into four, although a number of other approaches are also being tried (Table 11.3.3).

The cost associated for developing tidal power station can vary depending on the capacity. Project Severn Estuary in UK cost US $15 billion which produces about 8000 MW. The proposed 2200 MW tidal power station project in San Berandino cost about US $3 billion.

11.3.5 Wave power

Wave power can be used for electricity generation, as well as water for desalination and pumping of water into reservoirs. Wave power is distinct from the diurnal flux of tidal power and the steady flow of ocean currents.

Waves are generated by wind passing over the surface of the sea. As long as the waves propagate slower than the wind speed just above the waves, there is an energy transfer from the wind to the waves. Both air pressure differences between the upwind

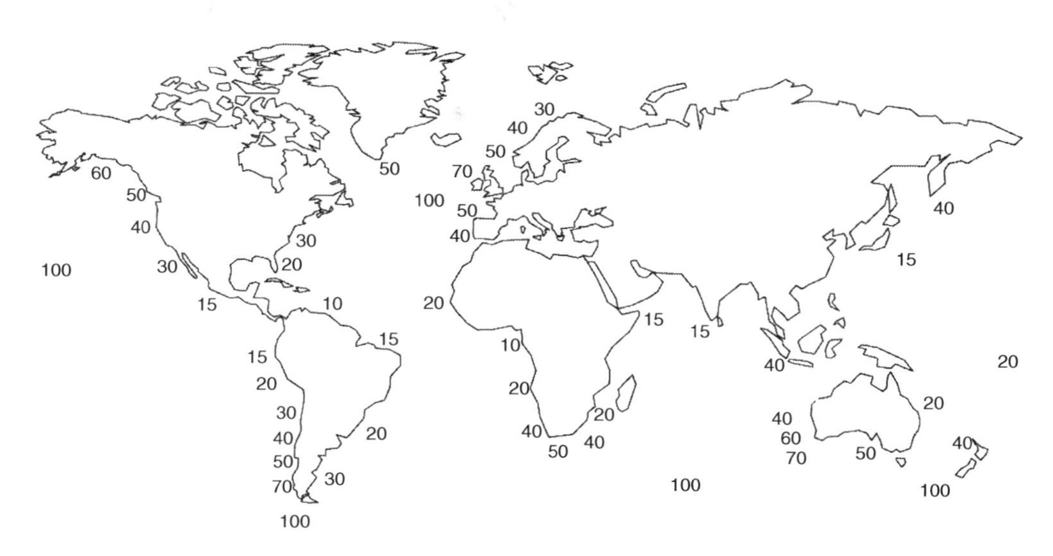

Figure 11.3.2 Approximate global distribution of wave power levels (kW/m of wave fuel)

and the lee side of a wave crest, as well as friction on the water surface by the wind causes the growth of the waves (Cruz, 2008).

Wave height is determined by wind speed, the duration of time the wind has been blowing, fetch or the distance over which the wind blows and by the depth and topography of the seafloor. The depth and topography of the sea floor can focus or disperse the energy of the waves. A given wind speed has a matching practical limit over which time or distance will not produce larger waves.

In general, larger waves are more powerful but wave power is also determined by wave speed, wavelength, and water density. When an object bobs up and down on a ripple in a pond, it experiences an elliptical trajectory. This oscillatory motion is highest at the surface and diminishes exponentially with depth.

The waves propagate on the ocean surface, and the wave energy is also transported horizontally with the group velocity. The group velocity of a wave is the velocity with which the overall shape of the wave's amplitudes propagates through space. The mean transport rate of the wave energy through a vertical plane of unit width, parallel to a wave crest, is called the wave energy flux or wave power (McCormick, 2007).

Wave energy can be considered as a concentrated form of solar energy. Winds, generated by the differential heating of the earth, pass over open bodies of water, transferring some of their energy to form waves. The amount of energy transferred, and hence the size of the resulting waves, depends on the wind speed, the length of time for which the wind blows and the distance over which it blows. The useful worldwide resource has been estimated at >2 TW (WEC, 1993). The approximate global distribution of wave power levels is given in Fig. 11.3.2.

Wave power generation is not currently a widely employed commercial technology although there have been attempts at using it since at least 1890. The world's first commercial wave farm is based in Portugal, at the Aguçadoura Wave Park, which consists of three 750 kilowatt Pelamis devices.

There is a large amount of ongoing work on wave energy schemes. The devices could be deployed on the shoreline, near the shore and offshore:

Shoreline Devices: These devices are fixed to or embedded in the shoreline itself. It has the advantage of easier maintenance and/or installation. These would not require deep water moorings or long lengths of underwater electrical cable. However, they would experience a much less powerful wave regime. This could be partially compensated by natural energy concentration. The deployment of such schemes could be limited by requirements for shoreline geology, tidal range and preservation of coastal scenery.

One major class of shoreline device is the oscillating water column (OWC). It consists of a partially submerged, hollow structure, which is open to the sea below the water line. This structure encloses a column of air on top of a column of water. As waves impinge upon the device they cause the water column to rise and fall, which alternatively compresses and depressurizes the air column. If this trapped air is allowed to flow to and from the atmosphere via a turbine, energy can be extracted from the system and used to generate electricity (Falnes, 2002).

Nearshore Devices: The main prototype device for moderate water depths (i.e. <20 m) is the OSPREY developed by Wavegen. This is a 2 MW OWC, with provision for inclusion of a 1.5 MW wind turbine. Since there could be environmental objections to large farms of wind or wave energy devices close to the shore, this system aims to maximize the amount of energy produced from a given amount of near shore area (Thorpe, 1999).

Offshore Devices: This class of device exploits the more powerful wave regimes available in deep water (>40 m depth) before energy dissipation mechanisms have had a significant effect. In order to extract the maximum amount of energy from the waves, the devices need to be at or near the surface and so they usually require flexible moorings and electrical transmission cables. More recent designs for offshore devices have also concentrated on small, modular devices. The McCabe wave pump, OPT wave energy converter, Pelamis and Archimedes wave swing are some of the examples.

Some examples of wave power systems are given in Table 11.3.4.

The major technical challenges in deploying wave power devices are:

- The device needs to capture a reasonable fraction of the wave energy in irregular waves, in a wide range of sea states.
- There is an extremely large fluctuation of power in the waves. The peak absorption capacity needs to be much (more than 10 times) larger than the mean power. For wave power the ratio is typically 4.
- The device has to efficiently convert wave motion into electricity. Wave power is available at low speed and high force, and the motion of forces is not in a single direction. Most readily-available electric generators operate at higher speeds, and most readily-available turbines require a constant, steady flow.
- The device has to be able to survive storm damage and saltwater corrosion.

At present, the main stumbling block to deployment of wave energy devices is funding. The capital costs are the problem, as it is hard to get companies to invest in technologies that have not yet been completely proved. The position is similar to other forms of renewable energy sources.

Table 11.3.4 Some examples of wave power systems

Country	*Technology used*	*Project/Location*	*Type*	*Capacity*
Portugal	Pelamis Wave Energy Converter	Aguçadoura Wave Park/Póvoa de Varzim	Offshore	2.25 MW
Denmark	Wave Dragon	Danish Wave Energy Test Center/ Nissum Bredning fjord	Offshore	4–11 MW
Portugal	AquaBuOY	Finavera Renewables	Offshore	–
Australia	CETO Wave Power	Biopower/Carnegie Corporation/ Fremantle, Western Australia	Offshore	–
Australia	Oceanlinx	Near Port Kembla, near Sydney	Offshore	2 MWe
UK	Wavebob	Galway Bay near Galway in Ireland	Offshore	–
UK	Pelamis Wave Energy Converter	European Marine Energy Centre/ Orkeny	Offshore	3 MW
UK	Anaconda Wave Energy Converter	Engineering and Physical Sciences Research Council (EPSRC)/ Checkmate SeaEnergy	Offshore	~1 MW
UK	Oyster wave energy converter	Aquamarine Power/European Marine Energy Centre/Orkney	Nearshore	100 MW or more
Sweden	WEC (wave energy converter) with a linear generator	Centre for Renewable Electric Energy Conversion, Uppsala University/ Lysekil Project/Lysekil	Offshore	10 kW
USA	EPAM	SRI International/Santa Cruz, Calif	Offshore	–
USA	PowerBuoy	Pacific Northwest Generating Cooperative/Reedsport, Oregon	Offshore	150 kW

Until the technology matures, estimates of the cost of power from wave energy devices represent a snapshot of the status and costs of the designs at the current stages of their development.

That review found support for this proposition, with the predicted generating costs of several devices being reduced by factors of two or more as part of the review activities.

The electricity costs of a number of devices have been evaluated more recently using the same peer-reviewed methodology developed for the last UK review of wave energy. These figures show that there have been significant improvements in the predicted generating costs of devices, so that there are now several with costs of about 5 p/kWh (US 8 ¢/kWh) or less at 8% discount rate (if the devices achieve their anticipated performance) (Thorpe, 1999).

Wave devices that are on-shore have social implications for the surrounding area. They can be integrated within harbour walls, which can affect shipping and cause noise pollution. They can create employment in the area and attract visitors.

Offshore devices have an effect on navigation and consultation with affected bodies must be undertaken. The experiences of other offshore industries, such as oil, should aid this part of planning for wave devices.

There can be environmental impacts resulting from wave powered devices. Devices that are on-shore can have environmental benefits, such as helping to reduce the erosion of the landscape. Any devices off shore can have an effect on the aquatic life in that area but this again is very site specific and hard to predict. But anchoring systems can become almost like artificial reefs, creating a place for new colonization.

11.3.6 Damless hydro

Low head hydro power applications use river current and tidal flows to produce energy. These applications do not need to dam or retain water to create head. Using the current of a river or the naturally occurring tidal flow to create electricity may provide a renewable energy source that will have a minimal impact on the environment (Harvey and Brown, 1992).

Orthogonal rotor turbines equipped with blades of a symmetric profile can be regarded as a prospective type of free-flow hydraulic machine which can be installed either in the free flow in a river channel and ocean or in the channels of chutes, spillways, and irrigation systems.

A low-head hydro project usually is an installation with a fall of water less than 5 m. Since no dam is required, low-head hydro has the following advantages:

- No safety risks of having a dam, avoiding the risk of a flash flood caused by a breached dam;
- Environmental and ecological complications such as submergence of large tracts of forested and inhabited areas, need for fish ladder, silt accumulation in basin;

Low-head units are necessarily much smaller in capacity that conventional large hydro turbines. So many units must be built for a given annual energy production. Some of the costs of small turbine – generator units are offset by lower civil construction cost (Curtis and Langley, 2004).

Not every site can be economically and ecologically developed. Sites may be too far from customers to be worth installation of a transmission line, or may lie in areas particularly sensitive for wildlife.

A hydrokinetic turbine is an integrated turbine generator to produce electricity in a free flow environment. In-stream Energy Generation Technology (IEGT) turbines could be used in rivers, man made channels, tidal waters, or ocean currents. These turbines use the flow of water to turn them, thus generating electricity for the power grid on nearby land.

A 35 kilowatt hydrokinetic turbine has been installed in the Mississippi River near Hastings, Minnesota. If the viable river and estuary turbine locations of the US are made into hydroelectric power sites it is estimated that up to 130 000 gigawatt-hours per year – about half the yearly production of the country's dams – could be produced.

The axial flow rotor turbine consists of a concentric hub with radial blades, resembling a wind mill. Either a built-in electrical generator or a hydraulic pump which turns an electrical generator on land provides the electricity. The open center fan turbine consists of two donut shape turbines which rotate in the opposite direction of the current. This in turn runs a hydraulic pump that in turn drives a standard electrical generator.

A helical turbine has hydrofoil sections that keep the turbine oriented to the flow of the water. The leader edge of the blades turns in the direction of the water. The cycloidic turbine resembles a paddle wheel, where the flow of the water turns the wheel with lift and drag being optimized. Hydroplane blades are made to oscillate by the flowing water, thus generating electricity. The FFP turbine generator uses a rim-mounted, permanent magnet, direct-drive generator with front and rear diffusers and one moving part (the rotor) to maximize efficiency.

The turbines can be installed in a variety of ways, multiple banks set on pilings driven into the river beds or mounted on existing river structures such as bridge piers. The turbine generators can be attached to bridge abutments or pilings, which minimize disruption to river beds.

Turbines are to be deployed in arrays of multiple units spaced no less than 15 m apart where the site conditions, depth, and needed infrastructure are suitable. Exact depth and spacing is determined based on site conditions, including current flows and water depth. Since the turbines do not block waterways, and the water passing through the device is not subject to high pressure, these systems are designed to not impede or damage fish or other wildlife.

Another approach is to suspend the turbines from a floating barge. The turbines suspended from the bottom of a floating barge can accommodate changes in flow. The barges can be deployed and have the generators come on line more quickly with fewer disturbances to the river bed. The obvious disadvantage to the barge system would be interference with navigation and recreational use of the waterway.

Concerns have been raised about the danger to marine animals, such as seals and fish, from wave and tidal devices. There is no evidence that this is a significant problem. Such devices may actually benefit the local fauna by creating non-fishing 'havens' and structures such as anchoring devices may create new reefs for fish colonization.

11.4 ENHANCED GEOTHERMAL SYSTEMS

Enhanced Geothermal Systems (EGS) are a new type of geothermal power technologies that do not require natural convective hydrothermal resources. Present geothermal power systems depend on naturally occurring water and rock porosity to carry heat to the surface. Majority of geothermal energy within drilling reach is in dry and non-porous rock. EGS technologies "enhance" and/or create geothermal resources in this hot dry rock (HDR) through hydraulic stimulation (Armstead, 1987).

EGS offer great potential for expanding the use of geothermal energy. Present geothermal power generation comes from hydrothermal reservoirs, and is somewhat limited in geographic application to specific ideal places.

EGS utilise new techniques to exploit resources that would have been uneconomical in the past. These systems are still in the research phase, and require additional research, development and deployment for new approaches and to improve conventional approaches, as well as to develop smaller modular units that will allow economies of scale on the manufacturing level.

Several technical issues need further government-funded research and close collaboration with industry in order to make exploitation of geothermal resources more economically attractive for investors. These are mainly related to exploration of reservoirs, drilling and power generation technology, particularly for the exploitation of low-temperature cycles.

When natural cracks and pores will not allow for economic flow rates, the permeability can be enhanced by pumping high pressure cold water down an injection well into the rock. The injection increases the fluid pressure in the naturally fractured granite which mobilizes shear events, enhancing the permeability of the fracture system.

Water travels through fractures in the rock, capturing the heat of the rock until it is forced out of a second borehole as very hot water, which is converted into electricity using either a steam turbine or a binary power plant system. All of the water, now cooled, is injected back into the ground to heat up again in a closed loop.

EGS technologies, like hydrothermal geothermal, are expected to be baseload resources which produce power 24 hours a day like a fossil plant. Distinct from hydrothermal, EGS may be feasible anywhere in the world, depending on the economic limits of drill depth.

EGS is one of the few renewable energy resources that can provide continuous base-load power with minimal visual and other environmental impacts. Geothermal systems have a small footprint and virtually no emissions, including carbon dioxide. Geothermal energy has significant base-load potential, requires no storage, and, thus, it complements other renewables – solar (CSP and PV), wind, hydropower – in a lower-carbon energy future.

The accessible geothermal resource, based on existing extractive technology, is large and contained in a continuum of grades ranging from today's hydrothermal, convective systems through high- and mid-grade EGS resources. Improvements to drilling and power conversion technologies, as well as better understanding of fractured rock structure and flow properties, benefit all geothermal energy development scenarios.

Field studies conducted worldwide for more than 30 years have shown that EGS is technically feasible in terms of producing net thermal energy by circulating water through stimulated regions of rock at depths ranging from 3 to 5 km.

EGS systems are versatile, inherently modular, and scalable from 1 to 50 MWe for distributed applications to large "power parks," which could provide thousands of MWe of base-load capacity. EGS also can be easily deployed in larger-scale district heating and combined heat and power (cogeneration) applications to service both electric power and heating and cooling for buildings without a need for storage on-site.

Favourable locations are over deep granite covered by a thick (3–5 km) layer of insulating sediments which slow heat loss. HDR wells are expected to have a useful life of 20 to 30 years before the outflow temperature drops about 10 degrees Celsius and the well becomes uneconomic. If left for 50 to 300 years the temperature will recover.

11.4.1 Technical considerations

The EGS concept is to extract heat by creating a subsurface fracture system to which water can be added through injection wells. Creating an enhanced or engineered, geothermal system requires improving the natural permeability of rock.

Geothermal energy consists of the thermal energy stored in the Earth's crust. Thermal energy in the earth is distributed between the constituent host rock and the natural fluid that is contained in its fractures and pores at temperatures above ambient levels. These fluids are mostly water with varying amounts of dissolved salts; typically, in their natural in situ state, they are present as a liquid phase but sometimes may consist of a saturated, liquid-vapor mixture or superheated steam vapor phase.

The source and transport mechanisms of geothermal heat are unique to this energy source. Heat flows through the crust of the Earth at an average rate of 59 mW/m^2. The intrusion of large masses of molten rock can increase this normal heat flow locally; but for most of the continental crust, the heat flow is due to two primary processes:

(i) Upward convection and conduction of heat from the Earth's mantle and core, and
(ii) Heat generated by the decay of radioactive elements in the crust, particularly isotopes of uranium, thorium, and potassium.

Local and regional geologic and tectonic phenomena play a major role in determining the location (depth and position) and quality (fluid chemistry and temperature) of a particular resource. For example, regions of higher than normal heat flow are associated with tectonic plate boundaries and with areas of geologically recent igneous activity and/or volcanic events (younger than about 1 million years).

Certain conditions must be met before one has a viable geothermal resource. The first requirement is accessibility. This is usually achieved by drilling to depths of interest, frequently using conventional methods similar to those used to extract oil and gas from underground reservoirs.

The second requirement is sufficient reservoir productivity. For hydrothermal systems, one normally needs to have large amounts of hot, natural fluids contained in an aquifer with high natural rock permeability and porosity to ensure long-term production at economically acceptable levels. When sufficient natural recharge to the hydrothermal system does not occur, which is often the case, a reinjection scheme is necessary to ensure production rates will be maintained.

Thermal energy is extracted from the reservoir by coupled transport processes (convective heat transfer in porous and/or fractured regions of rock and conduction through the rock itself). The heat extraction process must be designed with the constraints imposed by prevailing in situ hydrologic, lithologic, and geologic conditions. Typically, hot water or steam is produced and its energy is converted into electricity, process heat, or space heat.

Rocks are permeable due to minute fractures and pore spaces between mineral grains. Injected water is heated by contact with the rock and returns to the surface through production wells, as in naturally occurring hydrothermal systems (Fig 4.3.4.1). The main technological details are:

- Injection Well: A well drilled into hot basement rock that has limited permeability and fluid content.
- Injecting Water: Water is injected at sufficient pressure to ensure fracturing, or open existing fractures within the developing reservoir and hot basement rock.
- Hydro-fracture: Pumping of water is continued to extend fractures some distance from the injection wellbore and throughout the developing reservoir and hot basement rock. This is a crucial step in the EGS process.
- Doublet: A second production well is drilled with the intent to intersect the stimulated fracture system created in the previous step, and circulate water to extract the heat from the previously "dry" rock mass.
- Multiple Wells: Additional production-injection wells are drilled to extract heat from large volumes of rock mass to meet power generation requirements.

EGS technologies are being developed and tested in France, Australia, Japan, Germany, the U.S. and Switzerland (Table 11.4.1). The largest EGS project in the world is a

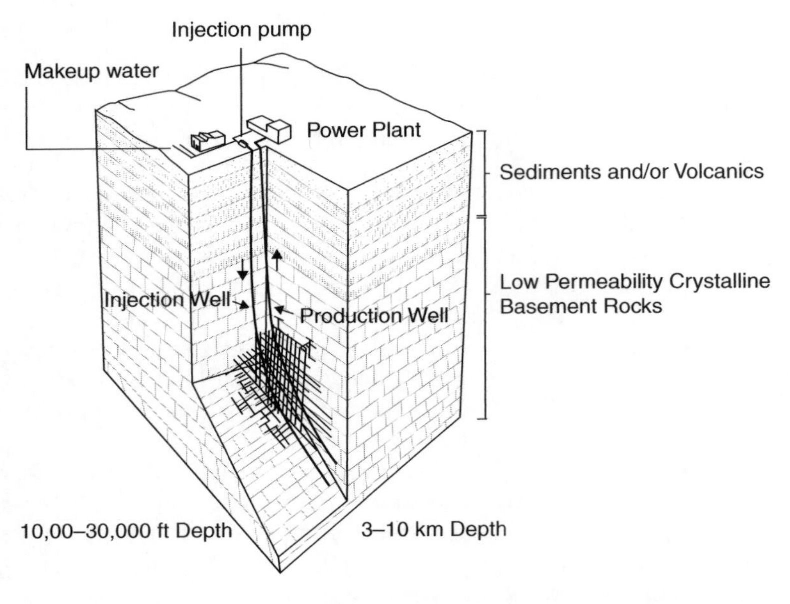

Figure 11.4.1 Two-well Enhanced Geothermal System

Table 11.4.1 Current EGS projects

Country	*Project*	*Type*	*Size (MW)*	*Depth (km)*	*Status*
France (EU)	Soultz/ENGINE	R&D	1.5	4.2	Operational
United States	Desert Peak/DOE, Ormat, GeothermEx	R&D	11–50		Development
Germany (EU)	Landau	Commercial	3	3.3	Operational
Australia	Paralana-Phase1/Petratherm	Commercial	7–30	4.1	Drilling
Australia	Cooper Basin/ Geodynamics	Commercial	250–500	4.3	Drilling
United States	Bend, Oregon/ AltaRock Energy, NCPA	Demonstration	–		Proposed
Japan	Ogachi	R&D	–	1.0–1.1	CO2 experiments
United Kingdom	United Downs, Redruth/ Geothermal Engineering Ltd	Commercial	10 MW	4.5	Proposed
United Kingdom	Eden Project/ EGS Energy Ltd.	Commercial	3 MW	3–4	Proposed

25 megawatt demonstration plant currently being developed in the Cooper Basin, Australia. The Cooper Basin has the potential to generate 5 000–10 000 MW.

It is reported that in the United States the total EGS resources from 3–10 km of depth is over 13 000 zetta joules. Out of this over 200 ZJ would be extractable, with the potential to increase this to over 2 000 ZJ with technology improvements — sufficient to provide all the world's current energy needs for several millennia.

With a modest R&D investment of $1 billion over 15 years, 100 GWe (gigawatts of electricity) or more could be installed by 2050 in the United States. The "recoverable" resource (that accessible with today's technology) is between 1.2–12.2 TW for the conservative and moderate recovery scenarios respectively (MIT, 2006).

11.4.2 Economic considerations

EGS could be capable of producing electricity at 3.9 cents/kWh. EGS costs were found to be sensitive to four main factors:

(i) Temperature of the resource;
(ii) Fluid flow through the system measured in liters/second;
(iii) Drilling costs; and
(iv) Power conversion efficiency.

EGS energy which is transformed into delivered energy (electricity or direct heat) – is an extremely capital-intensive and technology-dependent industry. The capital investment may be characterized in three distinct phases:

- Exploration and drilling of test and production wells
- Construction of power conversion facilities
- Discounted future redrilling and well stimulation.

Estimates of capital cost by the California Energy Commission (CEC, 2006), showed that capital reimbursement and interest charges accounted for 65% of the total cost of geothermal power. The remainder covers fuel (water), parasitic pumping loads, labor and access charges, and variable costs.

By way of contrast, the capital costs of combined-cycle natural gas plants are estimated to represent only about 22% of the levelized cost of energy produced, with fuel accounting for up to 75% of the delivered cost of energy.

Given the high initial capital cost, most EGS facilities will deliver base-load power to grid operations under a long-term power purchase agreement (typically greater than 10 years) in order to acquire funding for the capital investment.

There is a positive correlation between the development of new EGS fields and continued declines in delivered costs of energy. This reflects not only the economies from new techniques and access to higher value resources, but also the inevitable cost of competitive power sources.

For the US it is suggested that with significant initial investment, installed capacity of EGS could reach 100 000 MWe within 50 years, with levelized energy costs at parity with market prices after 11 years. It is projected that the total cost, including costs for research, development, demonstration, and deployment, required to reach this level of EGS generation capacity ranges from approximately $600–$900 million with an absorbed cost of $200–$350 million.

Center for Geothermal Energy Excellence at the University of Queensland, has been awarded $18.3 million (AUS) for EGS research, a large portion of which will be used to develop CO_2 EGS technologies. Research conducted at Los Alamos National Laboratories and Lawrence Berkeley National Laboratories examined the use of supercritical

CO_2, instead of water, as the geothermal working fluid with favorable results. CO_2 has numerous advantages for EGS:

- Greater power output
- Minimized parasitic losses from pumping and cooling
- Carbon sequestration
- Minimized water use

11.4.3 Further studies required

Further research is required in three areas:

- *Drilling technology* – both evolutionary improvements building on conventional approaches to drilling such as more robust drill bits, innovative casing methods, better cementing techniques for high temperatures, improved sensors, and electronics capable of operating at higher temperature in down-hole tools; and revolutionary improvements utilizing new methods of rock penetration will lower production costs. These improvements will enable access to deeper, hotter regions in high-grade formations or to economically acceptable temperatures in lower-grade formations.
- *Power conversion technology* – improving heat-transfer performance for lower-temperature fluids, and developing plant designs for higher resource temperatures to the supercritical water region would lead to an order of magnitude (or more) gain in both reservoir performance and heat-to power conversion efficiency.
- *Reservoir technology* – increasing production flow rates by targeting specific zones for stimulation and improving downhole lift systems for higher temperatures, and increasing swept areas and volumes to improve heat-removal efficiencies in fractured rock systems, will lead to immediate cost reductions by increasing output per well and extending reservoir lifetimes. For the longer term, using CO_2 as a reservoir heat-transfer fluid for EGS could lead to improved reservoir performance as a result of its low viscosity and high density at supercritical conditions. In addition, using CO_2 in EGS may provide an alternative means to sequester large amounts of carbon in stable formations.

11.4.4 Induced seismicity

Some seismicity is expected in EGS, which involves pumping fluids at pressure to enhance or create permeability through the use of hydraulic fracturing techniques. Depending on the rock properties, and on injection pressures and fluid volume, the reservoir rock may respond with tensile failure, as is common in the oil and gas industry, or with shear failure of the rock's existing joint set, as is thought to be the main mechanism of reservoir growth in EGS efforts.

Seismicity associated with hydraulic stimulation can be mitigated and controlled through predictive siting and other techniques. Based on substantial evidence collected so far, the probability of a damaging seismic event is low.

Chapter 12

Algal biofuels

Sabil Francis
University of Leipzig, Leipzig, Germany

12.1 INTRODUCTION

Algal biofuels or oilgae refer to a promising subcategory of liquid fuels produced from algae. The algae are *autotrophic* simple aquatic organisms that range from small unicellular organisms such as pond scum to complex multi-cellular ones such as kelp. Though *photosynthetic*, like *plants*, they are considered "simple" because they lack the many distinct organs found in *land plants* (Mousdale, 2008).

The oil productivity of microalgae surpasses that of the best oil seed producing terrestrial plants. Though both depend on sunlight for energy, microalgae are extremely fuel efficient when compared to land based plants. Microalgae are selected based on a number of factors, most notably high innate growth rates, favorable overall composition (lipids, carbohydrates, and proteins), and ability to grow in specific climatic conditions.

Fuel end products, such as biodiesel, ethanol, methane, hydrogen, jet fuel, bio crude and more via a wide range of processes can be produced using algae (Fig 12.1). Several by-products with wide ranging applications in the pharmaceutical and chemical industries are also created in the process of algal fuel extraction. Algal biofuels alone can replace all fossil fuel consumption on earth.

Two kinds of algae that have the potential for biofuels are the macro algae (with high oil content, but costly and cultivation intensive) and the micro algae (low oil yield but easy to cultivate and cheap). Newer processes, such as cellulosic fermentation (for deriving ethanol), gasification (for deriving biodiesel, ethanol and a wide range of hydrocarbons), or anaerobic digestion (for methane or electricity generation), have been developed to tap into the potential of macro algae (Table 12.1).

The major research effort in this area has been "Aquatic Species Program" (ASP), which ran from 1978 to 1996 under the US National Renewable Energy Laboratory

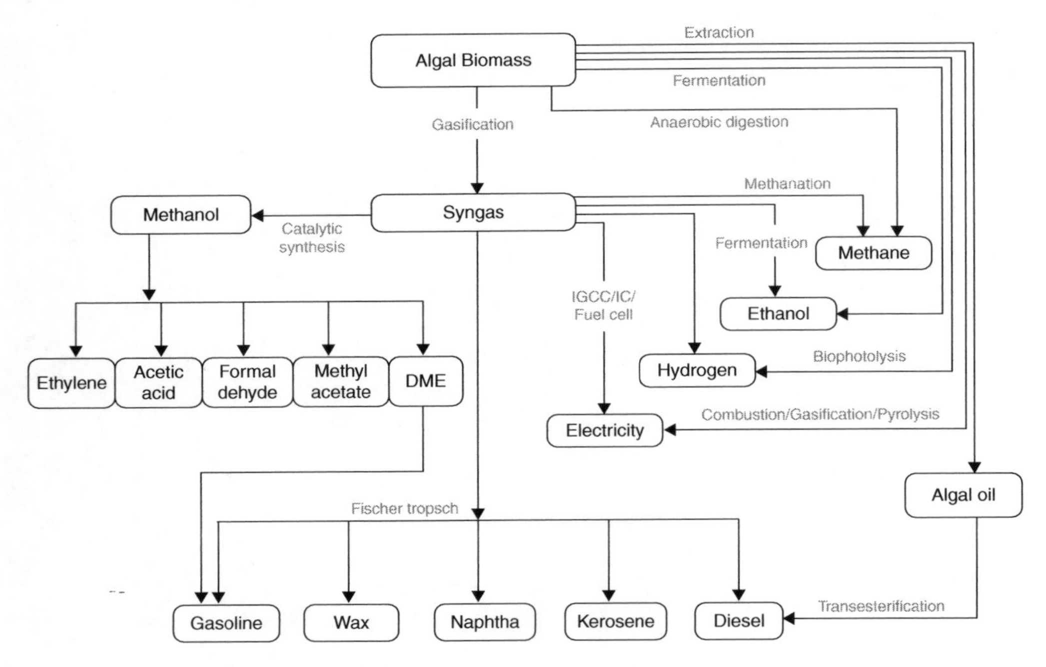

Figure 12.1 Paths to the various energy products from algae

Table 12.1 Summary of processes for converting algae to energy

Final Product	*Processes*
Biodiesel	Oil extraction and transesterification
Ethanol	Fermentation
Methane	Anaerobic digestion of biomass; methanation of syngas produced from biomass
Hydrogen	Triggering biochemical processes in algae; gasification/pyrolysis of biomass and processing of resulting syngas.
Heat & Electricity	Direct combustion of algal biomass; gasification of biomass
Other Hydrocarbon Fuels	Gasification/pyrolysis of biomass and processing of resulting gas.

(NREL), funded by the Office of Fuels Development, a division of the US Department of Energy (Biopact, 2007).

12.2 COMPARATIVE ADVANTAGES

From the perspective of bio fuel production algae they have enormous advantages over plants which can be summarized as (Exxon Mobil, 2010):

- *The use of otherwise unsuitable land*: Algae can be grown using land and water unsuitable for plant or food production, unlike some other first- and second-generation biofuel feedstocks.

Table 12.2 Flue gas composition from coal fired power plant that could be used for algae cultivation and bio fuel generation

Component	N_2	CO_2	O_2	SO_2	NOx	Soot dust
Concentration	82%	12%	5.5%	400 ppm	120 ppm	50 mg/m^3

Table 12.3 Fuel yield per acre of production per year

Algae	2000 gallons
Palm	650 gallons
Sugar Cane	450 gallons
Corn	250 gallons
Soy	50 gallons

- *Mass production*: Quantities of algae can be grown quickly, and the process of testing different strains of algae for their fuel-making potential can proceed more rapidly than for other crops with longer life cycles.
- *Energy efficient*: Some algae can produce bio-oils through the natural process of photosynthesis.
- *Greenhouse effect*: Growing algae consume carbon dioxide; this provides greenhouse gas mitigation benefits. Since algae flourish in high concentrations of carbon dioxide and nitrogen dioxide, the cultivation of algae in the vicinity of polluting industries such as cement plants, breweries, or steel plant is an ideal way to clean up the air, produce bio diesel, and cut down the amount of carbon dioxide in the atmosphere. One ton algae can absorb about 1.8 tons of CO2, and so the potential of this form of climate change control is enormous (Table 4.4.2.1) (Oilgae, 2010).
- *Similarity to petroleum*: Among the biofuels, bio-oil produced by photosynthetic algae have molecular structures that are the most similar to petroleum and refined products such as jet fuel. This would mean less transformative technology in the automotive and other industries.

12.3 PROBLEMS WITH ALGAL BIOFUELS

The extant problems with algal biofuels can be summarized as follows (Biopact, 2007):

- Lack of a constant and high lipid content that is key to the creation of biofuels;
- Lack of stability of algae cultures that leads to intermittent cultivation;
- Varying degrees of photosynthetic efficiency that does not result in high and constant biomass productivity;
- Inability to withstand seasonal climate changes, and fluctuating temperatures; and
- The physical nature of algae – the membranes have to be easily harvestable, and must be ideally done without too much loss and without the need for costly flocculants.

Biotechnology might offer a solution to these problems through genetic engineering of species. Because of its enormous potential, in comparison with other bio fuel sources, algae seem to be the best alternative (Table 12.3).

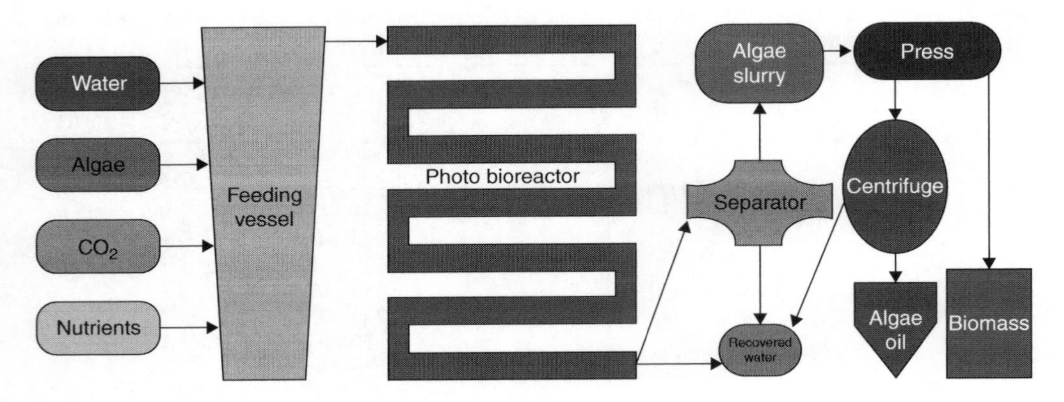

Figure 12.2 Working of a photo-bioreactor

12.4 TECHNOLOGIES

There are three stages in the creation of algal biofuels (a) the cultivation of algae (b) the harvesting of algae (c) the extraction of the oil.

12.4.1 Cultivation of algae

Open and closed methods of cultivation: One of the key advantages of algae is that it can be cultivated virtually anywhere. All that is needed is light, carbon dioxide and water. There are two types of algal cultivation. The first one is "open" cultivation, which has been the norm in the United States and in the NREL project. In this method, "raceway" ponds, which contain native strains of algae are stirred using a paddle wheel, while carbon dioxide is introduced. The water can be wastewater (treated sewerage) freshwater, brackish water, or salt water, depending on the strain of algae grown.

The Japanese prefer a "closed" system. One example of this is a photo-bioreactor (PBR) a closed translucent container. Depending on whether the heat – in the form of natural light or artificial light or both – is constant on intermittent, cultivation can be all the year around. Because PBR systems are closed, all essential nutrients must be introduced into the system (Fig 12.2).

A PBR can be operated in "batch mode", but it is also possible to introduce a continuous stream of sterilized water containing nutrients, air, and carbon dioxide and have continuous cultivation. There are two types of illumination that are used in PBRs—natural and artificial. Naturally illuminated *Algal Culture* systems with large illumination surface areas include flat-plate, horizontal/serpentine tubular airlift, and inclined tubular photo-bioreactors. Generally, laboratory-scale photo-bioreactors are artificially illuminated (either internally or externally) using fluorescent lamps or other light distributors.

Table 12.4 shows the comparative advantages and disadvantages of a PBR.

Table 12.4 Comparative advantages and disadvantages of photo-bioreactor cultivation

Advantages	*Disadvantages*
High Biomass Productivity and cell density	High capital cost associated with construction costs, circulation pumps, and nutrient-loading systems
Less contamination, water use, & CO_2 losses	Absence of evaporative cooling, which can lead to very high temperatures
Better light utilization & mixing	Accumulation of high concentration of photosynthetically generated O_2 leading to photo-oxidative damage
Controlled culture conditions	Absence of evaporative cooling, which can lead to very high temperatures

12.4.2 Harvesting of algae

The separation of algae from the growth medium, whether closed or open, can be defined as the harvesting of algae. One of the key aims in this step is the removal of the high water content, through technical processes such as flocculation, micro-screening and centrifugation. Again, the harvesting depends on the type of algae (Sheehan et al, 1998).

Harvesting of macro algae depends on the mode of cultivation. While macro algae that grows on a solid substrate has to be cut, free floating algae can be harvested merely by the raising of a net that has been installed in a pond, giving it a major cost advantage over micro-algae that has to be filtered and screened, for separation. While human harvesting was the norm earlier, currently petrol driven rotary cutters can be used to gather macro algae efficiently. The harvesting can be done 3 to 4 times, but the crop declines in yield.

Micro algae is usually cultivated in a thick algae paste, and the key step is the concentration of the algae so that its harvesting is viable—this can mean a one or two step process. This will depend on the strain of algae that is cultivated, especially its size and the particular properties of the strain. There are four ways in which microalgae can be harvested—floatation, centrifugation, filtration, and culture auto flocculation, which leads to a clustering of the microalgae. Fast growing algae are usually motile uni-cells, and therefore the hardest to harvest.

12.4.3 Extraction of various energy products from algae

Extraction can be broadly categorized into two methods: energy intensive mechanical methods that can be subdivided into (a) expression/expeller press (b) ultrasonic-assisted extraction. The second option is the use of environmentally hazardous chemical methods that can be further classified into (a) hexane solvent method (b) soxhlet extraction and (c) super critical fluid extraction that uses high pressure equipment that is expensive and energy intensive. Many manufacturers of *algae oil* use a combination of mechanical pressing and chemical solvents in extracting oil.

Rarer methods include enzymatic *extraction* that uses enzymes to degrade the cell walls with water acting as the solvent that makes fractionation of the oil much easier.

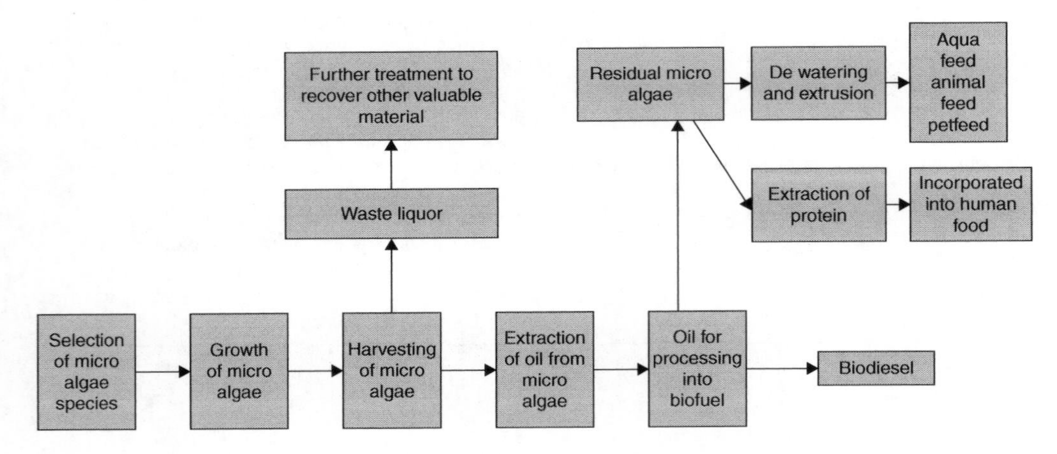

Figure 12.3 Detailed process of biodiesel from algae

The costs of this extraction process are estimated to be much greater than hexane extraction. While this method may become feasible in the future, but with the relatively long time enzymatic digestion requires and the developing of alternate methods, it does not hold much promise for the present (Whitcre et al, 2007).

The third method that can be used is that of osmotic shock which depends on a sudden reduction in osmotic pressure. An algae farm is designed to produce a number of products including algal oil, delipidated algal meal (DAM) and dried whole algae (DWA). The algal oil is suitable for conversion to biodiesel and can be substituted for any other vegetable oil (Soy, palm, Jatropha) in a commercial biodiesel production plant. The DAM and DWA are suitable for a wide variety of animal feed applications.

Biodiesel from algae: Bio diesel has been defined as any diesel-equivalent biofuel made from renewable biological materials such as vegetable oils or animal fats consisting of long-chain saturated hydrocarbons (Fig 12.2). It can be used in pure form (B100) or may be blended with petrodiesel at any concentration. Traditional sources such as bio diesels created from corn or soya take away from the food chain leading to higher food prices, and are extraction inefficient.

In the extraction process the first product is "green crude" which is similar to crude oil. It must be then refined, by mixing it with a catalyst, such as sodium hydroxide and an alcohol, such as methanol, resulting in biodiesel mixed with glycerol. The mixture is cleaned to remove the glycerol, a valuable by-product, leaving pure algal biodiesel fuel, which is similar to petrodiesel fuel.

Although algal biodiesel and petro diesel are similar, there are a few significant differences between their properties. However, low yields from naturally grown algae, the selection of high-oil content strains, devising cost effective methods of harvesting, oil extraction and conversion of oil to biodiesel options are some of the problems that remain to be overcome.

Ethanol from algae: Processes for the production of ethanol use food crops such as corn and sugar cane leading to a food crisis. One potential way to produce ethanol from algae is by converting starch (lipids) in algae into biodiesel and cellulose from the cell walls of algae (carbohydrate) into ethanol (Fig 12.4).

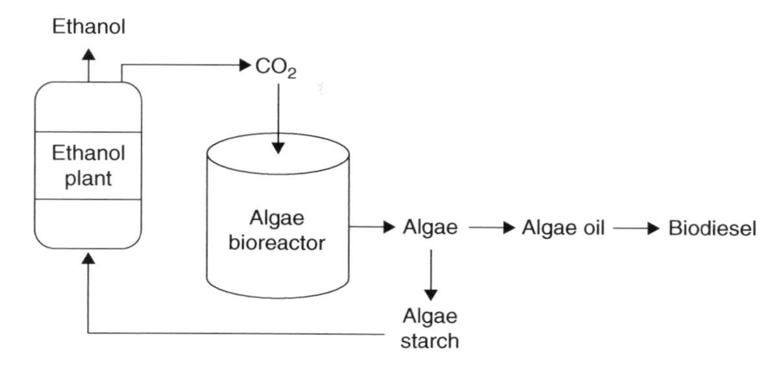

Figure 12.4 Extraction of ethanol using algae

Lipids are another generic name for TAGs (triglycerides), the primary storage form of natural oils. The advantage of algae in the creation of second generation bio ethanol is because they are high in carbohydrates/polysaccharides and have thin cellulose walls. Among the promising species of algae for ethanol production are *sargassum, glacilaria, prymnesium parvum*, and *euglena gracilis*.

Methane from algae: Methane can be used for the generation of electricity, by burning it as a fuel in a gas turbine or steam boiler. Advantage of methane is that it produces a relatively lesser amount of carbon dioxide for each unit of heat that is released. With a combustion heat of 891 kJ/mol, but with lesser molecular mass, it produces more heat per mass unit than other complex hydrocarbons. Known as natural gas, and is considered to have an *energy* content of 39 megajoules per cubic meter, or 1 000 *BTU* per standard cubic foot.

As a fuel for vehicles it is more environmentally friendly than petrol or diesel. Theoretically, methane can be produced from any of the three constituents of *algae* – carbohydrates, proteins and fats. Closed algal bioreactors offer a promising alternative route for biomass feedstock production for bio-methane. Using these systems, micro-algae can be grown in large amounts (150–300 tons per ha per year) using closed *Bioreactor* systems (lower yields are obtained with open pond systems). This quantity of biomass can theoretically *yield* 200 000–400 000 m of methane per ha per year.

Hydrogen from algae: One of the key products that have enormous potential to solve the "energy crisis" is hydrogen; especially it's potential to liberate large amounts of energy per unit weight in combustion, and the ease of its convertibility to electricity by fuel cells. The production of hydrogen through biological processes, such as the use of algae has an incomparable advantage over other chemical methods as they are, in comparison, incredibly cost efficient and cheap.

In contrast to algal production of hydrogen, electrochemical hydrogen production via solar battery-based water splitting on the hand requires the use of solar batteries with high energy requirements. On the other hand, for the production of hydrogen from algae all that is required is a transparent closed box, and a simple solar reactor, meaning low energy requirements. If such processes could be made to work on a large-scale, the world would have a renewable source of hydrogen (Fig 12.5).

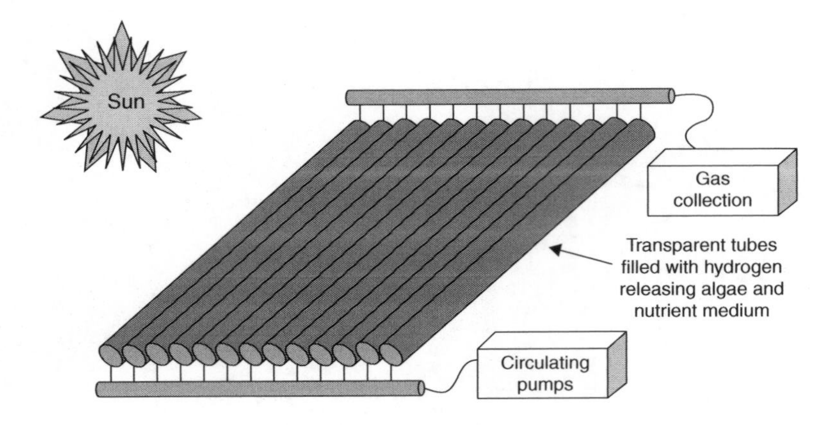

Figure 12.5 Hydrogen from algae

REFERENCES

Armstead, H. C. H. and Tester, J. W. (1987) *Heat Mining*. London: E and F. N. Spon.

Atzeni, S. and Meyer-ter-Vehn, J. (2004) *Nuclear Fusion Reactions*, The Physics of Internal Fusion. Oxford: University of Oxford Press.

Avery, W. H., Wu, C. (1994) *Renewable Energy From The Ocean – A Guide To OTEC*. New York: Oxford University Press.

Baker, A. C. (1991) *Tidal power*. London: Peter Peregrinus Ltd.

Beychok, M.R., (1974) Coal gasification and the Phenosolvan process, *Proc. American Chemical Society 168th National Meeting*, Atlantic City.

Biopact (2007) *An In-depth look at bio fuels from algae*. http://news.mongabay.com/bioenergy/2007/01/in-depth-look-at-biofuels-from-algae.html.

Bodansky, D. (2003) *Nuclear Energy: Principles, practices, and prospects*. Berlin: Springer.

BP (2009) *BP Statistical Review of World Energy* June 2009 (http://www.bp.com/statisticalreview)

Bradshaw, J. and T. Dance (2004) Mapping geological storage prospectivity of CO_2 for the world's sedimentary basins and regional source to sink matching. *Prodeedings of the 7th International Conference on Greenhouse Gas Technologies, Vol. I; peer reviewed Papers and Plenary Presentations*, pp 583–592. Eds. E.S. Rubins, D.W. Keith and C.F. Gilboy, Pergamon, 2005.

Brauns, E. (2008) Toward a worldwide sustainable and simultaneous large-scale production of renewable energy and potable water through salinity gradient power by combining reversed electrodialysis and solar power, *Environmental Process and Technology*, vol. 219, pp. 312–323.

Brogioli, D. (2009) Extracting renewable energy from a salinity difference using a capacitor, *Phys. Rev. Lett.*, Vol 103, pp 058501.1–058501.4.

CEC (2006) *California Energy Commission*. http://energy.ca.gov/geothermal/index.html

Charlier, R.H. and Justus, J.R. (1993) *Ocean Energies: Environmental, Economic and Technological Aspects of Alternative Power Sources*. Amsterdam: Elsevier Science Publishers.

Cohen, B.L. (1990) *The Nuclear Energy Option*. New York: Plenum Press

Cruz, J. (2008) *Ocean Wave Energy – Current Status and Future Prospects*. Berlin: Springer.

Curtis, D and Langley, B. (2004) *Going With the Flow: Small Scale Water Power*. Powys: Centre for Alternative Technology Publications.

DOE (1993) *Nuclear Physics and Reactor Theory DOE* Fundamentals Handbook Volume 1 and 2. Washington, D.C.: Department of Energy

DOE (2002) A Technology Roadmap for Generation IV Nuclear Energy Systems. Washington, D.C.: DOE Nuclear Energy Research Advisory Committee.

DOI (2006) *Technology White Paper on Ocean Current Energy Potential on the U.S. Outer Continental Shelf*. Washington D.C.:US Department of Interior.

Exxon Mobil (2010) *Algal biofuels.* http://www.exxonmobil.com/Corporate/energy_climate_con_vehicle_algae.aspx.

Falnes, J. (2002) *Ocean Waves and Oscillating Systems.* Cambridge: Cambridge University Press.

Hammons, T. J. (1993) Tidal power, *Proceedings of the IEEE*, 81(3).

Harvey, A. and Brown, A. (1992) *Micro-Hydro Design Manual.* Stockholm: ITDG Publishing.

IAEA (1992) INSAG-7, *The Chernobyl Accident: Updating of INSAG-1.* Vienna: International Atomic Energy Agency.

IAEA (2003) *Review of National Accelerator Driven System Programmes for Partitioning and Transmutation.* Vienna: International Atomic Energy Agency.

IAEA (2005a) *Country Nuclear Fuel Cycle Profiles* – Second Edition. Vienna: International Atomic Energy Agency.

IAEA (2005b) *Thorium Fuel Cycle – Potential Benefits and Challenges.* Vienna: International Atomic Energy Agency.

IAEA (2006) *Nuclear Technology Review 2006.* Vienna: International Atomic Energy Agency.

IAEA (2008a) *Advanced Applications of Water Cooled Nuclear Power Plants.* Vienna: International Atomic Energy Agency.

IAEA (2008b) *Spent Fuel Reprocessing Options.* Vienna: International Atomic Energy Agency.

IAEA (2008c) *Nuclear Safety Review for the year 2008.* Vienna: International Atomic Energy Agency.

IAEA (2008d) *Improving the International System for Operating Experience Feedback INSAG-23.* Vienna: International Atomic Energy Agency.

IAEA (2009a) *Nuclear Technology Review 2009.* Vienna: International Atomic Energy Agency.

IAEA (2009b) *Nuclear Fuel Cycle Information System: A Directory of Nuclear Fuel Cycle Facilities.* Vienna: International Atomic Energy Agency.

IAEA (2009c) *Design Features to Achieve Defence in Depth in Small and Medium Sized Reactors.* Vienna: International Atomic Energy Agency.

IAEA (2010) *Power Reactor Information System* http://www.iaea.org/programmes/a2/ Vienna: International Atomic Energy Agency.

IEA (2007) *Renewables in Global Energy Supply, An IEA Fact Sheet.* Paris: International Energy Agency.

Inderwildi, O.R., Jenkins, S.J. and King, D.A. (2008) Mechanistic Studies of Hydrocarbon Combustion and Synthesis on Noble Metals, *Angewandte Chemie International Edition*, 47 (5253).

IPCC (2005) *IPCC Special Report on Carbon Dioxide Capture and Storage Prepared by Working Group III of the Intergovernmental Panel on Climate Change* [Metz, B., O. Davidson, H.C. de Coninck, M. Loos and L. A. Meyer (eds.)]. Cambridge: Cambridge University Press.

Jones, A.T. and Rowley, W. (2003) Global Perspective: Economic Forecast for Renewable Ocean Energy Technology, *Marine Technology Society Journal*, 36.

Larson, E.D., Williams, R.H., Regis, M. and Leal, L.V. (2001) ?A review of biomass integrated-gasifier/gas turbine combined cycle technology and its analysis for Cuba, *Energy for Sustainable Development* 5 (1).

Lecomber, R. (1979) *The evaluation of tidal power projects, Tidal Power and Estuary Management.* Dorchester: Henry Ling Ltd.

Lindeberg, E., J.-F Vuillaume and A. Ghaderi (2009) Determination of the CO_2 storage capacity of the Utsira formation, *Energy Procedia*, 1, pp 2777–2784.

Marchetti, C. (1977) On Geoengineering and the CO_2 Problem, *Climatic Change*, 1, 59–68.

McCormick, M. (2007) *Ocean Wave Energy Conversion*. New York: Dover Publications.

Meshik, A. (2005) The working of an Ancient Nuclear Reactor *Scientific American*, vol. 293, pp. 82–91.

Michal, R. (2001) Fifty years ago in December: Atomic reactor EBR-I produced first electricity *Nuclear News* pp 28–29

MIT (2006) *The future of geothermal energy*. Cambridge, MA: Massachusetts Institute of Technology.

Moritsuka, H., M. Ashizawa and K. Ichikawa (2009) Electric power stable supply toward 2050 – Substitution fossile fuels for biomass fuel by co-firing. *Proceedings of the International Conference on Power Engineering-09* (ICOPE-09) November 16–20, 2009, Kobe, Japan.

Mousdale, D. M. (2008) *Biofuels: Biotechnology, Chemistry, and Sustainable Development*. Boca Raton: Taylor & Francis.

Nakashiki, N., T. Ohsumi and K. Shitashima (1991) *Sequestering of CO_2 in a deep-ocean – Fall velocity and dissolution rate of solid CO_2 in the Ocean*, CRIEPI Report EU91003.

NEI (2007) *Power Market Development: How much*? Kent: Nuclear Engineering International.

Nuttall, W.J. (2008) *Fusion as an Energy Source: Challenges and Opportunities*. Bristol: Institute of Physics.

OECD/NEA (2009) Nuclear energy and addressing climate change, *Nuclear Energy in Perspective*. Paris: OECD/Nuclear Energy Agency.

OECD/NEA-IAEA (2008) *Uranium 2007: Resources, Production and Demand*? (Red Book). Paris: OECD Publishing.

Oilgae (2010) *Flue gas and its part in global warming*. http://www.oilgae.com/algae/cult/cos/flu/flu.html.

Olsson, M., Wick, G. L. and Isaacs, J. D. (1979) Salinity Gradient Power: utilizing vapour pressure differences, *Science* 206.

Saling, J.H and Fentiman, A.W. (2002) *Radioactive Waste Management*. New York: Taylor & Francis.

Sheehan, J., Dunahay, T., Benemann, J. and Roessler, P. (1998) *A Look Back at the U.S. Department of Energy's Aquatic Species Program: Biodiesel from Algae*. Colorado: U.S. Department of Energy's Office of Fuels Development.

Takahashi, T., T. Ohsumi, K. Nakayama, K. Koide and H. Miida (2009) Estimation of CO_2 Aquifer Storage Potential in Japan, *Energy Procedia* 1, 2631–2638.

Thorpe, T. W. (1999) An Overview of Wave Energy Technologies: Status, Performance and Costs, *Proc Wave Power: Moving towards Commercial Viability*, IMECHE Seminar, London, pp 1–16.

Trimble, L.C. and Owens, W.L. (1980) Review of mini-OTEC performance, Energy to the 21st century, *Proceedings of the Fifteenth Intersociety Energy Conversion Engineering Conference*, New York.

Twidell, J., Weir, A. D. and Weir, T. (2006) *Renewable Energy Resources*. New York: Taylor & Francis.

Volk, T.A., Abrahamson, L.P., White, E.H., Neuhauser, E., Gray, E., Demeter, C., Lindsey, C., Jarnefeld, J., Aneshansley, D.J., Pellerin, R. and Edick, S. (2000) Developing a Willow Biomass Crop Enterprise for Bioenergy and Bioproducts in the United States, *Proceedings of Bioenergy 2000*, New York: North East Regional Biomass Program.

Walker, S.J. (2004) *Three Mile Island: A Nuclear Crisis in Historical Perspective*. Berkely: University of California Press.

WEC (1993) *Renewable Energy Resources: Opportunities and Constraints 1990–2020*. London: World Energy Council.

WEC (2001) *Survey of World Energy Resources*. London: World Energy Council.
Whitcre, G., Ware, W. and David, M. (2007) *Reviews of Environmental Contamination and Toxicology : Continuation of Residue Reviews*, V189. New York: Springer.
Wick, G.L. and Schmitt, W.R. (1977) Prospects for Renewable Energy from the Sea, *Marine Technology Society Journal*, 11.
WNA (2009) *World Uranium Mining*. London: World Nuclear Association.
WNA (2010) *The Economics of Nuclear Power*. London: World Nuclear Association.

Section 4

Demand-side energy technologies

U. Aswathanarayana (India)

Chapter 13

Industry

U. Aswathanarayana

13.1 INDUSTRIAL ENERGY USE AND CO_2 EMISSIONS PROFILE

Key industries are listed in Table 13.1 in the order of decreasing magnitude of CO_2 emissions, worldwide. Iron and Steel industry emit the largest amounts of CO_2, and Chemicals and Petrochemicals industry account for the largest energy use.

Industry-caused CO_2 emissions (6.7 Gt in 2005) constitute about 25% of the total worldwide emissions. Iron and steel industry accounts for about 30% of the CO_2 emissions, followed by 27% from non-metallic minerals (mainly cement), and 16% from chemicals and petrochemicals production. If the Best Available Technologies (BATs) are applied worldwide, current CO_2 emissions can be reduced by about 19% to 32%. Improvements in steam supply systems and motor systems have the potential to raise efficiencies from 15% to 30%. If CHP is included in the process designs, it will reduce heat demand per unit of output.

Table 13.1 Energy and CO_2 emissions of key industries

	Energy	*CO_2 emissions*
Iron and Steel	Second largest; 20% of the world's	Largest; 30% of the world's
Non-metallic minerals	Third largest; 10% of the world	Second largest; 27 % of the world's
Chemicals and Petrochemicals	Largest; 28 % of the world's	Third largest; 16 % of the world's
Pulp and Paper	Fourth largest; 6% of the world's	Fourth largest/3% of the world's
Non-ferrous metals	Fifth largest; 3% of the world's	Fifth largest; 2% of the world's

Industrial emissions can be drastically reduced by the application of CCS in the production of chemicals, iron and steel, cement, paper and pulp. This option is still in the process of development.

Technological improvements to bring about savings in CO_2 emissions vary from one industry to another. For instance, the energy savings potential of the petrochemical industry depends upon the development of alternative feedstocks. Life cycle CO_2 emissions can be reduced substantially by using biomass feedstocks and recycling more plastic waste.

There are three main ways in which CO_2 emissions from the industries can be reduced:

(i) Through improvements in efficiency that can be brought about through the recycling of waste materials, and changes in the product design, (ii) feedstock substitution, such as the greater use of biomass, and (iii) CO_2 capture and storage (CCS).

Industrial sector has been generally efficient. Improving the energy efficiency in industries has the effect of reducing the industrial emissions. Commercially available technologies are capable of bringing about efficiencies of the order of 10% to 20% in energy-intensive industries, such as, chemicals, paper, steel and cement manufacture.

Where the energy prices are lower, there is no incentive to improve the energy efficiency and consequently, energy efficiency tends to be less in such situations. New improvements in motor and steam system technologies have the potential to bring about energy efficiency in all industries, with energy savings in the range of 15% to 30%, and payback period of about 2 years.

In some favourable situations, there can be financial savings as high as 30% to 50% over the operating life of the improved systems (ETP, 2008, p. 472).

A number of technologies, such as, smelt reduction, near net-shape casting of steel, new separation membranes, black liquor gasification, advanced co-generation, are in the process of development. It may take 10–15 years before these technologies become commercially viable. CCS technologies are most economical in the case of large CO_2 emitters, such as blast furnaces, cement kilns, ammonia plants and black liquor boilers.

Table 13.2 gives the CO_2 emissions sector-wise. The direct energy and process emissions in 2050 will be 66% above the 2005 level in the ACT Map scenario, and 22% below the 2005 level in the case of the BLUE Map scenario. The differences in the

Table 13.2 Industrial CO_2 reductions by sector in the ACT and BLUE Map scenarios

Reference	*ACT Map (%) Baseline 2050*	*BLUE Map Baseline 2050 (%)*	*ACT Map 2005 (%)*	*BLUE Map 2005 (%)*
Iron and Steel	−20	−65	71	−26
Cement	−22	−68	38	−44
Chemicals and Petrochemicals	−2	−53	101	−5
Pulp and Paper	−36	−97	83	−91
Non-ferrous metals	−9	−24	258	200
Other	−11	−48	54	−10
Total	−16	−61	66	−22

(Source: ETP, 2008, p. 474)

reduction of emissions in the case of different sectors is attributed to the projected levels of economic activity under ETP scenarios.

Total final use by industry (2763 Mtoe in 2005) is nearly one-third of the total global energy use. These figures do not include either about 1 000 million tonnes of wood and biomass feedstock used (eq. 380 to 430 Mtoe) or the energy used for the transportation of raw materials and finished industrial products. Two-thirds of the final energy use is accounted for by the G-8 nations (Canada, France, Germany, Italy, Japan, U.K., USA) plus the five (+5) emerging economies (Brazil, China, India, Mexico and South Africa) (Table 13.3).

Raw materials production accounts for most of the industrial energy use. Out of the 68% of the final industrial energy use, chemical and petrochemical industries account for 29%, and iron and steel industry 20%. During the last three decades, virtually all the manufacturing industries recorded an improvement in the industrial energy intensity defined as energy use per unit of industrial output. During the period, 1971 and 2005, energy use and CO_2 emissions increased by 65% , i.e. at the annual growth rate of about 1.5%. But the growth rates vary widely among different sectors – energy and feedstock use has doubled in the case of chemicals and petrochemicals sub-sector, whereas the growth was flat in the case of iron and steel sub-sector.

China is currently the largest producer of aluminium, ammonia, cement, and iron and steel in the world. The energy efficiency of industrial production in China is less than that of the OECD countries, as the energy production in China is largely coal-based, with consequent higher CO_2 emissions. The position is, however, changing fast. There was about 100 GW of additional power generation in 2006 alone. China has now 30 GW of super-critical units, and about 100 GW of supercritical plants are in

Table 13.3 Industrial direct energy and process CO_2 emissions in world regions (2005 Mt CO_2/yr)

	OECD	*Transition economies*	*Developing countries*	*World*
Chemical and Petrochemical	491	113	482	1 086
of which petrochemical feedstocks	172	76	191	439
Iron and Steel	626	266	1 101	1 992
of which process emissions	45	12	54	111
Non-metallic minerals	471	82	1218	1770
of which process emissions	244	40	656	940
Paper, pulp and print	128	2	59	189
Food and tobacco	133	10	100	243
Non-ferrous metals	48	2	59	110
Machinery	64	5	59	129
Textile and Leather	31	1	63	96
Mining and quarrying	37	14	47	98
Construction	51	9	37	96
Transport equipment	28	3	19	49
Wood and wood products	16	2	9	27
Non-specified	191	55	528	775
Total	**2315**	**563**	**3 782**	**6 660**
of which process emissions	289	52	710	1 051

(Source: ETP, 2008, p. 481)

Table 13.4 Final energy use by energy carrier and direct CO_2 emissions related to energy use, 2005

	Mtoe/yr	*Gt CO_2/yr*
Coal and coal products	714	3.10
Natural Gas	561	1.28
of which petrochemical feedstocks	129	0.30
Oil and oil products	666	1.24
of which petrochemical feedstocks	338	0.13
Combustible renewables and waste	180	
Electricity	532	
Heat	110	
Other	0	
Total Direct Energy Emissions	**2 763**	**5.61**
Process emissions (cement & steel)		1.05
Total Direct Energy and Process Emissions		**6.66**
Electricity Generation emissions		3.19
Total Direct and Indirect Emissions		**9.86**

(Source: ETP 2008, p. 479)

order. China has now some of the most efficient steel and paper plants in the world. Africa has some efficient aluminium smelters, and the cement kilns in India are highly energy efficient.

Table 13.4 gives the breakdown of the industrial energy use by fuel and energy carrier. Biomass in the form of combustible renewables and waste is largely used in pulp and paper industry. Most of the CO_2 is emitted by iron and steel, cement, chemicals and petrochemicals, paper and pulp industries, but this pattern of distribution is different in the case of energy demand. This is so because (i) the petrochemical products contain large quantities of fossil carbon, (ii) the emission of CO_2 in the case of sectors such as cement production, is unrelated to their energy use, and (iii) Different industrial sub-sectors have different fuel mixes.

Total direct and indirect CO_2 emissions from industry (9.9 Gt in 2005) is equivalent to 37% of the global total emissions. The G-8+5 countries account for 89% of global CO_2 emissions.

13.2 IRON AND STEEL

The Iron and steel industry is the world's second largest consumer of energy, accounting for 20% of the world's energy use. It is also the largest emitter of energy and process CO_2, accounting for 30% of the global emissions. The four leading producers of steel, namely, China, the European Union, Japan and USA, account for 67% of the CO_2 emissions.

There are three principal ways of producing steel:

(i) EAF method, based on iron scrap being melted in an electric arc furnace,
(ii) DRI/EAF method, whereby iron ore and some scrap are melted in an arc furnace,
(iii) BF (Blast furnace) and BOF (Basic Oxygen Furnace) method, whereby 70–100% iron ore and the reminder scrap are used to produce steel.

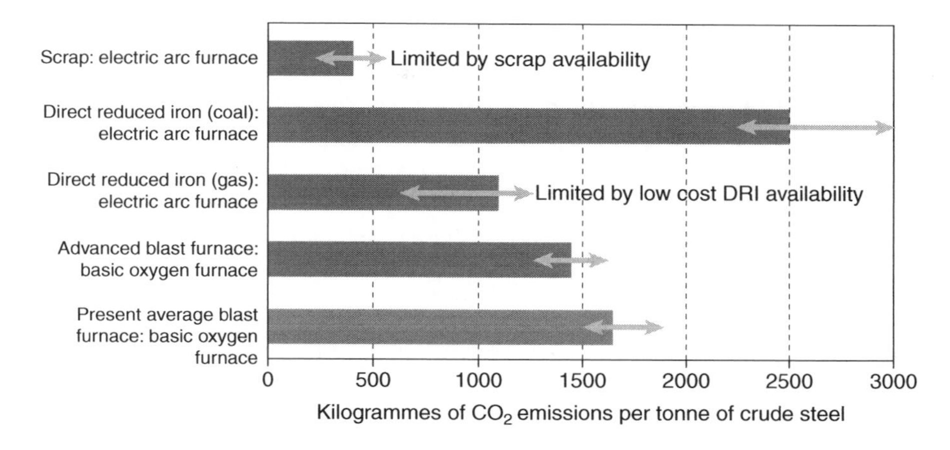

Figure 13.1 CO_2 emissions per tonne of steel produced
(Source: ETP 2008, p. 484, © IEA-OECD) shows the potential of CO_2 reductions in respect of the three key processes of steel making

The Open Hearth Furnace route is similar to BF – BOF route. It is, however, considered outmoded, and accounts for just 3% of the world steel production. Where scrap is in short supply, and gas supply is available, Direct Reduced Iron (DRI) can be used in place of scrap. In India, DRI is coal-based, as gas use is not available.

In USA where scrap is plentiful, EAF is the preferred method of steel production. In countries like India, where the availability of scrap is limited, steel is produced by the BF – BOF method. The scrap – EAF route uses much less energy (4–6 GJ/t) relative to BF – BOF route (13–14 GJ/t). This is so because the scrap – EAF route avoids the need for the expenditure of energy for the preparation of iron ore and washing of coal, reduction of iron ore to iron, and for the conversion of coal to coke. There would hence be significant energy saving if it possible to switch to scrap/EAF route from BF-BOF route.

The energy efficiency of primary steel production varies greatly between countries and even between different plants in a country, depending upon the size of the steel plant, the extent of waste energy recovery, quality of the iron ore, process technology used, and so on. Use of scrap in steel making reduces the CO_2 emissions.

Best Available Technologies (BAT) to reduce CO_2 emissions are: Steel finishing improvements, efficient power generation from BF gas, switch from OHF to BOF, increased BOF gas recovery, blast furnace improvements, COG recovery, and CDQ (advanced wet quenching). Improvements in the blast furnace are the most important way to increase efficiency. The average global potential is 0.30 t CO_2/t steel produced. The total global potential is ~340 Mt CO_2. China accounts for nearly half of this, as it is the largest steel producer in the world.

Energy exists in the gas streams in the steel making in the form of heat, pressure and combustible content. If the gas streams are made use of effectively, energy efficiency increases and CO_2 emissions decrease. If these approaches are widely adopted world wide, there would be reductions of 100 Mt CO_2 per year.

Efficiency in steel making can be improved by reducing the number of steps involved and the amount of materials processed in any step, in the following ways: (i) Injecting pulverized coal in place of coke in the blast furnace. Before coking, ROM coal is invariably washed in order to reduce the ash content. Direct injection of pulverized coal avoids the need for coal washing as well as burning coal in coke ovens to convert it into coke, (ii) New technologies such as COREX can use coal instead of coke, and (iii) New reactor designs (FINEX and cyclone converter furnaces) can use coal and ore fines. Coal injection has the potential to save up to half of the coke presently used, thus saving the energy needed in coke production (2–4 GJ/t). The potential for coal savings globally would be 12 Mtoe per year, equivalent to 50 Mt of CO_2/yr.

Improvement in energy efficiency and reduction in CO_2 emissions can be achieved through process streamlining.

Smelt reduction and efficient blast furnaces: Smelt reduction involves the development of a single process in the place of ore preparation, coke making and conversion to iron in the blast furnace. Small and medium-scale steel plants stand to benefit from this approach. In the COREX plant design, coal fines, iron ore fines and limestone fines are palletized into self-fluxing sinter. Such plants are in use in South Africa, Korea and India. The new kinds of smelt-reduction plants generate about 9 GJ/t of surplus off-gas, whose reuse could bring about significant additional CO_2 reductions. By blowing oxygen, instead of air, into blast furnaces, and by recycling top gases, it is possible to achieve a 20–25% reduction of CO_2. Japan and the European Union are developing the ULCOS (Ultra Low CO_2 Steel-making) process. The combination of smelt reduction and nitrogen-free blast furnaces may bring about 200 to 500 Mt of CO_2 by 2050.

Direct casting: Customarily, steel is continuously cast into slabs, billets and blooms. They are later reheated and rolled into desired shapes. Near-net casting and thin-strip casting integrates the casting and hot rolling processes into one step, and saves considerable energy (typically, 1–3 GJ/t of steel). Material losses are also reduced. (Table 13.5).

Fuel and feedstock substitution: Iron ore is reduced to iron through the use of coal and coke. Where available, natural gas is used for the production of DRI. In South America, particularly Brazil, wood is used, in small-scale plants. In south India, Mysore Iron and Steel works at Bhadravati used wood for many years. Japan has been using 0.5 Mt (20 PJ) of plastic waste as a coal substitute in the blast furnaces. Hydrogen and electricity could also be used in steel making, but as the CO_2 reduction benefits cost more than USD 50/t CO_2, they are not much favoured.

Table 13.5 Global technology prospects for direct casting

Direct casting	*2008–2015*	*2015–2030*	*2030–2050*
Technology stage	R&D, Demonstration	Commercial	Commercial
Investment costs (USD/t)	200	150–200	150–200
Energy reduction (%)	80%	90%	90%
CO_2 reduction (Gt/yr)	0–0.01	0–0.03	0–0.1

(Source: ETP 2008, p. 488)

CCS for the current blast furnaces would cost USD 40–50/t CO_2, excluding the expenses in the furnace redesign. DRI production would allow CCS at a much lower cost of USD 25/t CO_2 (Borlée, 2007) When the DRI production picks up in the Middle East, this would contribute significantly to the reduction of CO_2 emissions.

CCS in iron and steel production could save around 0.5–1.5 Gt of CO_2 per year.

13.3 NON-METALLIC MINERALS

Non-metallic minerals are used for the production of cement, bricks, glass, ceramics and other building materials. This sector is the third largest consumer of energy (10% of the global energy use), and second largest emitter of CO_2 (27% of the global energy and process CO_2 emissions). China, India, the European Union and USA account for 75% of the CO_2 emissions. Out of the global cement production of 2310 Mt in 2005, the developed countries accounted for 563 Mt (24% of the world output), transition economies 98 Mt (4% of the world output) and the developing countries 1649 Mt (72% of the world output).

Yates et al (2004) described ways and means of reducing the emission of greenhouse gases in the cement industry.

China is the world's largest producer of cement.

Cement industry accounts for 83% of the total energy use and 94% of the total CO_2 emissions pertaining to the non-metallic minerals sector.

Limestone is the principal raw material for making cement. Clinker is produced by heating limestone and chalk to temperatures above 950°C. Clinker production accounts for most of the energy consumed in making cement. Large amounts of electricity are also used in grinding of the raw materials, and in the production of finished cement. The calcination of limestone leads to the emission of CO_2, and these emissions are unrelated to energy use. CO_2 emissions in the course of calcination cannot be reduced through energy efficiency measures – they can only be reduced through appropriate raw material selection.

Improvements in cement-making have the potential to reduce CO_2 emissions by 290 Mt. If clinker substitutes are included, the potential saving could rise to 450 Mt of CO_2 . The world average potential is 0.18 t CO_2/t of cement.

The following Best Available Technologies (BAT) has the potential to reduce the CO_2 emissions: BF slag clinker substitutes, other clinker substitutes, alternative fuel, electricity savings and fossil fuel savings.

Heat efficiency and management: Large-scale rotary kilns, which are used in the industrialized countries, are more efficient than small-scale vertical shaft kilns that are used in developing countries, such as China and India, but these countries are also switching to rotary kilns. All over the world, the wet process of making Portland cement is being replaced by dry process, because of two benefits: saving of water to make the slurry, and saving of energy as drying will not be needed. Dry-process kilns use about half of the energy as wet-process kilns. The most efficient arrangement is the dry kiln, with six-stage preheating and pre-calcining.

Grinding is necessary to produce cements. Cements with high fly ash content reduce energy use and CO_2 emissions. The energy efficiency of grinding is low, typically 5–10%, as the remainder is converted to heat. Grinding is done more efficiently by

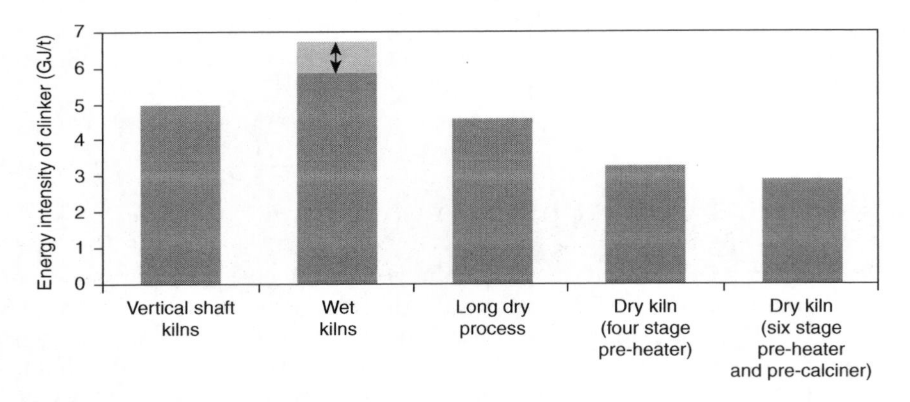

Figure 13.2 Energy efficiency of various cement clinker production technologies (Source: ETP, 2008, p. 492, © IEA-OECD)

using roller presses and high-efficiency classifiers. High-strength cements used in building skyscrapers involve superior grinding technologies and the use of additives. Such cements are expensive, and require sophisticated knowledge for using them.

Fuel and feedstock substitution: The use of wastes and biomass (including tyres, wood, plastics, etc.) in the place of fossil fuels in the cement industry not only brings about saving in fuels but also reduction in CO_2 emissions. A number of cement plants in Europe, wastes are co-combusted in the cement kilns to the extent of 35% to more than 70%. Some individual plants have achieved even 100% substitution. The cement industry in USA burns 53 million used tyres per year. Another potential source of energy in USA is carpets. Instead of dumping them in the landfill, as is the usual practice, they can be burned in the cement kilns. Their fuel value is estimated at 100 PJ.

There is potential for alternative fuels to be raised from 24 Mtoe to 48 Mtoe. When that happens, there would be CO_2 reductions of the order of 100–200 Mt per year.

Clinker substitutes and blended cements: Increasing the proportion of non-clinker feedstocks, such as volcanic ash, granulated blast furnace slag , and fly-ash from coal-fired power generation, is an effective way to reduce energy and process emissions. The CO_2 savings from blended cements could be 300–500 Mt by 2050.

Blast furnace slag which has been cooled with water is more suitable than the slag cooled with air. If all the water-cooled blast furnace slag is used, there will be CO_2 reduction of approximately 100 Mt . The setting time of cement is a critically important consideration in cement use. When fly ash from the coal-fired power plants is used as a non-clinker feedstock, its carbon content may adversely affect the setting time of cement. The pre-treatment of fly ash will allow it to be substituted to the extent of 70%. If half of the fly ash is used in the cement industry, instead of dumping it in the landfill, there will be saving of 75 Mt of CO_2. If the EAF and BOF steel slag resource of 100–200 Mt per year, were used in the cement manufacture, there would be CO_2 savings of 50 to 100 Mt per year.

Other feedstocks possible are volcanic ash, ground limestone and broken glass. They can bring about reduction in the use of energy and CO_2 emissions. When limestone

Table 13.6 Global technology prospects for CCS for cement kilns

CCS	*2008–2015*	*2015–2030*	*2030–2050*
Technology stage	R&D	R&D Demonstration	Demonstration, Commercialization
Costs (USD/t CO_2)	150	100	75
Emission reduction (%)	95	95	95
CO_2 reduction (Gt CO_2/yr)	0	0–0.25	0.4–1.4

(Source: ETP 2008, p. 495)

is calcined in the cement kilns, the off gas will have a high content of CO2. If oxygen instead of air were used in the cement kilns, the off gas would be pure CO2. The use of CCS in the cement kilns will raise the production cost by 40 to 90% (Table 13.6).

13.4 CHEMICALS AND PETROCHEMICALS

The chemicals and petrochemicals industry is the largest consumer of energy (28% of the world's industrial energy) and the third largest emitter of energy and process CO_2 emissions (16% of the world's emissions). The industry is highly complex, both in terms of processes (distillation, evaporation, direct heating, refrigeration, electrolytic and biochemical), the number of final products, and size (ranging from a few kgs. to thousands of tonnes). However, three processes account for 537 Mtoe of energy use (which is 70% of the energy use in the sector):

High-value chemicals (HVC), such as, olefins (ethylene and propylene) and aromatics (benzene, toluene, and xylene) are produced by the steam-cracking of naphtha, ethane and other feedstocks. This process accounts for more than 39% of the final energy use in the chemicals and petrochemicals industry. Out of the total of 318 Mtoe, about 50 Mtoe is used for energy purposes, and 268 Mtoe is locked up in the cracking products. The energy used in steam cracking is determined principally by the nature of the feedstock, and secondarily by the furnace design and process technology. For instance, 1.25 tonnes of ethane, 2.2 t of propane or 3.2 t of naphtha are needed to produce 1 t of ethylene. Naphtha cracking is more in use in Asia-Pacific, and Western Europe, whereas ethane cracking is more prevalent in North America, Middle East and Africa. This difference is evidently attributable to feedstock availability. Improvements in steam cracking design have led to 50% reduction in the energy consumption since 1970s.

Methanol: Methanol is used as anti-freeze, solvent and fuel. About 80% of ethanol production is natural gas-based, and so there is spurt in methanol production in Middle East and Russia. About 30 GJ of natural gas is needed to produce one tonne of methanol. China uses coal for methanol production. In 2006, the global production of methanol was 36 Mt, of which 40 % was used for the production of formaldehyde, 19% was used to make methyl tertiary butyl ether (MBTE), which is a gasoline additive, and 10% for the production of acetic acid.

Ammonia: Almost all the synthetic nitrogen fertilizers are based on anhydrous ammonia. Ammonia is made by combining nitrogen from air with hydrogen from natural gas or naphtha, coke-oven gas, refinery gases and heavy oil. Global ammonia production was 145.4 Mt in 2005. East and West Asia account for 40% of the global production. About 77% of world ammonia production is based on natural gas-steam reforming, 14% on coal gasification (mostly in China), and 9% on the oxidation of heavy hydrocarbon fractions (mostly in India). Coal-based process uses 1.7 times more energy, and heavy oil-based process uses 1.3 times more energy than the gas-based process. The cost of natural gas accounts for 70–90% of the cost of ammonia production.

USA, European Union, Japan and China are the largest producers of HVCs, and account for 62% of the CO_2 emissions. China, the European Union, India and Russia are the largest producers of ammonia, and account for 72% of the energy use in the production of this chemical.

Oil, natural gas and coal feedstock provide more than half of the energy (469 Mtoe/yr) consumed in the sector. Products, such as plastics, solvents and methanol hold most of the carbon input of the feedstock, but some of the carbon gets released at a later stage, say, for instance when the product is incinerated. During the complete life cycle, chemicals and petrochemicals emit far more CO_2 than indicated by the industrial CO_2 emissions.

Energy and Materials efficiency: New Process technologies

Steam cracking per tonne of ethylene cracked needs 18–25 GJ of energy. Energy efficiency of steam cracking is being improved through the use of higher temperature (>1100°C) furnaces, gas – turbine integration (by which process heat is provided to the cracking furnace), advanced distillation columns and combined refrigeration plants. These improvements could result in the savings of 3 GJ per tonne of ethylene. The adoption of BAT would lead to an energy saving of 24 Mtoe.

Bowen (2006) gave an account of the development trends in ethylene cracking.

The amount of energy used for producing ammonia ranges from 28 GJ to 53 GJ/t, averaging 36.9 GJ/t. High-capacity, modern plants are about 10% more energy efficient than smaller and older plants. CO_2 emissions range from 1.5 to 3.1 Mt , with an average of 2.1 Mt, per one Mt of ammonia produced. Two-thirds of CO_2 is process related, and one-third is from fuel combustion. If all the production of ammonia is based on the natural gas feedstock, it has the energy saving potential of 48 Mtoe, which represents reduction in CO_2 emissions of 75 Mt. If CO_2 is separated from hydrogen using high-efficiency solvents, there would be two benefits: there would be an energy saving of 1.4 GJ/t of ammonia produced, and the CO_2 separated could be used for the production of urea fertilizer for which there is good demand.

Biomass feedstock: There will be considerable saving of energy when the biomass feedstock is substituted in place of petroleum feedstock. ETP, 2008, p.500, lists four principal ways of producing polymers and organic chemicals from biomass:

- Direct use of several naturally occurring polymers after subjecting them to thermal treatment, chemical derivatisation or blending.
- Thermochemical conversions, such as Fischer-Tropsch process of converting coal to oil, and methanol-to-olefins (MTO) via pyrolysis or gasification. There is tremendous potential for using low-cost coal and stranded gas feedstocks for MTO

Table 13.7 Global technology prospects for biomass feedstocks and biopolymers

	2008–2015	*2015–2030*	*2030–2050*
Technology stage	R&D	R&D Demonstration	Demonstration, Commercialization
Investment costs (USD/t)	5 000–15 000	2 000–10 000	1 000–5 000
Life-cycle CO_2 reductions	50%	70%	80%
CO_2 reduction (Gt/yr)	0–0.05	0.05–0.1	0.1–0.3

(Source: ETP 2008, p. 501)

process. Since the Second World War, South Africa which is deficient in oil, has been using coal to produce oil by Fischer-Tropsch process.

- "Green" biotechnology whereby genetically-modified potatoes or miscanthus are used to produce biopolymers. Sapphire Energy, San Diego, California, uses single-cell algae to produce an organic mix which is chemically identical to low-sulphur, sweet crude (vide image on the cover). The Company plans to produce one million gallons (3.8 Million litres) of biodiesel and jet fuel by 2011.
- "White" biotechnology which makes use of fermentation processes and enzymatic conversions to produce some specialty and fine chemicals.

Bio-ethylene can be used to produce polyethylene and a wide range of chemical derivatives. The production of biobased chemicals involve not only saving of energy but also reduction in greenhouse gases. For instance, there would be energy saving of as much as 60% through the substitution of cellulosic fibre in place of synthetic fibre.

The production of ethylene from bio-ethanol leads to a saving of energy and reduction in the emission of greenhouse gases by about one-third, relative to petrochemical ethylene. If advanced fermentation and separation technologies are used, the saving can be as high as 50%. The large amount of biomass waste produced after the production of bio-ethanol from sugarcane, can be used to generate electricity, thereby saving fossil fuels.

Carbon credits and higher oil prices will make biomass feedstocks competitive. Though theoretically, whatever is producible from petroleum can be produced from bio-based feedstocks, the market penetration of bio-based products would depend upon the relative prices, technological developments, government support and synergies with biofuel production (Table 13.7).

Plastic waste recycling and energy recovery

Only 20 to 30% of the plastic waste can be mechanically recycled, and the rest can be used for energy recovery. Considering that the energy recovery per tonne of plastic waste is 30 to 40 GJ/t, the primary energy saving potential is in the range of 48–96 Mtoe/yr. The quantity of plastic waste produced worldwide is about 100 Mt. Out of this, only 10 Mt is recycled. About 30 Mt is incinerated. Energy recovery from the plastics is estimated at 17.9 Mtoe which is 3 % of the energy used in its production.

Table 13.8 Global technology prospects for membranes

Membranes	*2003–2015*	*2015–2030*	*2030–2050*
Technology stage	R&D, Demonstration	Demonstration, Commercial	Commercial
Internal rate of return	8%	10%	15%
Energy savings (%)	15%	17%	20%
CO_2 reductions (Gt/yr)	0–0.03	0.1	0.2

(Source: ETP 2008, p. 502)

Membranes:

Separation technologies involve processes such as distillation, fractionation and extraction. Their primacy in the chemical industry could be judged from the fact that they account for 40% of the energy used and 50% of the operating costs of the chemical industry. Customized membranes are being increasingly used to replace the energy-intensive separation processes not only in chemical industry but also in food processing, water purification, paper, petroleum refining, etc. industries. The market penetration of membranes is impeded because of the higher costs of the membranes and their susceptibility to fouling . Global technology prospects for membranes are given in Table 13.8.

Innovations in process technology and equipment in the petrochemical sector have the potential to increase the energy efficiency in the petrochemical sector by 5% in the next 10–20 years, and by 20% in the next 30–40 years. The main barrier is the large-scale demonstration.

13.5 PULP AND PAPER

The pulp and paper industry accounts for 6% of the world's industrial use and 3% of the energy and process CO_2 emissions. About 80% of the paper in the world is produced in European Union, USA, China and Japan. The paper and pulp industry generates about 50% of the energy needs from its own biomass residues. This explains the lower intensity of CO_2 emissions of the industry. Greater efficiencies are still possible through the use of lesser amounts of bioenergy resources which can replace fossil fuels. Berntsson et al (2007) gave a vision of future possibilities of biorefining. Hector and Berntsson (2007) described the ways and means of reducing greenhouse gases in pulp and paper mills.

As the pulp and paper industry uses large quantities of steam, Combined Heat and Power (CHP) is an attractive technology for the pulp and paper industry. Chemical pulp mills produce large quantities of black liquor, which is used to produce electricity through the boiler system. But the efficiency of this process is low. Higher efficiencies are achievable by the gasification of the black liquor (syngas), and using the gas to produce electricity through the operation of gas turbines. The total cost of the gasifier-gas turbine system is 60 to 90% higher than the standard boiler system. USA, Sweden

Table 13.9 Global technology prospects for black liquor gasification

Black liquor gasification	*2003–2015*	*2015–2030*	*2030–2050*
Technology stage	R&D, Demonstration	Demonstration, Commercial	Commercial
Investment costs (USD/t)	300–400	300–350	300
Energy reduction (%)	10–15%	10–20%	15–23%
CO_2 reduction (Gt/yr)	0–0.01	0.01–0.03	0.1–0.2

(Source: ETP 2008, p. 507)

and Finland are collaborating in this effort with the goal of producing electricity at US cents 4/kWh.

A highly attractive proposition is to use the black liquid gasifiers to produce dimethyl ether (DME) which can serve as a substitute for diesel fuel.

Carbon dioxide is produced when black liquor is combusted for energy and production of chemicals. The total black liquor production worldwide is 73 Mtoe, which has a CCS potential of 300 Mt of CO_2 per year.

Black liquor production is expected to grow to 79 Mtoe by 2025. This could yield an additional 8 Mtoe of electricity per year. The consequent savings of primary energy is estimated to be 12 to 19 Mtoe, and the CO_2 savings potential may be 30 to 75 Mt per year. Global technology prospects for black liquor gasification are given in Table 13.9.

Best Available Technologies (BAT)

If all waste paper is used for energy recovery, it is theoretically possible to have paper and pulp industry without CO_2 emissions. This is not, however, the most sensible option. As much waste paper as possible should be recycled in order to avoid cutting trees to make pulp. More than 90% of the electricity used in mechanical pulping ends up as heat. If this heat is recovered and used in paper drying, energy will be saved. Paper mills which integrate mechanical, chemical, recycled paper and pulp operations are 10–50% more efficient than stand-alone paper mills. In the industrialized countries, more paper is recycled than produced. Recycling of paper is a common practice in most countries – for instance, China recycles 64% of its paper. There can be saving of 10 GJ to 20 GJ of energy per tonne of paper recycled, depending upon the kind of paper waste. Canada and USA are rich in wood resources. Canada is the largest producer of mechanical pulp, and USA is the largest producer of chemical pulp, in the world.

Paper production involves the drying of process fibres. Paper drying consumes 25 to 30% of the energy used in the pulp and paper industry. There could be energy saving of at least 15–20%, if not 30%, if this is done efficiently. Improved forming technologies, increased pressing and thermal drying could be made use of to remove water efficiently. Super-critical CO_2 use and nanotechnology have great potential to manage the role of water and fibre orientation process. Table 13.10 gives the global technology prospects for drying.

The Best Available Technologies (BAT) for pulp and paper industry recommended by the European Union are given in Table 13.11.

Table 13.10 Global technology prospects for energy-efficient drying technologies

Efficient drying	*2003–2015*	*2015–2030*	*2030–2050*
Technology stage	R&D, Demonstration	Demonstration, Commercial	Commercial
Investment costs (USD/t)	800–1100	700–1000	600–700
Energy reduction (%)	20–30%	20–30%	20–30%
CO_2 reduction (Gt/yr)	0–0.01	0.01–0.02	0.02–0.05

(Source: ETP 2008, p. 507)

Table 13.11 Best Available Technology (BAT) for the paper and pulp industry

	Heat GJ/t	*Electricity GJ/t*
Mechanical pulping		7.5
Chemical pulping	12.25	2.08
Waste paper pulp	0.20	0.50
De-inked waste paper pulp	1.00	2.00
Coated papers	5.25	2.34
Folding boxboard	5.13	2.88
Household and Sanitary Paper	5.13	3.60
Newsprint	3.78	2.16
Printing and writing paper	5.25	1.80
Wrapping and packaging paper and board	4.32	1.80
Paper and paperboard not elsewhere specified	4.88	2.88

(Source: ETP 2008, p. 505)

13.6 NON-FERROUS METALS

The non-ferrous metals sector comprises of aluminium, copper, lead, zinc and cadmium. Copper, lead and zinc are called base metals. In 2005, the non-ferrous metals accounted for 3% of the industrial energy, and 2% of the energy and process CO_2 emissions.

World Aluminium (2007) gave a detailed account of the role of electricity in aluminium industry. European Commission (2001) reviewed the Best Available Techniques in the non-ferrous metal industry.

Bauxite is the principal ore of aluminium. It is composed of the minerals, gibbsite – $Al(OH)_3$, boehmite – γ AlO(OH), and diaspore – α AlO(OH). There are two kinds of bauxite: karst bauxite (carbonate bauxite) and lateritic bauxite (silicate bauxite). Bauxite formation involves desilication and separation of aluminium from iron, under conditions of tropical weathering characterized by warm temperatures, high rainfall and vegetation, and good drainage. Australia is the largest producer of bauxite in the world, accounting for one-third of the bauxite production. China, Brazil, Guinea, India and Jamaica are important producers. World production of bauxite in 2008 was 205 Mt. Reserves are 27 billion tonnes. Bauxite is a kind of soil, and hence it is recovered by surface mining.

Bauxite is treated with sodium hydroxide in pressure vessels at temperatures of 150–200°C, to separate the aluminous part from the ferruginous part (red mud) (Bayer

Table 13.12 Global technology prospects for inert anodes and bipolar cell design in primary aluminium production

Inert anodes	*2003–2015*	*2015–2030*	*2030–2050*
Technology stage	R&D	Demonstration	Commercial
Investment costs (USD/t)	N/A	Cost savings	Cost savings
Energy reduction (%)	N/A	5–15%	10–20%
CO_2 reduction (Gt/yr)	N/A	0–0.05	0.05–0.2

(Source: ETP 2008, p. 512)

process). Most of the energy consumed in alumina production (about 12 GJ/t of alumina) is in the form of steam. Integration of alumina plants with CHP units, can bring down the energy consumption to around 9 GJ/t. The world alumina production is 60 Mt, involving the use of 16 Mtoe of energy. Two kg. of alumina is needed to produce one kg. of aluminium metal.

The calcined alumina is molten with cryolite at a temperature of 1000°C, and aluminium metal is produced electrolytically by Hall- Héroult process.

The conversion of alumina to aluminium metal is highly energy intensive. The amount of electricity used to produce one tonne of aluminium metal varies from 14 622 to 15 387 kWh, with a weighted average of 15 194 kWh/t. New generation smelters use much less energy of 13 000 kWh/t. About 18 GJ of pitch and petroleum coke is needed for the production of anodes per tonne of aluminium.

Since electricity cost constitutes the bulk of the cost of aluminium metal production, aluminium smelters are invariably located not where bauxite is, but where large quantities of cheap electricity are available. For instance, alumina is shipped all the way from Guiana in South America for being smelted in Ghana in West Africa where cheap hydropower ($\sim$1 000 MW) from Volta dam is available (the Aksombo dam on the Volta river created the Volta Lake, the fourth largest man-made lake in the world). Aluminium smelters have come up in countries like Norway, Iceland, Canada, Russia and the Middle East where low-cost electricity is available.

Most of the growth in the aluminium industry has taken place in China. China's production has doubled from 7 Mt in 2005 to 14 Mt in 2008.

The primary aluminium production requires twenty times more energy than recycling.

The use of inert cathodes in place of carbon anodes not only reduces the energy consumption by 10–20%, but also eliminates the CO_2 emissions. But this technology has yet to achieve market penetration. The global technology prospects for inert anodes are given in Table 13.12.

13.7 RESEARCH & DEVELOPMENT, DEMONSTRATION AND DEPLOYMENT

Much R&D, Demonstration and Deployment work is needed to reduce costs, improve energy efficiency and reduce CO_2 emissions, in order to achieve ACT targets. Table 13.13 gives the RD&D breakthroughs needed, technology wise (source: EPP, 2008, p. 586–589

Table 13.13 RD&D breakthroughs needed

RD&D breakthroughs, technology-wise	*Stage*	*ACT target*
Biorefineries: Pulp and paper Black liquor to methanol pilot plants	Demonstration	
Biorefineries: Biomass for various industries: Lower-cost biomass collection system for large-scale plants	Applied R&D	
CCS overall: Reduce capture cost and improve overall system efficiencies; and storage integrity and monitoring	Basic science/ Applied R&D	
CCS for blast furnace (iron/steel): Development of new blast furnace with oxygen and high temperature CO_2 mixture		195 Mt CCS (2050)
CCS for cement kilns (cement): Use of physical absorption systems (Selexol or other absorbents); use of oxygen instead of air; and process design to accommodate potentially higher process temperatures		400 Mt (2050) (energy + process)
CCS for black liquor (paper): Integration with IGCC + CCS and maximized production of biofuels for other use.		
Feedstock substitution – cement: Clinker substitute (reduction of carbon contents by upgrading of high carbon fly ash through froth flotation); triboelectrostatic separation, or carbon burnout in a fluidized bed; special grinding to increase the pozzolanic reaction rate of fly ash, and use of steel slag.	Applied R&D	
Feedstock substitution – Chemical & Petrochemical: Biopolymer (e.g., polyactic acid; polytrimethylene terephthalate fibres; polyhydroxyalkanoates; monomers from biomass and more advanced fermentation and separation technology, e.g. butanol; and naphtha products from biomass FT process	Applied R&D	26 Mtoe biomass feedstocks by 2050
Fuel substitution: Electric heating technologies; and development of suitable heating and drying technologies	Applied R&D	
Fuel substitution – heat pump: Higher temperature application; larger system; and higher coefficient of performance (COP)		
Plastics recycling/ energy recovery – Chemicals and Petrochemicals: Better low-cost separation technologies; and dedicated high efficiency energy recovery technologies	Applied R&D	
Process innovation in basic materials production processes – Aluminium: Development of inert anodes; fundamental materials research; bipolar cell design; and anode wear of less than 5 mm per year	Basic science	5–10% energy reduction (2030)
Cement : Development of high performance cement using admixtures	Basic science	
Chemicals and petrochemicals: Increased nitrogen fixation (new nitrogen fertilizer formation and understanding of steps that lead from recognition of signals exchanged between plant and bacteria to the differentiation and operation of root nodules; the genes responsible for rhizobia and legumes; the structural chemical bases of rhizobia/legume communication; and the signal transduction pathways responsible in respect of the symbiosis-specific genes involved in nodule development and nitrogen fixation); and use of membranes (performance improvement of various membranes for specific gases; liquid and gas membranes for liquid-liquid extraction and cryogenic air separation; and development of membrane reactors).	Basic science/ Applied R&D	

(Continued)

Table 13.13 Continued

RD&D breakthroughs, technology-wise	*Stage*	*ACT target*
Iron/Steel: Smelt reduction (reduction of surplus gas); and direct casting – i.e. near-net shape casting and thin-slip casting (increased reliability, control and adoption of the technology to larger-scale production units; product quality improvement; and usability improvement by steel processors and users).	Demonstration/ Deployment/ Commercialisation	195 Mt/yr (2050)
Pulp and Paper: Black liquor gasification; increased reliability of gasifier	Applied R&D	
Gasifier with a gas turbine	Demonstration	

Chapter 14

Buildings & Appliances

U. Aswathanarayana

14.1 INTRODUCTION

Buildings are large consumers of energy – in 2005, they consumed 2 914 Mtoe of energy. The residential and service sectors account for two-thirds and one-third of the energy use respectively. About 25% of the energy consumed is in the form of electricity. Thus the buildings constitute the largest user of electricity.

Globally, space and water heating account for two-thirds of the final energy use. About 10–13% of the energy is used in cooking. Rest of the energy is used for lighting, cooling and appliances. The end-uses dominated by electricity consumption are important from CO_2 abatement perspective, in the context of the CO_2 emissions related to electricity production.

CO_2 emissions can be reduced significantly through the use of Best Available Technologies in the building envelope, HVAC (heating, ventilation and air conditioning), lighting, appliances and cooking. Heat pumps and solar heating are the key technologies to reduce emissions from space and water heating. New designs of energy-efficient houses can reduce the heating demand by as much as a factor of ten, without much additional expense. Through a combination of compact design, careful orientation towards sunlight, proper insulation, high air tightness, and heat recovery from the insulation system, it is possible to have houses with virtually no heat loss. Government policies in respect of passive housing may be so framed as to promote, demonstrate and deploy new technologies, in the construction of new houses and refurbishment of old houses.

The buildings sector employs a variety of technologies for various segments, such as building envelope and its insulation, space heating and cooling, water heating systems, lighting, appliances and consumer products. Local climates and cultures have a profound effect on energy consumption, apart from the life styles of individual users.

The economic lifetimes of the individual segments of the buildings sector have an enormous range. Building shells can last for decades, even for centuries. So buildings tend to be renewed, rather than replaced. It is likely that more than half of the existing buildings will be standing in 2050. As against this, HVAC (heating, ventilation, air conditioning) systems are changed once in 10–15 years. Household appliances are changed over a period of 5 to 15 years. At the other end of the scale, incandescent light bulbs are changed yearly. The economic lifetimes of the various segments of the buildings have a large range – from a few years for light bulbs, to a few tens of years for electric transmission equipment, to hundred years or more for building stocks (ETP, 2008, p. 522). The magnificent *Brihadeswara* (Lord of the Universe) temple in Thanjavur, South India, built in 1010 A.D., is an excellent shape after a thousand years.

Under the circumstances, government policies and standards should be such as to promote the deployment of Best Available Technology for the infrastructure components at the time of refurbishing a building.

Many a time a person living in a residential building may not be the owner of the building. Though the tenant would replace on his own, items of infrastructure with short economic lifetimes (such as lights and fans), major improvements, such as, fitting the building with solar panels, have to be undertaken by the owner. Evidently, the owner and tenant need to coordinate their efforts to improve efficiency and reduce CO_2 emissions.

Through the application of integrated, intelligent building systems, it is possible to achieve about 80% reduction in energy consumption and emission of greenhouse gases. This involves integrated passive solar design with structural components of advanced design, such as, high-performance windows, vacuum-insulated panels, and high-performance reversible heat pumps. Research and Development and professional training activities have to be undertaken to realize the goal.

Now, France has launched a massive greening programme, starting with the construction business. About 25% the country's greenhouse-gas emissions come from energy consumption in the buildings. About 200,000 to 500,000 jobs are expected to be created in the process of bringing about a 40% drop in the energy consumption in the construction sector by 2020, involving investment of hundreds of millions of euros.

In about twenty years' time, France is expected to have houses which are off-grid (i.e. self-sufficient in electricity). Some may even generate more energy than they consume.

The heating requirements of a building are very much dependent on the age of the building. The data obtained from Germany show that pre-1970 buildings require 55 to 130% more energy than modern buildings. The turnover of building stocks in the developing countries is much faster, typically 25–30 years (this situation is reflected in the regulation that banks in India do not give loans to buy flats/buildings older than 20 years).

The long economic life of the buildings in the OECD countries act as a constraint in reducing the energy requirements and greenhouse gas emissions.

About 38% of the global total final energy consumption is attributed to buildings (this includes structures used in agriculture and fisheries). Buildings in OECD countries account for 45% of this consumption, transition countries account for 10%, and the developing countries account for 46%.

Table 14.1 Final Energy consumption in the services and residential sectors in different regions in 2005

	OECD countries	*Economies in Transition*	*Developing countries*
Electricity	38%	14%	15%
Oil products	19%	17%	17%
Natural Gas	33%	37%	5%
District heat	4%	31%	1%
Coal products	1%	3%	6%
Renewables and waste	5%	3%	56%

(Source: ETP, 2008, p. 525)

Most of the OECD countries which lie in the temperate and cold regions, use natural gas and oil for space heating. District heating is important in transitional economies (like Russia). Developing countries are largely dependent upon renewables and wastes.

Though reliable information is not available for non-OECD countries, IEA estimates that two-thirds of energy is used for space and water heating, 10–13% for cooking, and the rest for lighting, cooling and other appliances. There is much variation among the developing countries in regard to the fraction of energy use in the buildings sector for space and water heating – about two-thirds in the case of China and about a quarter in the case of Mexico (Table 14.1).

Energy consumption in the buildings sector varies greatly among countries and even parts of countries, depending upon the size of the household, heating and cooling load, lighting, number and types of appliances, and the pattern of their use. For instance, the number of light bulbs used in a household in China is 6.7, as against 40 for a household in Sweden. Senior citizens in USA prefer to live (say) in Florida, because of low heating bills.

Global population which was about 6.5 billion in 2005, is expected to reach 9.2 billion in 2050, i.e. by about 1.4 times. The demand for energy in the residential buildings sector is, however, expected to rise much faster, for two reasons: (i) with the decrease in the number of persons per household, the number of households globally is expected to rise 50% faster than the population growth, and (ii) the household floor area and the appliances used in a household are expected to increase, thus requiring more energy. Service sector floor area is projected to rise by 195% during the period, 2005–2050.

As a consequence of the rapid growth in the use of appliances, electricity consumption in the buildings sector is expected to increase by 180% during 2005 and 2050. Consequently, CO_2 emissions (produced in the process of electricity generation) related to the building sector are expected to increase by 129% during this period. In the residential sector, energy demand is projected to grow by 1.2% per year according to the Baseline scenario. The growth is projected to be 1.7% during 2005–2015, falling to 0.9% during 2030 to 2050.

The growth of the energy consumption in the services sector varies greatly among the regions: Latin America: 3.2%, Middle East: 3.1%, Africa: 2.7%, developing Asia: 2,6%, transition economies: 2.3%, and OECD countries: 1.0–0.8%.

In the OECD countries, the pattern of consumption of energy in 2004, was as follows: Space heating: 54%; Water heating: 17%; Appliances: 20%, Lighting: 5%, and cooking: 4%.

The energy consumption in the buildings sector in the non-OECD countries is expected to grow by 98% during 2005 and 2050. During 2005 and 2050, the energy demand in the service sector is expected to grow at a much faster rate (227%) than in the case of the residential sector (84%).

The rise in the middle class incomes triggered an urban construction boom in China and India. This is manifested in increase in the number of households, and the housing floor area per person. While China is switching from solid fuels (biomass and coals), India has shifted from fuel wood, cow dung and agricultural waste to kerosene and LPG. The number of Indians who use biomass for cooking is expected to drop from 668 million in 2005 to 300 million in 2050. It is also projected that by 2050, virtually all Indians will have access to electricity.

The Baseline scenario makes the following projections upto 2050 in the case of China: increase in the urban residential floor area at the rate of about 530 million sq, ft. per year; proportion of the population living in cities from 40 to 60% by 2030 to 73% in 2050; reduction in the size of the average household from 3.5 persons in 2005 to 2.9 persons in 2050. In the case of India, the residential floor area is expected to increase by 3.2 times during the period, 2005 to 2050.

China has prescribed energy-efficiency standards for the buildings sector, but compliance is not satisfactory – it varies from 60% in the northern region to 8% in the southern region.

In China and India, the use of household appliances has soared, as incomes rose, and the prices of the appliances fell. In some cases, they reached saturation levels. Improvements in the energy efficiency of appliances have partly offset the increasing energy demand due to larger numbers.

The share in the economy and the energy consumption in the services sector are expected to grow five-fold in the case of China, and even at a greater rate in the case of India.

Ageing population is the characteristic of Russia, which is the most important member of the transition economies. Consequently, there will hardly be any increase in the number of households. However, as incomes rise, Russia is experiencing a building boom – the average size of the new apartments (83 m^2) is 63% larger than the stock average.

Because of the cold climate of the transition economies, space heating accounts for two-thirds of residential sector energy consumption.

Residential Energy demand is projected to be reduced by 31% below the Baseline in 2050 in the ACT Map, and by 41% under the BLUE Map scenario. There would be a decline in all fuel sources, with the exception of non-biomass renewables, which are expected to increase by 128% under the ACT Map scenario and by 270% in the case of BLUE Map scenario (Table 14.2).

14.1.1 The building shell, heating and cooling

The energy efficiency of a building shell is critically dependent upon the insulation and the thermal properties of the building shell (walls, ceiling, and ground or basement floor). It therefore follows that improvement in insulation can reduce the heating requirement by a factor of two to four compared to the standard practice. *It should be mentioned here that this improvement in insulation can be brought about at a few*

Table 14.2 Reduction below the baseline scenarios in 2050 by scenario

	Residential		*Services*	
	ACT Map	*BLUE Map*	*ACT Map*	*BLUE Map*
Coal	−58%	−90%	−56%	−68%
Oil	−46%	−74%	−61%	−82%
Gas	−31%	−61%	−48%	−75%
Electricity	−30%	−27%	−39%	−45%
Heat	−37%	−45%	−31%	−17%
Biomass	−34%	−42%	−28%	−27%
Other/solar	128%	270%	328%	538%
Total	**−31%**	**−38%**	**−41%**	**−50%**

(Source, ETP 2008, p. 537)

percent of extra cost in the case of residential buildings, and little or no incremental cost in the case of service sector buildings (Florides et al, 2002). In the case of countries that have mild winters but still require heating, improved insulation could reduce the heating requirements by a factor of two or more, besides substantially reducing the indoor temperatures during summer (the comfortable temperature round the year within the magnificent government buildings in Lutyens-designed New Delhi, capital of India, is attributed to the insulation provided by the massive sandstone walls that were used in their construction).

While it could be demonstrated that the owner of a residential building can achieve cost savings through the improved insulation of the building shell, the catch in it is that while the costs will have to be incurred straightaway, the benefits will accrue over a period of time. The problem can only be solved through financing arrangements. Retrofitting of high-rise residential buildings with energy-efficiency improvements can result in energy savings as high as 80%, but experience in retrofitting in the case of detached and terraced houses in some countries (say, UK) has shown that such retrofitting is not economical.

Advantage may be taken to install energy-efficient structures after demolition/reconstruction, as the difference in the costs between renovation as against demolition and reconstruction, may some times be insignificant (Palmer et al, 2006). That said, the problem here is the higher CO_2 emissions from the construction work, whose effects may last for 30 years.

The economic viability of retrofitting varies greatly among countries:

Canada: USD 368 to 203/t CO_2 saved, to keep to Canadian R 2000 standard.

USA: USD 42/t CO_2 abated.

14.1.2 Windows

A number of technologies, such as, multiple glazing layers, low-conductivity gases (such as, argon) between glazing layers, low-emissivity coatings on one or more glazing surfaces, and the use of very low conductivity forming materials, such as fiberglass, have been used to improve the thermal performance of the windows. Windows with heat losses of 25 to 35% relative to the standard design are available in the market.

When glazing is achieved using low-conductivity gases, maintenance is the key to achieve good results – the performance may deteriorate by 60% if the maintenance is poor.

Glazings are available which while maximizing the transmission of visible sunlight, reflect or absorb a large fraction of the incident solar radiation. This will reduce the need for cooling.

The improvement in the technological performance of the windows does not entail any extra expense. In some cases, the costs have gone down (Jakob and Madlener, 2004).

When windows need to be replaced, replacing single glazing with more efficient glazing, need not be expensive, in the range of USD 57–490/t CO_2 saved.

14.1.3 Hot water

There are two principal ways to improve the efficiency of hot water systems: installing hot water cylinder insulation, and installing condensing boilers and heat pumps. It is possible to have solar water heating, at the cost of USD 1 to 2 per kW, depending upon the location and the number of sunshine hours. It is possible to have solar power to provide 60% to 70% of domestic hot water needs in residential buildings. Even in the case of the service sector buildings which may require higher temperatures (~250°C), solar power could provide 50% of heat requirements. Research is on to develop solar hot-water system which can deliver 2500 kWh per year at the cost US cents 4/kWh.

Experience has shown that it is highly economic to have hot-water cylinder insulation and to switch from an inefficient boiler to a condensing gas boiler. In Canada, hot water insulation and water saving devices showed negative costs of USD 209 to 360/t CO_2 saved (Seeline Group, 2005).

14.1.4 Cooling systems: Air conditioning

The efficiency of air conditioners varies a great deal. The least efficient are portable air conditions with energy ratio of less than 1.5 W/W (watts of cooling output per watts of power input). Modern split-room air conditioners have energy efficiency of 6.5 W/W.

Further improvements are being made in the energy efficiency through using "variable speed drive compressors, improving heat transfer at heat exchangers, optimizing the refrigerant, utilizing more efficient compressors, and optimizing controls" (ETP, 2008, p. 542).

The energy efficiency of the air conditioners is critically dependent upon the energy efficiency of heat pumps. The Coefficient of Performance (COP) of heat pump air conditioners has increased from 4.3 in 1997 to 6.6 in 2006. Some heat pumps have been able to achieve COP of 9.0.

Evaporative coolers work well in hot, dry climates. In these units, the outdoor air is cooled by evaporation, and the cool air is blown inside. Evaporative coolers cost about half of air-conditioners, and use a quarter of as much energy.

A room may be hot or cold in different seasons, or at different times in a day. Reversible heat pumps may be used if a room is hot and cold in different seasons or in different parts of the day. The efficiency of these systems depends upon the COP of the unit, and ventilation and thermal efficiency of the building.

The number of air conditioners is increasing rapidly. Air conditioning has become a major energy consumer, often accounting for 50% of the cost of the running of the buildings. Though the initial costs of some types of air conditioners are on the high side, they may have lower life-cycle costs. The use of programmable thermostat controls is recommended, as they could save energy and money.

In USA, use of advanced unitary compressors for air conditioning units is characterized by an abatement cost of USD 95/t CO_2. In the case of service sector, an advanced rooftop air conditioner unit could save over 4000 kWh per year, at a negative cost of USD 72/t CO_2 saved. In India today, electricity consumption due to air conditioning can be saved from its present to 10 to 11% (for USD 14 to 65/t CO_2 saved) to about 30 % (for USD 120 to 170 / t CO_2 saved). Use of split system heat pump air conditioning can bring down China's air conditioner electricity consumption by 27% (at the cost of USD 20/t CO_2 saved).

14.1.5 Appliances

Households are using more and more appliances. Some appliances like refrigerators have become more efficient. As against this, higher penetration of electronic home entertainment has raised the electricity consumption by 20%. On one hand, there is considerable improvement in the energy efficiency of home entertainment appliances. For instance, the switch from CRT televisions to more efficient LCD screens did reduce electricity consumption. But against this, expansion in the size of the screens, increasing in viewing hours and increase in the number of televisions per household, have tended to increase energy consumption.

Availability of low-cost electronic control technologies, improved materials, reduced manufacturing costs, and economies of scale have helped the Developed countries to improve energy efficiencies by 10 to 60%. Technical potential exists for further improvements in energy efficiency by 30 to 60%. Potentials for improvements in energy efficiency in developing countries and transition economies are even greater as they can leap-frog to more efficient technologies.

As BATs achieve greater market penetration, their costs will be greatly reduced (Table 14.3).

In most developed countries, rising family incomes meant that the overall running costs of appliances constitute only a small proportion of the household income, and is therefore ignored. Often, the consumers are unaware of the energy consumption of current TV technologies, and do not realize that a saving of 50% energy consumption is possible with LCD televisions with backlight modulation, or organic LEDS.

Governments should promulgate policy measures, such as regulatory and voluntary approaches, fiscal measures and procurement policies, to bring efficient energy technologies to the market. Manufacturers should be instructed to design all their devices with the ability to move to lowest power needed for their required functionality.

14.1.6 Lighting

Daylight is the largest, cleanest and highest quality source of light. Poor architecture some times creates need for lighting, where it is not necessary. Each of these areas in buildings has great potential to reduce lighting energy needs, without compromising on lighting services.

Table 14.3 Comparison of BAT for cold appliances and energy efficiency options in the European Union, India and China

	BAT/Improvement characteristics		*CO_2 abatement cost*	
	Incremental Cost today (USD)	*Energy Savings KWh/year*	*Before deployment USD t/CO_2*	*After Deployment USD t/CO_2*
European Union				
Upright freezer	394	110	465 to 1356	−151 to 461
Chest freezer	400	146	198 to 905	−339 to 223
Average for freezers	397	129	349 to 1097	−259 to 324
Fridge freezers	367	139	171 to 871	−364 to 190
India				
Refrigerator: direct cool	32	180		−38
Refrigerator: frost freeze	54	440		−46
China				
Refrigerators	96	261		−28

(Source: ETP, 2008, p. 5445; McNeil et al, 2005)

Lighting accounts for 1 900 Mt of CO_2 per year. This is equivalent to 70% of the emissions from the world's light passenger cars. The demand for lighting varies greatly – from 101 megalumen-hours per year per capita in USA to 3 megalumen-hours per year per capita in India.

Many new efficient lighting systems are highly cost-effective – they have internal rates of return on investment as high as 20%. It therefore makes economic sense to retire the old, inefficient systems and retrofit them with the new. A worldwide market shift from incandescent lamps to compact fluorescent lamps (CFLs) would cut the global electricity demand by 18%. If all end-users were to install efficient lamps, ballasts and controls, the global electricity demand in 2030 would be the same as in 2005. The average negative cost of such a development would be USD 161 per t of CO_2 saved. But this would need a stupendous government effort.

Solar-powered, solid-state lighting is a robust, low-energy, less expensive option.

All lighting is not interchangeable. What is recommended is substituting lower energy versions of a given lamp and ballast technology with the higher efficiency equivalent of the same technology. Also, potential for improvement in the case of semi-conductor (e.g. LED) and metal halide lamps is much greater than in the case of other kinds of lighting.

14.1.7 Heat pumps

Heat pumps find their most important use in cooling, space heating, hot water and industrial heat. They are capable of transforming low temperature heat from sources such as air, water, soil or bedrock into high temperature heat. Most heat pumps operate on vapour-compression cycle and can be driven by gas or waste heat. Absorption heat pumps used in space air conditioning may be driven by gas, or high-pressure steam or waste heat. Heating-only heat pumps have a good market in some countries (e.g.

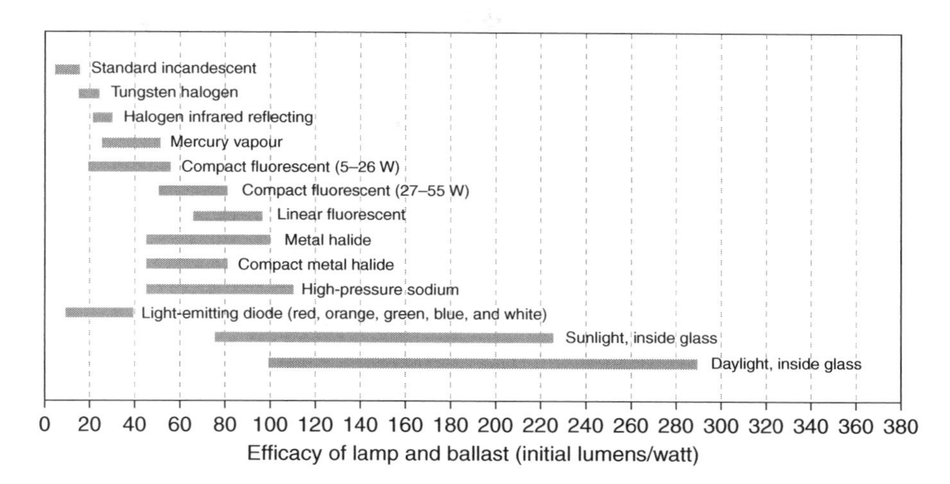

Figure 14.1 System efficacy of various light sources
(Source: ETP 2008, p. 548, © OECD – IEA)

Sweden, Switzerland, Canada). Reversible heat pumps which could be used both for heating and cooling are popular in countries where heating loads are moderate and cooling is required during summer.

Electric heat pumps account for 20% to 50% of electricity used for space and water heating. According to US EPA, the electricity consumption of ground-source heat pumps is 44% less than air-source heat pumps. Though ground-source heat pumps are more efficient than air-to-air source heat pumps, they are characterized by higher installation costs (Sachs et al, 2004).

Heat pumps can also be used for water heating. Some designs (e.g. ECO Cute Heat Pump) are far more efficient than conventional hot water heaters, but they are 2.5 times more expensive to install than conventional options. Their prices are, however, coming down. These new pumps may soon present a significant CO_2 abatement opportunity.

It is interesting to compare the installation and running costs of heat pumps versus boilers. A typical gas boiler costs about USD 1 500, as against USD 5 000 for a heat pump. A gas boiler would need about 50 GJ of gas per year, whereas a heat pump would need only 15 GJ of electricity per year. When a gas boiler is replaced by a heat pump, there will be a saving of CO_2 emissions of 2.8 t per year (assuming that electricity was produced CO_2-free), at a lifetime cost of USD 160/t CO_2 saved.

Presently, heat pumps are not a viable option of CO_2 abatement. In China, the average water heater with a tank capacity of 8–10 litres, costs around USD 100. The comparable ECO Cute heat pump in Japan with much greater capacity, costs about USD 5000/-. There is a good possibility that when high-efficiency, less expensive, reversible heat pumps enter the market, the developing countries may find them an attractive proposition to be used in regions characterized by moderate heating loads and significant summer cooling requirements.

14.1.8 Solar thermal heating

Solar thermal heating is making rapid progress – there has been over 15 GW_{th} new capacity in 2006 alone, increasing the total capacity by 16% globally. China leads the world in solar thermal heating. The total capacity of glazed flat plate and evacuated tube water collectors in China in Dec. 2005, was 52 500 GW_{th}, which is eight times more than that of Turkey (6 300 GW_{th}) which has the next largest installed capacity. Solar thermal heating is making good progress in Japan, European Union countries, India, etc.

There are two kinds of solar thermal heating: passive systems which use windows directed mainly towards the sun, and active systems involving collectors, heat storage and controls. In the active systems, solar radiation on the collector panel warms water or some other heat transfer fluid when they are circulated through a duct. The amount of heat energy that could thus be collected varies from 300 to 800 $kWh/m^2/yr$. In some designs, the warmed heat transfer fluid transfers the heat to be used as hot water or for space heating.

Solar thermal heating has achieved good market penetration and is used for crop drying, heating of buildings, and for industrial heat. In China, simple, cheap systems of solar heating have penetrated the market deeply, in the context of the expensive conventional hot water systems (Philibert, 2006). These systems do not, however, have freeze protection which is not necessary in China. Without freeze protection, the Chinese-type solar thermal heating could not be used in Europe, and hence did not receive adequate market penetration.

In latitudes below 40°, the substitution of solar heaters in the place of conventional inputs of gas or electricity could lead to saving of 50% in cases that require temperatures upto 250°C, and saving of 60% to 70% for domestic water heating at temperatures of around 60°C. In such a situation, the coefficient of performance (COP) of solar water heaters and heat pumps become comparable.

"Combi" designs which can be employed both for water and space heating, are becoming popular as they are more economically viable. As solar energy is intermittent, active solar space and water heating need back-up energy in the form of (say) electricity, bioenergy, or fossil fuels, which add to the costs. New technologies which integrate solar-assisted water heating with heat pump, are far more efficient than condensing boilers, have good potential.

Solar hot-water heating can result in modest CO_2 abatement costs. In South Africa, the abatement cost works out around USD 30/t CO_2.

Solar thermal systems have achieved good market penetration not only for water heating, but also heating swimming pools, and residential buildings, in countries, such as Cyprus, China, Germany, Turkey, Austria, etc. Their market penetration in cold countries has been insignificant, because of the high costs involved in freeze protection.

There has been a 20% reduction in the initial costs of solar thermal systems. As Combi systems are becoming cheaper, their market penetration is increasing. Prices of the solar thermal systems could come down further, through advances in materials and components, such as the development of effective optical coatings on surfaces, and anti-reflective and self-cleaning glazing materials. The ability of the materials to withstand higher temperatures, will prolong the lifetime of, and reduce the need for servicing of, the devices. Efforts are under way to develop designs whereby new flat-plate collectors are integrated into building facades and roofs. There is good market for photovoltaic-thermal collectors which can heat water as also generate electricity.

Currently, the solar thermal market largely involves small-scale units installed in single-family houses. There is need to enlarge the size of the units to serve larger establishments like schools, hotels and commercial buildings. In the place of solar thermal units with back-up energy in the form of electricity, bioenergy, etc., stand-alone units, i.e. without back-up, with high-efficiency storage applications and well-insulated buildings, are coming up, and may soon become competitive.

Improvements are needed in the efficiency of Concentration of Solar Heating (CSH) technologies in order for CSH to be used in district heating or large industrial applications of megawatt scale. Several collector and component designs need to be optimized for industrial applications.

There is little doubt that the more the architectural designs are integrated with standardized elements, the more will be the market penetration. Governments could give a thrust to market penetration by prescribing standards. In the absence of such standards and guidelines, the consumer has difficulty in making choices.

14.2 PASSIVE HOUSES AND ZERO-ENERGY BUILDINGS

The energy system of a building is complex, and involves a large number of components which interact with one another in complex ways. Building-energy simulations allow the design of "zero-energy" buildings (ZEB), in which the energy demand is balanced by the energy produced in the building. Similarly, it is possible to design "zero-carbon" buildings where the net CO_2 emissions over a year will be zero. Where a building satisfies the requirements of both zero-energy and zero-carbon, it is called zero-squared.

Off-the – grid buildings are stand-alone ZEBs that are not connected to an off-site energy facility. They require energy storage capacity to take care of the situation when the sun is not shining or the wind is not blowing.

Zero-energy buildings achieve that state in two ways:

(i) By using solar cells and wind turbines for electricity generation and biofuels and solar collectors for other forms of energy generation. The buildings are connected to the grid – excess power from solar cells and wind turbines is fed to the grid during (say) the daytime. During the nights when solar cells cannot generate electricity or during times when there is no wind, energy requirements for the house are drawn from the grid.
(ii) By reducing the heating and cooling loads drastically through the use of high-efficiency equipment, improved insulation, high-efficiency windows, natural ventilation, etc. Skylites are designed to provide 100% of the day-time illumination. Fluorescent and LED lamps which require one-third less energy are used for illumination in the nights. Exhaust from refrigerators and heat recovered from wastewater can be used to warm domestic hot water.

Where supply-side options (e.g. generating more power) are expensive, recourse could be taken to demand-side possibilities. The development and deployment of smart grids, smart appliances and advanced metering could be made use of to manage the peak demand. The back-up power in the case of intermittent renewables (e.g. solar PV in the nights) could be substantially reduced using this approach.

Energy consumption in the case of passive houses can be brought down by 70% to 90% through "intelligent" design using 3-D simulation.

Structures have come up in different parts of the world to demonstrate the techno-economic feasibility of zero-energy buildings. For instance, the 71-storey Pearl River Tower which houses the headquarters of the Guangdong Company, gets its energy from solar and wind power. Z-Squared Design Facility has been set up in San Jose, California, USA. Canada has set up R-2000 standard for energy-efficient housing. Germany has set up a self-sufficient, solar house in Fraunhofer Institute.

There is increasing public interest in passive/zero-energy/zero-carbon houses. The design and construction of a building have to be done in such a way that the energy efficiency of the individual parts of a building as well as that of HVAC (houses, heating, ventilation and air conditioning) is optimized. The passive house design should achieve a level of 15 kWh/m^2 for heating and cooling. In some European countries, such as Germany and Austria, passive housing has gone beyond the demonstration stage. The market penetration of passive housing will continue to be limited until there are cost reductions of individual components. Considering that bulk of the 1200 GW solar PV to be deployed under BLUE scenario will be mounted on buildings, there is little doubt that prices for passive housing will come down.

The typical additional investments needed for passive houses are in the range of 6–8%. The payback period for passive houses can be long – it is 30 years in Belgium. The abatement cost to refurbish passive houses in Germany is estimated to be USD 800/t CO_2. Existing multi-family houses can save as much 90% of the heating costs, if at the time of refurbishment passive housing technologies are adopted.

In India, green buildings are estimated to cost 8% to 10% more than basic buildings. The pay-back period is 5 to 7 years (Srinivas, 2006).

There are barriers impeding the market penetration of passive houses. People are concerned that the initial costs are high, and the long-term benefits and resale values are uncertain.

There is need to design ultra-passive housing with heating demand of 7 kWh/m per year or less. Passive buildings require to be integrated with renewable energy resources. When the new technologies are installed in schools, and public offices, more people can convince themselves of the viability of passive housing.

14.3 BIOENERGY TECHNOLOGIES

Biomass and waste account for about 10% of the global energy supply. More than 80% of the biomass is used for cooking and heating. For about 2.5 billion people, i.e. 40% of the world population who live in developing countries, biomass in the form of fuelwood, charcoal, agricultural waste and animal dung, is the only affordable fuel for cooking. About 668 million people in India continue to use animal dung, agricultural waste and fuelwood as fuel for cooking. Because of the low fuel efficiency of the cook stoves (~8%), the particulate matter in the Indian households burning biomass is 2 000 μg/m^3 (as against the allowable 150 μg/m^3), leading to 400 000 premature deaths.

According to WEO projections, the percentage of the population in the developing countries that rely on biomass for fuel is expected to decrease from 52% to 42% by 2050, but this would still be a substantial number.

Liquified Petroleum Gas (LPG), ethanol gel and Dimethyl Ether (DME) are important bioenergy sources that could substitute or supplement biomass for heating purposes. They are projected to contribute 50 Mtoe according to ACT Map scenario and 150 Mtoe according to BLUE Map scenario. Biogas has considerable potential for rural communities.

DME is a versatile fuel – it can be used in power generation turbines, diesel engines, or as a replacement for LPG for household cooking. DME is non-toxic, and could be produced from a variety of feed-stocks, such as, coal, natural gas and biomass. Currently DME production involves two steps – methanol is produced from syngas, which is then dehydrated to form DME. New technologies allow the production of DME from syngas in one step. Other possibilities that are being developed are co-production of methanol and DME and co-generation of DME and electricity. China leads the word in coal-based DME production, which is expected to reach 1 Mt/yr (0.03 EJ/yr) by 2009. Gas-based DME production is picking up in the Middle East.

Investment costs for conventional DME production vary from USD 11 to 15/GJ per year. The conversion efficiencies for biomass-based DME production range from 45 to 65%. The capital investment cost for DME production varies from USD 450 to 1050 per tonne of biomass. Sweden has reported a DME production cost of USD 11.6 to 14.5/GJ. The price is less in the developing countries.

By the use of homogenous dry fuel, such as saw dust pellets and wood chips, and by operating the plant continuously, some developed countries have been able to achieve high efficiencies in small-scale heating plants. Smaller CHP (Combined Heat and Power) plants of the capacity of <5 MW, which use solid biomass, operate at steam pressures of 50 to 60 bars. Investment costs for such systems are typically in the range of USD 615/kW_{th}. Research is going on to develop small-scale (1 to 5 MW) power plants with lower investment costs and better heat-to-power ratios. Such plants will revolutionise the energy consumption not only in the services sector but also in the residential sector. Future designs are expected to have provision for flue gas cleaning and particle separation to protect air quality in densely populated areas.

Table 14.4 (source: ETP, 2008, p. 559) provides particulars about biomass conversion technology in terms of heat output.

Straw-fired, district heating plants are used in some countries. They have maximum boiler temperature of 120°C, maximum steam pressure of 6 bars, overall plant efficiencies of 85%, power-to-heat ratio of 0.25, energy cost of USD 68/t of straw, or USD 17/MWh.

Table 14.4 Biomass conversion technologies

Combustion technology	*Minimum output (MW)*	*Typical Output (MW)*
Mechanical grate	1	2–30
Fluidised bed	2	10
Circulating fluidized bed	7	20
Gasification	0.5	2–10

Table 14.5 RD&D Targets

	Technical targets (Index)		
Technologies	*Current*	*2030 ACT Map/BLUE Map*	*2050 ACT Map/BLUE Map*
Electric appliances	1	1.15–1.5	1.25–2
Heating and cooling Technologies			
Heat pump	1	1.4/1.6	1.8/2
Air conditioning efficiency	1	1.3/1.5	1.4/17
Lighting systems			
Light Emitting Diodes (LED)	1	4/7	6/10

Table 14.6 RD&D breakthroughs needed

Technologies	*RD&D breakthroughs*	*Stage*
Heating and Cooling technologies	Reasonable-cost, high-temperature heat pump systems (new and retrofit applications); system integration and optimization with geothermal heat pumps	Applied R&D
Lighting systems	Improvement of semiconductors; Modification and optimization of known light-emitting substances for LED	Applied R&D
	New materials for LED; and stability of organic LED	Basic science

Woody biomass is used to heat homes and service centres. When heat-retaining fire places are used, heat is stored in the retaining structure and released into the room over a period of time, with efficiency of 80% to 85%. Well-designed fire places in which wood burns cleanly, could keep the whole building warm.

Fuel pellets can be fabricated by compressing dry sawdust or straw, so as to have less than 15% moisture content. Freshly produced saw dust must will have high moisture content of more than 50%, and therefore has lower fuel value. Pellet production costs in Austria and Sweden show that wet feedstock costs USD 10 to 12.9/GJ, whereas dry feedstock costs USD 5–7.8/GJ.

Pellet production cost is largely dependent upon the cost of the raw biomass material, and the amount of drying needed to bring down the moisture content to 15%. The pellet production cost breakdown is as follows: raw material: 35%, drying: 23%; pelleting: 17% personnel costs: 11%, crushing: 6%, general investment: 4%, storage and conveyor: 3%, and cooling: 1%.

14.4 RESEARCH & DEVELOPMENT, DEMONSTRATION AND DEPLOYMENT

Research, Development & Demonstration (RD&D) targets for existing technologies in buildings and appliances are given in Table 14.5 (source: ETP 2008, p. 589).

Table 14.6 gives the RD&D breakthroughs required for key technologies in buildings and appliances (source: ETP, 2008, p. 589).

Chapter 15

Transport

U. Aswathanarayana

15.1 OVERVIEW

The production of oil in the world is 30 billion barrels a year. The transport sector consumes half of the oil produced in the world, and accounts for a quarter of CO_2 emissions. IEA estimates that the oil use and CO_2 emissions will increase by 50% by 2030, and more than double by 2050, through greater air travel, road freight and light-duty vehicles, mostly in developing countries, such as China and India.

As against the projected increases in the use of oil, there is the question of peak oil. Most studies estimate that the oil production will peak between now and 2040, but some optimistic estimates (e.g. Exxon Mobil) extend it to the next century. Currently, there is much interest in shale gas. Shale is a common sedimentary rock, and most countries have it. Organic shales of Palaeozoic – Mesozoic age, with high natural gamma radiation, have the best potential. Recent advances in hydraulic fracturing and horizontal drilling have made shales gas economically viable. Wellhead gas price of about USD 4.25 per thousand cu.ft. is the minimum price required to make shale gas economical. Some estimates put the shale gas reserves to be 50% to 160% of the present known natural gas reserves in the world. Both China and India which are expected to consume large quantities of oil, have large areas where shale occurs. The European countries where organic shale occurs near some cities, find it a highly attractive idea to tap this shale gas, and thereby reduce dependence on Russian imports of natural gas.

Attempts are being made, especially in China, to produce oil from coal. As these fuels are more carbon intensive than oil, such an approach is unsustainable from climate change point of view.

A sustainable low-carbon transport system is a long way off. Technologies such as fuel cells and vehicle on-board energy storage (e.g. batteries, ultra-capacitors and H_2

storage) are still in the stage of development, and it may be many years before they become cost-effective.

In 2005, transport accounted for 23% of the global energy-related CO_2 emissions. If emissions from feedstocks, fuel production and distribution to vehicles are taken into account, transport GHG emissions constitute 27% of total emissions.

A low-carbon transport sector can be promoted by the following measures:

(i) *Fuel-efficient light-duty vehicles (LDVs)*: A fuel savings of 30% may be realized in the next 15–20 years through technologies such as, engine/drive-train efficiency, tyres, aerodynamics, light-weight materials and accessories (e.g. air-conditioning). Toyota's Prius automobile is a good example of what could be achieved with already known technology. It is the most fuel-efficient (50 mpg), fully hybrid, electric, mid-size car which is the cleanest vehicle (based on non-CO_2 toxic emissions). Governments could facilitate the market penetration of fuel-efficient, light-duty vehicles by prescribing efficiency standards and tightening their use.
Nano is the rear-engined, four-passenger, city car built by Tata Motors of India. It is very fuel efficient – 26 km/L on the high way and 22 km/L in the city. It is the cheapest car in the world (INR 100 000, eq. ~USD 2 000). Tatas have announced that they will be releasing eco-friendly versions based on compressed air as fuel, and electric versions.

(ii) *Advanced public transport system*: such as "bus rapid transit", which is very much in vogue in European cities. This would provide a low-cost, high quality mobility option.

(iii) *Freight movement*: Hybridization, engine efficiency, improvements in cab and trailer (weight and aerodynamics), and improved routing and logistics systems can bring down costs of medium-duty, urban-use trucks by 40% by 2030. Government regulations could promote this kind of development, but to date, only Japan has moved actively in this area. If 25% of all trucking for distances of about 500 km is switched to rail transport, there will be a saving of about 0.4 Gt of CO_2 per year.

(iv) *Rail transport*: energy and the relative GHG emissions presently account for about 3% of the transport sector. However, rail transport has tremendous future, particularly in the developing countries. If 25% of air travel for distances of about 750 km switch to high-speed rail travel, about 0.5 Gt of CO_2 per year could be saved.

(v) *Maritime transport*: Most of the maritime energy (~80%) is concerned with international shipping, the rest 20% being accounted for domestic commercial shipping, and recreational boating. As countries make progress economically, global trade and international shipping are bound to increase. By outfitting ships with energy-saving devices, and improving the operational efficiency through the use of high-tech parachute-type sails, it may be possible to bring down the energy use by 30% by 2050. Use of biofuels and hydrogen fuel may help, but technical problems require to be overcome before this is possible.

(vi) *Air transport*: As fuel costs constitute a significant part of the operational expenses of the airlines, new aircraft models incorporate many fuel-efficiency devices and practices. Average fleet efficiency is likely to increase by 20% by 2050, beyond the 30% increase expected under the Baseline scenario.

(vii) *Alternative fuels*, such as sugarcane ethanol in Brazil, ethanol from ligno-cellulosic feedstocks, biodiesel, biomass gasification via Fischer-Tropsch process, algae-based "green" crude, etc. are likely to have major impact on GHG reductions during the next 10–15 years.

(viii) *Battery-operated vehicles*: Scooters which run on Li-ion batteries and have a range of about 200 km per charge, have entered the market. Their fuel costs are about 25% of gasoline, they are silent, and do not have any emissions. In some countries, they do not require any registration. Germany is developing a system which wirelessly charge the cars as they are driven along. Roads are fitted with an electric system that create a magnetic field that will charge your battery-powered vehicle as you drive along. Radio chips will identify your vehicle, and bill you appropriately. The conductors are resistant to weather and mechanical wear.

The critical element here is the battery. Battery costs need to be reduced to USD 300/kWh, if electric vehicles are to have an impact on CO_2 emissions. If, as expected, vehicle electrification makes good progress, the transport electricity demand may reach 650 Mtoe, or 20% of the world electricity demand, necessitating a 2000 GW additional electricity capacity.

In sum, advances in RD&D into energy storage systems, fuel cell systems and advanced biofuel systems hold the key for bringing down CO_2 costs in the transport sector (ETP 2008, p. 425).

Japan is promoting two kinds of "green" cars. Toyota is concentrating on gasoline-electric hybrids. It sold 1.4 million Prius cars. It is now launching a new model, Sai (meaning talent) which is bigger and costlier than Prius, and gives 54 mpg (double the mileage of comparable gasoline engine). They expect to sell 500 000 to 600 000 hybrid cars per year. According to Toyota, plug-in hybrids would be the dominant automobile technology in future. Plug-in hybrids avoid expensive batteries, and will also increase the range of plug-in hybrids. Prius cars are in trouble because of brake problem.

Honda experimented with hydrogen-fuel vehicles which are totally pollution-free (in these vehicles, hydrogen combines with oxygen to produce water and electricity). The exorbitant cost of technology, and non-availability of adequate number of hydrogen filling stations, constrain the development of hydrogen-fuel vehicles.

Nissan is concentrating on purely electric car. They expect the electric car will capture 10% of the global market. As the battery is the key to the success of the electric car, Nissan is investing heavily in RD&D of batteries.

China has overtaken Japan as the largest producer of cars. It has also overtaken USA as the largest market for cars.

Table 15.1 (source: ETP, 2008, p. 428) summarizes the targets for the transport sector under ACT and BLUE Map scenarios.

15.2 ALTERNATIVE FUELS

The realization of low-carbon transport sector critically depends upon switching to low GHG fuels, such as biofuels, electricity and hydrogen.

Table 15.1 Targets for the transport sector

	ACT Map	*BLUE Map*
LDVs		
New LDV fuel economy improvement	50% reduction in LDV fuel/km by 2050 (includes hybrids), but no EVs & FCVsf	70% reduction in new LDV fuel/km by 2050 from FCVs and EVs.
Gasoline and diesel hybrids	75–95% market share in 2050, depending on region	About 70% market share in 2030, dropping to 35% in 2050 due to electric and fuel cell vehicles
Electric, plug-in hybrids	Beginning in 2020, hybrid vehicles reach 20% travel on electricity by 2050	Beginning in 2015, hybrid vehicles reach 60% electric share by 2050
Electric vehicles	None	Reach 20% of LDV sales in 2050
Fuel cell vehicles	None	Reach 40% of LDV sales in 2050
Travel Baseline: Total worldwide LDV travel triples between 2000 and 2050.	ACT is 15% lower in 2050 than baseline due to modal switch and telematic substitution	Same as ACT Map
Trucks	Average 35% efficiency improvement, including 50% hybridization by 2050	FCVs and EVs each may reach up to 25% stock by 2050
Buses	40% improvement by 2050, including 50% hybrids	50% improvement by 2050, including u74% hybrids
Rail	15% more efficient in 2050	30% more efficient in 2050
Air	Aircraft stock 35% more efficient in 2050, and 5% routing improvement	Aircraft stock 45% more efficient in 2050, and 10% routing improvement
Travel (non-LDV) Baseline travel (more than doubles for most modes).	Up to 10% reduction for air, trucking in 2050; up to 25% increase for buses, rail due to mode switching	Up to 15% reduction for air, trucking in 2050; up to 35% increase for buses, rail due to mode switching
Biofuels	About 570 Mtoe in 2050, mostly second generation	About 700 Mtoe in 2050, all second generation, mostly BTL
Low GHG hydrogen	No H_2	260 Mtoe in 2050
Low GHG Electricity	130 Mtoe mostly for plug-ins	320 Mtoe in 2050 for plug-ins and pure EVs

EV = Electric vehicles; FCV = Fuel Cell vehicle

15.2.1 Biofuels for transport

Transport fuel may be liquid or gas. It is generally agreed that production of transport fuel from biomass would reduce the need for fossil energy and thereby bring down the GHG emissions. But this is easier said than done, as there are formidable impediments in realizing this objective, such as, food security and land competition with biofuels, potential impact of biofuels on water resources, biodiversity and other environmental issues. Advanced biofuel technologies which make use of non-food biomass feedstocks, hold promise to achieve sustainable, low CO_2, and cost-effective biofuels.

Transport fuel would require the consumption of about 700 Mtoe of biomass. Equal amount of biomass will be needed to produce 2 450 TWh/yr of power. Biomass could be co-fired with coal, and could also used in Combined Heat and Power (CHP) systems. IEA projects that 2 200 Mtoe will be used to produce biochemicals, heating and cooking (including DME production) and in industry (e.g. CHP plants and black liquor).

Through the use of advanced biofuels for heavy goods vehicles, marine vessels and aeroplanes, it is possible to achieve very low GHG emissions in the transport sector by 2050. In due course, the first generations biofuels produced from grains and vegetable oil feedstocks will be replaced by the second generation biofuels produced from non-food biomass. Though the latter kind of biofuels may bring about large CO_2 reductions compared to presently used petroleum fuels, the question remains whether their production could be achieved without jeopardizing food security and environment.

15.2.2 Electricity in transport

Electricity presently accounts for about 1% of the transport fuel. That is so because electricity is presently used only in rail transport. Electricity is not a practical option for aircraft. Also, it is unlikely to be used in the near future for long-haul trucking and international shipping because of the limitations of batteries. In future, electricity will be used to power most cars and LDVs, some types of trucks and most rail systems. In hybrid vehicles, electricity is generated on board. Plug-in cars are charged from the electricity grid. Such vehicles will run on electricity completely or primarily.

Decarbonisation of the transport sector will be facilitated when it is possible to produce electricity with low GHG emissions, when there is significant improvement in the energy storage systems, and when there is shift to vehicles which use electricity as a fuel.

Electricity costs per kilometre of vehicle travel will be low.

15.2.3 Hydrogen in transport

Hydrogen is an emerging fuel for transport. But unlike in the case of electricity, there are no established hydrogen fuel production and distribution systems in the world. Its strong points are: very high fuel efficiency, near-zero pollution from vehicle operation, and near-zero GHG emissions if hydrogen is produced from GHG sources. Its disadvantages are shorter range than liquid-fuelled vehicles, long fuelling time and most importantly, higher costs. Hydrogen fuel is likely to be two to three times costlier than gasoline or diesel, though this will be partly offset by higher efficiency.

In the medium term, hydrogen – fuel cell combinations have the highest potential to be used in cars, buses and urban duty trucks. Use of hydrogen in trains and marine transport should await the development of affordable, heavy-duty fuel cells, and this is long way off.

Currently, there are three well-established modes of hydrogen production, namely, natural gas forming, coal gasification and water electrolysis. They annually produce 40 million tonnes of H_2, which is less than 0.5% of the world's energy. In order to produce large volumes of hydrogen economically for the transport sector, not only are the efficiencies of these technologies require to be improved greatly, but also new technologies, such as, solar and nuclear heat to split water, biomass gasification and photo-biological processes, have to be developed. This would require considerable RD&D effort.

Decentralised hydrogen units do not require much infrastructure for the transportation and distribution of hydrogen, but they are inefficient and expensive. Hydrogen production in decentralized units costs USD 50/GJ (or USD 1.60/Lge), compared to large units USD 10–15/GJ (or USD 0.35–0.50/Lge). Hydrogen production from natural gas reforming may go down to less than USD 15/GJ by 2030, and by electrolysis to less than USD 20. GJ, by coal gasification to less than USD 10/GJ, by water splitting by nuclear heat USD 10/GJ, and by solar heat to USD 20/GJ.

Apart from production, infrastructure need to be developed for the distribution, storage and delivery of hydrogen to vehicles. The global cost of this will be in trillions of USD. In any event, the retail price for hydrogen users will be well above USD 1.00 per litre of gasoline equivalent in foreseeable future. The demand for hydrogen fuel will depend upon the availability of the infrastructure, and infrastructure will develop when there is demand for the fuel. So unless there is a massive effort on the part of the governments, hydrogen fuel has a limited potential in the transport sector.

15.3 LIGHT-DUTY VEHICLES

As incomes rise, the ownership of cars, sport utility vehicles and light-trucks is expected rise pro rata. The total stock is projected to rise from 700 million in 2005 to about 2 billion in 2050. There is a possibility that this may rise even to 3 billion. A consequence of this growth will be pro rata increase in the fuel use, unless vehicles become far more efficient, and fewer cars are used because of profound improvement in the mass transit systems, and long-distance bus and rail transport. Government policies should promote a 50% in fuel use per kilometer, incremental technology improvements, and hybridization. Vehicle registration fees should reflect fuel economy standards and CO_2 emissions. This regulation which is being followed in OECD should be the global norm.

One way to achieve fuel economy is to lower the vehicle mass, through the increased use of aluminium and magnesium alloys (in place of steel), plastics and other lightweight material in engine and vehicle components. The use of some substitute materials may require significant design revisions, and the price of the vehicle may go up. The higher cost of the vehicle made up of lighter composite materials may be offset by reduced fuel costs, or due to large-scale production. If the problems associated with carbon-fibre vehicles (such as, their sensitivity to transversal loads) are overcome in due course, they will have profound influence on LDVs, because fuel consumption and GHG emissions may be reduced by half.

Under the BLUE map scenario, by 2050, the use of light materials in LDVs may bring out 25% weight reduction of LDVs at an estimated additional cost of USD 1 000 per vehicle. Such vehicles could save 10% fuel.

Powertrains

A number of powertrains are being developed to improve fuel efficiency. Advanced gasoline powertains have the potential to approach the fuel efficiency of the diesel powertrains, and both can approach the fuel efficiency of hybridized systems.

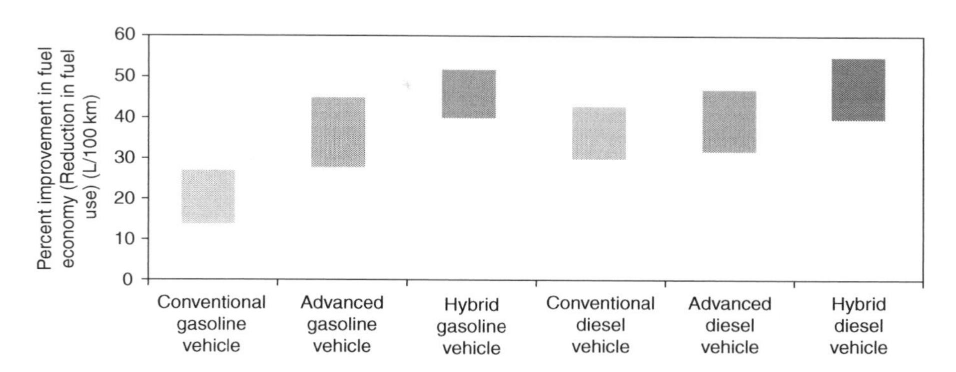

Figure 15.1 Fuel efficiency improvements of different powertrains, ETP, 2008, p. 440 © OECD-IEA

Vehicle hybridization involves the addition of electric motor, a controller and energy storage system (typically, a battery) to the existing internal combustion engine. Toyota's Prius is a good example of this. Hybrid cars are more efficient than the ordinary internal combustion engine, as the motor/battery system handles some of the peak power requirement, and engine power is made use of recharge batteries during the periods of low load. Computers are the key to the efficiency of the hybrid car. The on-board computer manages the use of the electric motor, the loads on the combustion engine and batteries, engine shutdowns, and the use of regenerative braking. Improved use of computers will thus play an important role in improving the efficiency of hybrid cars (Passier et al, 2007).

IEA projects that these small cars (such as, Nano of Tatas, India) will achieve a market share of 2% in the OECD countries, and 10% in non-OECD countries by 2015, and will remain constant thereafter. Though these small cars will use less fuel, there may be no reduction in the gross fuel use, as their lower fuel will be counterbalanced by larger numbers of vehicles.

Energy Storage

Energy storage is the key to hybridization. Lithium-ion batteries are the best batteries in the market. They are extensively used in small electronic devices. Attempts are being made to improve them to make use them in vehicles, but they suffer from the disadvantages of high cost and inadequate performance. As more batteries are used, the vehicle becomes heavier, and this puts practical limits on the benefits that batteries can offer.

Fig. 15.2 shows the energy density of batteries and liquid fuels (ETP, 2008, p. 442; © OECD-IEA).

Batteries are expensive per unit of energy. For instance, nickel-metal-hydride (NiMH) batteries cost USD 1 000/kWh of storage capacity. For Toyota's Prius car, the cost of the battery comes to USD 1 000 per vehicle. If batteries are to play an important part in carbon-free transport, their cost should come down by half to two-thirds. Also, they should be to endure up to 15 years recharge-discharge cycles.

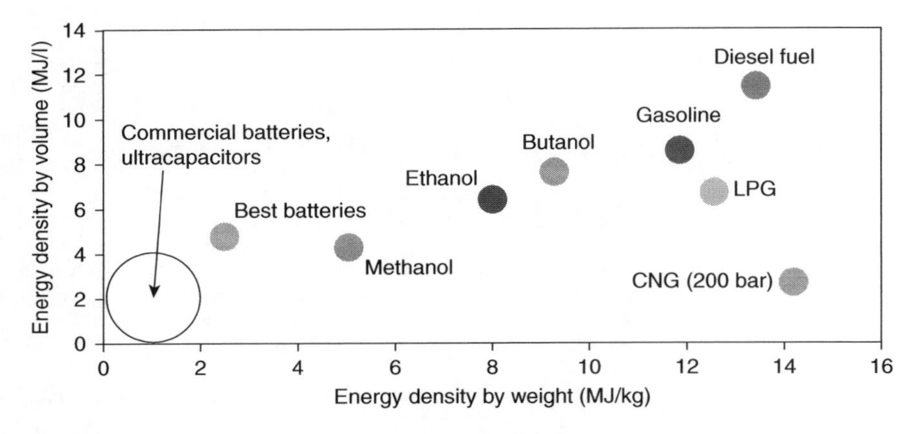

Figure 15.2 Energy densities of batteries and liquid fuels, ETP, 2008, p. 442 © OECD-IEA

In the case of Li-ion batteries, safety has emerged as a consideration. There have been reports of Li-ion batteries (used in, say, computers) exploding. There is a good possibility that the ongoing R&D will be able to overcome this problem.

Ultra-capacitors are analogous to batteries as electricity storage devices, with the difference that they store energy in charged electrodes, rather in electrolytes. Their strong point is that they can be steadily charged by the battery pack and then discharged rapidly when the peak power is needed. The cost and performance of the ultracapacitors has been improving dramatically from USD 2 per farad to US ten cents per farad. Their price should go down to one cent a farad if they could be affordable to mass market automotive applications.

Plug-in hybrids

Plug-in hybrids have much potential for the reduction of oil use and CO_2 emissions, as they can travel part-time on the electricity provided by the grid, rather than through the vehicle's internal recharging system. Thus they rely more on the electricity sector which is less expensive to decarbonise. Under the BLUE Map scenario, the transport sector is expected to be fully decarbonised by 2050. Plug-in hybrids, however, need several times (~10 to 20) battery capacity than today's non-plug-in hybrids. If this problem can be overcome, plug-in hybrids could play a significant role in the decarbonisation process (Simpson, 2006).

The benefit of fuel savings in the case of hybrid vehicles has to be weighed against the additional cost of buying such a vehicle.

The economics of plug-in hybrids are summarized in Table 15.2 (source: ETP, 2008, p. 444).

The above calculations assume, (i) vehicle efficiency of batteries at 0.16 kWh/km; (ii) system configured to discharge up to 66% maximum, (iii) future battery costs may drop to USD 300/kWh, (iv) Percentage of daily driving on batteries are based on US figures, may be higher for other countries.

Table 15.2 Costs of plug-in hybrids in terms of driving range

Plug-in vehicle Battery capacity	*Vehicle driving range on batteries (km)*	*Battery storage needed (kWh)*	*Vehicle battery cost (USD)*		*Percentage of average of daily driving on batteries*
			Current (USD 1 000/kWh)	*Future (USD 300/kWh)*	
Low	20 km	5	5 000	1 500	20–40%
Medium	50 km	12.5	12 500	3 750	40–60%
High	80 km	20	20 000	6 000	60–80%

The market penetration of plug-in cars would strongly depend upon battery costs. As battery costs go down, and as electric recharging infrastructure spreads and the number of opportunities to recharge the battery during the day improves, the economics of plug-in cars improves. According to BLUE scenario, a 50-km range, plug-in hybrid is expected to achieve good market penetration after 2030.

Electric Vehicles

Electric vehicles do not require the entire internal combustion engine, the drive train and fuel tank. This saves USD 4 000 compared to the hybrids. As against this, electric vehicles require a more powerful motor/battery system. For a 500-km driving range which the drivers expect, the battery capacity needed is 50 kWh, which at present prices will cost USD 50 000 per vehicle. Even when the battery prices come down to USD 300/kWh, the battery price component of the car will still be USD 15 000. Thus, pure electric vehicles will be at least USD 10 000 more expensive than comparable hybrids.

In due course, the electric vehicles may benefit from the more efficient batteries developed for hybrid vehicles. The fuel cost of electric cars is 25% of gasoline cars. They are silent, environment-friendly, with no GHG emissions. Some countries do not require registration for such vehicles.

Fuel Cell vehicles

Fuel cell consists of two electrodes sandwiched around an electrolyte. (Fig. 15.3 Fuel cell concept. Source: ETP, 2008, p. 267; © OECD-IEA).

It is a device that can convert hydrogen or natural gas into electricity. It can be operated at temperatures ranging from 80 to 1000°C. The efficiency of fuel cells range from 40% to 60%. In a fuel cell vehicle, the power plant is a fuel-cell stack. The fuel-cell vehicles could store H_2 on board or carry a liquid fuel like ethanol from which H_2 fuel can be drawn into the fuel-cell stack.

There are several varieties of fuel cells. Phosphoric acid fuel cells are used in large scale stationary market. Molten carbonate fuel cells (MCFC) and solid-oxide fuel cells (SOFC) are used as back-up in the case of remote power generation. Direct Methanol fuel cells (DMFC) are entering the market for portable devices. Transportation sector uses polymer electrolyte membrane fuel cells (PEMFC). Much R&D work is being performed to improve the efficiency of PEMFC.

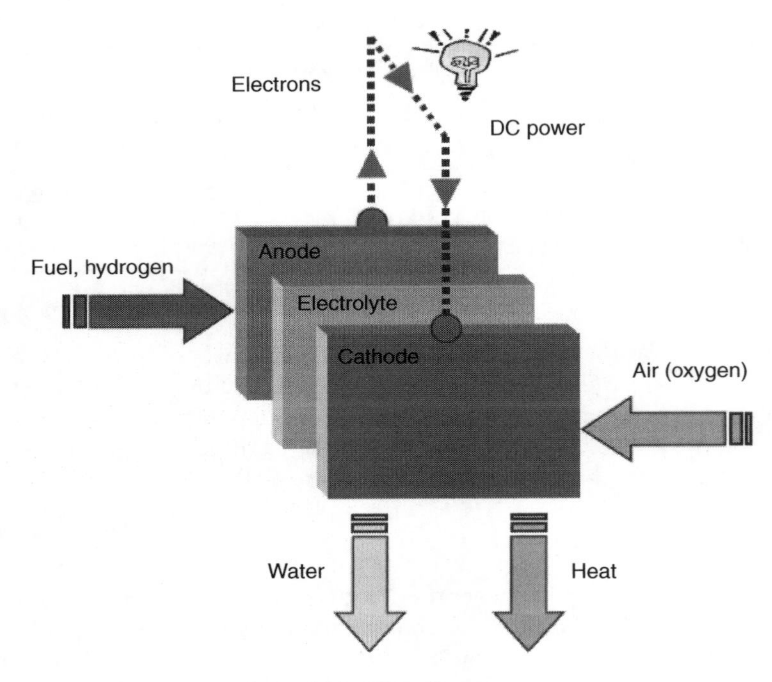

Figure 15.3 Fuel cell concept
(Source: ETP, 2008, p. 267, © OECD – IEA)

Table 15.3 Performance and use of different fuel cells

	PEMFC	*SOFC*	*MCFC*	*DMFC*
Operating temp. (°C)	80–150	800–1 000	>650	80–100
Fuel	H_2	H_2, Hydrocarbons	Natural gas, hydrocarbons	Methanol
Electrical efficiency (%)	35–40	<45	44–50	15–30
Applications	Vehicles, power	Stationary power	Stationary power	Portable power
Lifetime (h) Vehicles	2 000	6 000	8 000	Not known
Power	30 000	20 000	20 000	
Target lifetime (h)				
Vehicles	4 000	40 000	40 000	Not known
Power	25 000	60 000	60 000	

Table 15.3 lists the performance and use of different fuel cells. (source: ETP, 2008, p. 268).

A fuel-cell car requires 75 kW of power, which at USD 500 per kW of power would cost USD 37 500. Besides, a fuel-cell vehicle requires at least 5 kg of hydrogen stored on board to achieve a range of 500 km. At the rate of USD 1 000/kg of high-pressure cylinder systems, this would cost USD 5 000 per vehicle. Demonstration fuel cell cars are priced at USD 100 000 per vehicle. Thus fuel-cell vehicle can penetrate the market only after there is a drastic reduction (of the order of magnitude) in terms of fuel-cell stack system and energy storage system.

Future

Progress is being made at such a rapid pace that Toyota is launching gasoline-electric hybrids (successor to Prius), and Nissan and Honda are launching fully electric vehicles. Honda is launching hydrogen fuel-cell vehicle (*New York Times*, Oct. 21, 2009).

Advanced technology vehicles are expected to play a key role, particularly after 2020.

Governments need to promote simultaneously the development of EVs, PHEVs and FCVs, batteries, recharging infrastructure, while providing incentives for the market promotion of such vehicles. A practical way will be for governments is to choose regions and metropolitan areas which have shown enthusiasm to implement the new approaches.

Biofuels may find increasing use in LDVs. Currently biofuels production is dominated by ethanol from grain crops and biodiesel from oil-seed crops. This should be phased out. Governments should provide incentives to shift to second generation biofuels from non-food feedstocks. Such fuels have to be sustainable, low GHG and cost-efficient, with minimum adverse land-use impacts.

It is possible to reduce CO_2 emissions by shifting the passenger travel to more efficient modes such as mass transit systems (as Singapore has done successfully). Such a modal shift brings other benefits such as lower traffic congestion, lower pollutant emissions and more livable cities. Also, citizens may be encouraged to make short trips on foot or by bicycle (as Paris has done).

Fig. 15.4 (source: *Transport, Energy and CO_2: Moving towards sustainability, 2009*; © OECD-IEA) shows the extent different technologies and fuels contribute to CO_2 reductions from LDVs in the BLUE Map scenario by 2050.

These projections are no doubt uncertain, but the curves do tell a story. It is possible to bring about reductions of the order of 5 Gt in CO_2 equivalent emissions from LDVs, at a marginal cost of about USD 200/tonne with oil at USD 60/bbl. If a higher price of USD 120/bbl is assumed, the emission reductions can be realized at a marginal cost of about USD 130/tonne. There is a good possibility that most of the emission reductions could be achieved at costs far below this. It is expected that most reductions, particularly up to 2030, could come about from incremental improvements in internal combustion engine vehicles and hybrid vehicles, at very low average cost.

15.4 TRUCKING AND FREIGHT MOVEMENT

Trucks come in many shapes and sizes – ranging from small delivery vans to heavy duty tractor-trailers which can carry loads of about 300 tonnes. For most vehicles, fuel costs represent a significant part of the operating costs. Fuel efficiency gains may be achieved in the following ways: (i) Downsizing and downweighting, (ii) Improvements in the engine/drivetrain efficiency through turbo-charging, advanced higher compression diesel engines, and computer controls, (iii) Hybrid drivetrains – they improve the efficiency of urban delivery trucks and short-haul vehicles, by 25 to 45%,

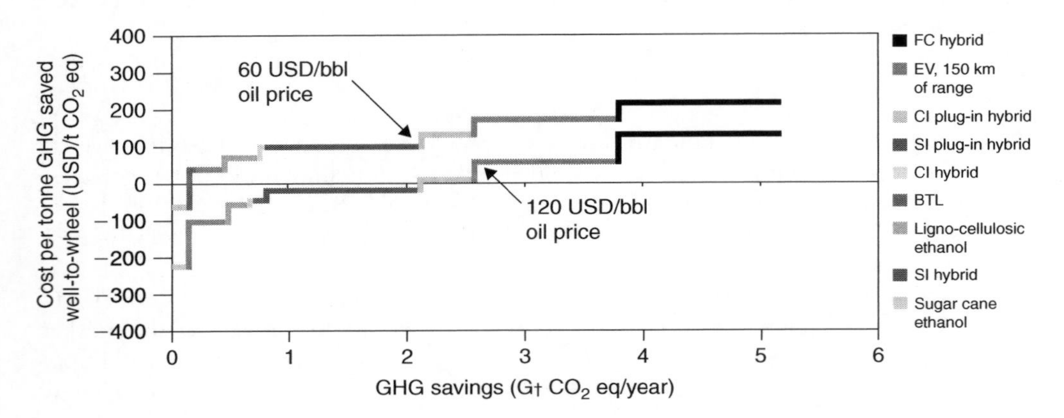

Figure 15.4 Projected GHG reduction of light duty vehicles and fuels.
Transport, Energy and CO_2: Moving towards Sustainability, 2009, Executive Summary, p. 37
SI = Spark Ignition (gasoline) vehicle; CI = Compressed Ignition (diesel) vehicle;
ICE = Internal Combustion Engine (ICE) vehicle; Hybrid = Hybrid vehicle;
BtL = Biomass-to-Liquids (Biodiesel); FC = Fuel Cell; EV = Electrical vehicle

(iv)Aerodynamic improvements, particularly for long-haul trucks, through better integration of tractor-trailer integration, (v) low-rolling, second generation resistance tyres, (vi) More efficient auxiliary improvement., such as cabin heating/cooling systems and lighting – long haul trucks use substantial amount of fuel while stationary.

Technology improvements in trucks pay back their costs in fuel savings over the life of the trucks.

As has happened in the case of LDVs, hybrid propulsion systems are being used with medium-duty delivery trucks (Duleep, 2007). Electric and fuel-cell powered delivery trucks and buses used in urban setting, have a good future, as they are often centrally fuelled. It is unlikely that electric and fuel cell-powered long haul trucks will be viable in the near future, because of the problems of fueling and durability (long-haul trucks need to travel 100 000 km/yr).

Truck operational efficiency can be improved in the following ways: (i) On-board diagnostic systems (real-time, fuel economy computers, data loggers help the drivers and companies to ensure that they are optimally driven and maintained, (ii) Speed governors and advanced cruise-control systems helps the drivers to drive safely and efficiently, (iii) Driver training programmes and good vehicle maintenance system help to improve trucking efficiency, (iv) Logistical improvements, such as, computerized truck dispatching and routing, and use of terminals and warehouses.

As the Canadian experience has shown, regular training of drivers in fuel-efficient driving techniques can yield fuel saving of up to 20% per vehicle kilometer.

Trucking has been growing rapidly during the last two decades, and this is expected to continue. Trucks can be made 30% to 40% more efficient by 2030 through technological measures, operational measures and logistical improvements in handling and routing of goods. In order to optimize the process, governments need to work with the trucking companies to regulate the driver training programmes and create incentives for better efficiency. Japan is a pioneer in this effort,

Biodiesel produced from biomass gasification and liquefaction can be readily used in trucks. Shifting to electricity or hydrogen is not a viable option in the case of trucks due to constraints of range and energy storage limitations. Thus, second generation, non-food based biofuels is effectively the only way to decarbonise the trucking fuel.

Shifting to rail transport constitutes an attractive option to save energy and cut CO_2 emissions. Rail transport in the OECD countries costs one-fifth of the truck transport. Bulk raw materials like coal are often transported by rail. China moves a billion tonnes of coal per year, using dedicated rail links and trains with payloads of 25 000 tonnes.

High speed rail

Trains with cruise speed of more than 200 km/hour exist in Japan, Europe, and western USA. High speed rail (HSR) trips of about three hours (700–800 kms.) constitute an attractive alternative to air travel, as they avoid the hassles of traveling to the airport, checking-in and security checks. Since electricity used in HSR trips will be generated primarily by zero-carbon sources after 2030, there will be saving in energy and CO_2 emissions.

Studies made in Europe and Japan show that the energy consumption per line-km in HSR is about one-third to one-fifth of the aeroplane and car energy use per passenger line-km (ENN, 2008). The total CO_2 emissions of rail systems are near zero (ignoring possible fossil-fuel use to heat the rail stations).

The cost of HSR construction varies from country to country, ranging from USD 10 million to 100 million per line-km, depending upon the land costs, labour costs, financing methods and topography. Europe has 2000 km of HSR in operations, and plans to add 4000 km by 2020. China is expected to build 3 000 km of HSR in the next 15 years. IEA estimates that HSR travel will save 0.5 Gt of CO_2 per year by 2050.

15.5 AVIATION

Commercial travel has been growing at the rate of about 5% per year in terms of passenger-kilometers. There has been a steep drop in air travel after September 11, 2001 attack in US, but it picked up later. In 2006, global average growth rate has been 5% in passenger air traffic and 6% in cargo traffic. As air traffic increases, there would be increase in fuel use and CO_2 emissions. IEA estimates that the technical potential for efficiency improvement (in terms of energy – intensity reduction) of aviation will be 0.5% to 1% on an average, i.e. 25% to 50% by 2050. Load factor improvement in energy efficiency may be 0.1–0.3% per annum. The total potential annual rate of change may be 0.7 to 1.2%.

Large aircraft burn up to a billion litres of jet fuel over their life times. So reducing fuel use could provide enormous fuel cost savings. So improvements in the aircraft design and operation are cost-effective, definitely in the long-term.

Apart from CO_2, aircraft emissions include nitrogen oxide, methane, and water vapour which are capable of radiative forcing (i.e. climate warming). More work is needed to understand the impact of GHG emissions due to aviation.

Improvements in aviation fuel efficiency can be brought about through increasing engine efficiencies, lowering weight, and lift-to-drag ratio (Karagozian et al, 2006).

Potential for improved aerodynamics

The higher the lift-to-drag ratio, the less the fuel consumption. The lift-to-drag ratio can be increased in the following ways: (i) Wing modifications- retrofitting the aircraft with winglets has improved the lift-to-drag ratio by 4% to 7%, (ii) Hybrid laminar flow control: when hybrid laminar control processes are applied to fin, tail-plane and nacelles as well as to the wings, fuel consumption has been found to be reduced by 15%. Improvement of 2 to 5% efficiency are more typical, (iii) Flying wing/blended wind-body configuration: In this design, the entire aeroplane generates lift, and the body is streamlined to minimize drag, leading to a high lift-to-drag ratio, and 20% to 25% less fuel consumption. The commercialization of flying wing aircraft may be possible by 2025.

Structure/materials-related technology potential

Fuel efficiency can be improved and GHG emissions reduced by making the aircraft lighter through the use of new materials and composites.

(i) *Carbon-fibre reinforced plastic*: Carbon fibre – reinforced plastic (CRPF) has many merits: it is stronger and more rigid than metals such as aluminium, titanium and steel. Its density is half of that aluminium, and one-fifth that of steel. It is corrosion-resistant, and fatigue-resistant. If aluminium is fully replaced by CRPF, the weight of the aircraft will be reduced by. 10–15%. Boeing 787 uses CRPF for 50% of the body (on a weight basis) and one-third of the fuel efficiency gain of 20% in this kind of aircraft is attributed to this substitution. As CRPF technology matures, it will be used for wings, wing boxes and fuselages.
(ii) Fibre-metal-laminate (FML): FML is made up of a central layer of fibre sandwiched between thick layers of aluminium. It is stronger than CRPF. About 3% of the fuselage skin of Airbus A 380 is made up of FML. It is also finding increasing use in the construction of aircraft wings.
(iii) Reduction in the weight of engines: New composites not only reduce the weight of the engines, but they also allow higher operating temperatures and greater combustion efficiency, which have the consequence of reduced fuel consumption.

Baseline scenario envisages 25% technical efficiency improvement. BLUE Map scenario projects 35% technical efficiency improvement by 2050.

Operational system improvement potential

Fuel consumption can be reduced in the following ways:

(i) Continuous Descent Approach (CDA): Computerised CDA systems ensures smoother descent that reduces changes in the engine thrust, and thereby saves fuel and reduces noise.
(ii) Improvements in CNS/ATM system: Improvement in communications, navigation and surveillance (CNS) and air traffic management (ATM) systems would enable the optimization of flight paths, with resulting fuel economy. The International Civil Aviation Organization (ICAO) projects fuel savings of about 5% by 2015 in USA and Europe by this approach (ICAO, 2004).

(iii) Multi-stage long distance travel: Today's technology is standard for a range of 4000 km. Fuel efficiencies may be improved by developing fleets with ranges of 5000 to 7500 km. This may not be acceptable to all travelers, however.

Alternate Aviation Fuels

Aviation fuel needs to satisfy a number of stringent requirements: it should have large energy content per unit mass and volume; it should be thermally stable (in order to avoid freezing at low temperatures; and it should have the prescribed viscosity, surface tension, and ignition properties. Synthetic jet fuels, derived from coal, natural gas or biomass, have characteristics similar to conventional jet fuel, and could serve as alternate aviation fuels. Also, their use reduces GHG emissions. Liquid hydrogen is another possibility, as it delivers a large amount of energy per unit mass. Its use as fuel require major modifications in aircraft design (Daggett et al, 2006). Other alternatives, such as methane, methanol and ethanol, do not make the grade because of their low energy density.

Thus, high-quality, high energy-density aviation biofuels hold great potential as low-GHG aviation fuels in future. Their sustainability is dependent upon production from non-food sources. In the BLUE map scenario, second-generation biofuel, such as biomass-to-liquid (BTL) fuel, will be providing 30% of the aircraft fuel by 2050.

In the BLUE scenario, air travel growth can be tripled rather than quadrupled by 2050, through alternatives such as high-speed rail systems, and substituting teleconferencing for long-distance trips. Governments and businesses are urged to promote these developments through appropriate policy actions.

15.6 MARITIME TRANSPORT

International water-borne shipping has grown very rapidly in the recent years due to the high economic growth of countries like China and India. It now represents about 90% of all shipping use, the rest 10% being used through in-country river and coastal shipping. The average DWT of the ships is increasing, and so are tonne-kilometres of goods moved.

The structure of the shipping industry continues to be heavily fragmented, in terms of ownership, operation and registration. This has constrained optimizing the ship efficiency. It is not uncommon for a ship to be owned by the Greeks, registered in Panama and operated by Philippinos. There will be endless legal problems when the ship runs into trouble (e.g. oil leak).

The world shipping fleet made use of 200 Mtoe of fuel in 2005, which is about 10% of the total transport fuel consumption. During the last decade, the shipping fuel consumption and CO_2 emissions have been growing at the rate of 3% per annum. International shipping involves three types of freight movement: dry bulk cargo, container traffic, crude oil and other hydrocarbons such as liquefied petroleum gas. Among these, the container traffic has been growing at the fastest rate of about 9% (Kieran, 2003). It is projected that the container shipping will increase eight-fold by 2050.

Efficiency technologies

There are a number of ways to improve energy efficiency and reduce GHG emissions of maritime transport. The fuel consumption of ocean-going ships can be reduced by 30% through the optimization of the propulsion plant configuration, such as, operating one engine instead of two per shaft at moderate speeds, reducing auxiliary electricity demand through greater use of thermostats to regulate ship-board temperatures, and use of secondary propulsion systems, such towing sail. Towing sails can be retrofitted to existing ships. It has been claimed that the computerized operation of the towing sails, can bring down average fuel costs by 10% to 35% (SkySails, 2006).

Changes in hull design by tailoring the stern flaps and wedges to reduce energy consumption, and increasing ship speed, can reduce the fuel consumption and related CO_2 emissions by 4% to 8%. Using advanced light-weight materials in ship design can reduce the hull weight by 25 to 30%, resulting in significant reduction in fuel consumption.

It has been found that if the ship speed is reduced from 25 knots to 20 knots, there will be fuel savings of 40 to 50%. So slowing down is a cost-effective approach to reduce CO_2 emissions. Even if a 10% reduction in speed may require 10% more ships, that would still be worth it.

Use of high-efficiency, inter-cooled, recuperative (ICR) gas turbine engines can reduce fuel consumption by 25% to 30%.

Alternative Fuels

Ships presently use heavy fuel oils (HFO). Significant reductions can be achieved if the ships shift to new carbon-free fuels. Some of the large ship engines with output exceeding 50 MW have dual-fuel configuration involving natural gas (NG) and HFO and have thermal efficiencies of over 50%. It is feasible to introduce other liquid and gaseous fuels (H_2) in such a set-up. Carbon-free "Green" crude produced from algae has the potential to be used as a fuel in ships. It may presently be more expensive than heavy fuel oil. Some kinds of bio-crude are not as stable as petroleum fuel. Catalytic cracking or hydro-treating of bio-crude could upgrade it to the acceptable level, but that will add costs to the bio-crude.

Despite these constraints, bio-crude or its derivative products have good potential as low-carbon fuels usable in ships. Liquid hydrogen (LH_2) has high gravimetric energy density, as it is 2.8 times lighter than HFO. It increases useful payload, and hence brings higher economic returns. Most importantly, it is extremely clean. Much R&D effort is needed to develop LH_2 based fuel-cell systems for ship propulsion (Velduis et al, 2007). In BLUE map scenario, biofuels share is expected to go up by 30% of overall fuel use by 2050.

International agreements are needed to bring about improvement in international shipping efficiency and CO_2 reduction. CO_2 cap-and-trade system may be made applicable to shipping. A standard ship efficiency index to which all new registration of ships have to adhere (and old ships need to be retrofitted), may be designed and be brought into existence through institutions such as UN International Maritime Organization (IMO).

15.7 RESEARCH & DEVELOPMENT BREAKTHROUGHS REQUIRED FOR TECHNOLOGIES IN TRANSPORT

Table 15.4 Technology breakthroughs in transport sector

Technologies	*RD&D Breakthroughs*	*Stage*
Vehicles		
Hydrogen fuel cell vehicles	Material investigation for solid storage; Cost reduction and improvements in durability and reliability of hydrogen on-board gaseous and liquid storage; cost reduction for fuel-cell system; durability improvement of fuel cell stack and balance of system components (system controller, electronics, motor, and various synergistic fuel economy improvements, etc.)	Basic science/Applied R&D/Demonstration
Plug-in Hybrid/ Electric vehicles	Energy storage capacity and longer life for deep discharge (further development of Li-ion batteries, e.g. Li-polymer, Li-sulphur, etc.) ultracapacitors and fly-wheels; systems that combine storage technologies, (such as batteries with ultracapacitors); and optimization of materials characteristics and components for batteries	Basic science/Applied R&D/Demonstration
Fuels		
Advanced biodiesel (BtL with FT process)	Feedstock handling; gasification/treatment; co-firing of biomass and fossil fuels; syngas production/treatment; better understanding of cost trade-offs between plant scale and feedstock transport logistics	Applied R&D/ Demonstration
Ethanol (cellulosic)	Feedstock research; enzyme research (cost and efficiency); system efficiency; better data on feedstock availability and cost per region; land use change analysis; and co-products and biorefinery opportunities	Applied R&D/ Demonstration
Hydrogen	Development of hydrogen production; distribution and storage systems	Applied R&D

(Source: ETP, 2008, p. 590)

Chapter 16

Electricity systems

U. Aswathanarayana

16.1 OVERVIEW

About one-seventh of the electricity produced worldwide is lost. Out of this, Transmission and Distribution (T&D) losses account for 8.8%. In developing countries, considerable amount of electricity is lost through pilferage, often with the connivance of the local employees of electricity corporations. The total Transmission and Distribution losses are the highest in India (31.9%), and the lowest in Japan (8.7%).

Transmission and Distribution losses as a percentage of gross electricity production in various countries are given in Table 16.1 (source: ETP, 2008, p. 402).

Unlike other energy carriers, such as coal or oil, it is not possible to store electricity in large quantities (except in the form of other types of energy, such as pumped storage or compressed air).

Electricity demand varies according to the time of the day (lower demand in the night) and climate and season (air conditioning demand during the summer, and heating demand in the cold countries during winter). Consequently, peak national grid demand may be two to three times more than the minimum demand. In an electricity grid, it is imperative that electricity production should keep pace with consumption. If this condition is not ensured, there would be instability in the grid with severe voltage fluctuations.

In order to cope with this variability in electricity demand, grids make use of three types of power generating stations:

(i) *Base-load plants*, that can provide consistent supply of electricity over long periods, such as coal-fired thermal power stations and nuclear power stations. Though both capital and operating costs of coal-fired stations are low, moves

Table 16.1 Transmission and distribution losses

Country	*Direct use in plant (%)*	*T&D losses (%)*	*Pumped storage (%)*	*Total (%)*
India	6.9	25.0	0.0	31.9
Mexico	5.0	16.2	0.0	21.2
Brazil	3.4	16.6	0.0	20.0
Russia	6.9	11.8	−0.6	18.1
China	8.0	6.7	0.0	14.7
EU-27	5.3	6.7	0.4	12.5
USA	4.8	6.2	0.2	11.2
Canada	3.2	7.3	0.0	10.5
Japan	3.7	4.6	0.3	8.7
World	**5.3**	**8.8**	**0.2**	**14.3**

are afoot to phase them out because of their environmental and climate change impacts. The capital costs of nuclear power are high, but the operating costs are low. As they have no carbon footprint, they are being favoured, even though the problems of disposal of nuclear waste, safety and proliferation continue to be troublesome issues.

(ii) *Shoulder-load plants*, that can provide electricity during periods of extended high demand, such as, a natural gas combined cycle plant (NGCC) plant or gas turbine which has lower capital and operating costs. Such plants can also serve as base-load plants.

(iii) *Peak-load plants*, which can provide highly flexible power supply of short duration, in order to meet the fluctuations in demand, such as, pumped (hydroelectric) storage.

Variable renewables like wind and solar PV need to have back-up systems based on storable fuels, like coal or biomass.

The load duration curves have significant impact on CO_2 mitigation costs. In Europe and USA, the peak demand is double that of minimum demand. Irrespective of whether a power station is used as a base-load plant or peak-load plant, they will require the same capital investment. The base-load plant is likely to be coal-fired, whereas the peaking plant is likely to be gas-fired. CCS (CO_2 capture and storage) of an NGCC plant costs twice as much as coal-fired plant. At USD 50/t CO_2, the costs of mitigating CO_2 may turn out to be much higher for shoulder-load and peak-load plants than for base-load plants.

16.2 TRANSMISSION TECHNOLOGIES

Power generating units supply electricity to the consumers through a network of transmission and distribution (T&D) grids. Through an intelligent use of the grid system, France is able to cater to a total supply capacity with one-quarter of the total demand potential. This is possible because not all consumers will draw the maximum potential demand at the same time.

Table 16.2 Cost performance of transmission systems

Parameter	*Unit*	*HVAC*		*HVDC*	
Operation voltage	kV	760	1160	±600	±800
Overhead line losses	%/1000 km	8	6	3	2.5
Sea cable losses	%/1000 km	60	50	0.33	0.25
Terminal losses	%/station	0.2	0.2	0.2	0.6
Overhead line cost	M Eur/1000 km	400–750	1000	400–450	250–300
Sea Cable cost	M Eur/1000 km	3200	5900	2500	1800
Terminal cost	M Eur/1000 km	80	80	250–350	250–350

Customarily, electricity is transmitted over long distances on Alternating Current (A.C.). The higher the A.C. transmission voltage, the lower would be the transmission losses – the transmission losses would be 8% for 1 000 km at 750 KV, and 15% for 1 000 km at 380 KV. Residences use 220 V A.C. in most countries, and 110 A.C. in some countries, notably USA. As many as five step-downs may be involved between generation and actual use. T&D may cost USD 5.5 to 8/MWh, and may constitute 5 to 10% of the delivered cost of the electricity.

The development of high-voltage valves has enabled the transmission of DC power at high voltages for long distances with lower transmission losses. DC transmission losses are typically 3% for 1000 km. Most sub-sea cables use DC supply, as losses by AC cable will be excessive. 800 KV High voltage DC (HVDC) transmission lines are being increasingly used, as they are more economical than AC lines for longer distances (>500 kms.). Also, HVDC systems are easier to control, and occupy less space (Rudervaal et al, 2000).

HVDC has some disadvantages – failure in one line cannot receive help from elsewhere, as synchronization is not possible.

Because of the public resistance to new overhead HVDC lines, attempts are being made to lay the HVDC lines underground. This is technically feasible, but the costs are a deterrent - an underground DC line is 5 to 25 times more expensive than the overhead line. Advances in new technologies in respect of cables and insulation are bringing down the costs of the underground cables. This will improve the viability of underground cables.

The cost performance of transmission systems is summarized on Table 16.2 (source: ETP, 2008, p. 405).

16.3 DISTRIBUTION

Transformers are used to step-down the voltage from high to medium and then to low, in the process of supplying electricity to the consumer. In some cases, as many as five step-downs may be involved. Power transformers are very highly efficient – losses are usually less than 0.25% in large units, and do not exceed 2% even in the case of small units. In a power network, the losses due to transformers can exceed 3% of total electricity. Replacement of conventional steel cores by amorphous iron cores can reduce the losses by 30%. In rural India, where there are a large number of lower-capacity sub-stations, and where conversion of single-phase supply to three-phase supply is resorted to, the distribution losses may exceed 30%. During periods of peak load, the losses may even exceed 45%.

There has been rapid increase in the use of AC/DC transformers in the electronic equipment. These transformers are switched on permanently, but the device concerned is used intermittently. Such losses beyond the meter may amount to 5 to 10% of total electricity.

In the case of wind turbines, transportation over 2 000 km. would add 50% to the production cost (US cents 2 to 3/kWh).

The development of regional interconnections would reduce the need for storage and backup facilities, and therefore should be promoted.

Transmission and Distribution (T&D) losses are most serious in developing countries. It is possible to reduce the global T&D losses from the present 18% to 10%, through the application of new technologies, and policies.

16.4 ELECTRICITY STORAGE SYSTEMS

Electricity cannot be stored, except in a small way in the form of capacitors. It can, however, be converted to other forms of energy and stored. In batteries, it is converted to chemical energy. In pumped storage, it is stored as potential energy. Electricity can also be stored in the form of compressed air or in fly wheels.

The cost of storage or backup capacity typically adds US cents 1 to 2/kWh.

Fig. 16.1 (source: Thijssen, 2002, quoted by ETP, 2008, p. 407, © OECD-IEA) depicts the capital cost of different storage options.

Battery electricity storage is efficient, but expensive. For instance, Lithium-ion battery typically costs USD 500/kWh. Delivered costs are around USD 0.20/kWh.

The discharge times and system ratings of different storage options are given in Fig. 16.2 (source: Thijssen, 2002, quoted by ETP, 2008, p. 408, © OECD-IEA).

Pumped storage is the preferred option. It has an efficiency ranging from 55 to 90%, system rating of about 100 MW, and discharge times of hours. Pumped storage plants can respond to load changes almost instantly (less than 60 seconds). Compressed air energy systems (CAES) have efficiencies of about 70%. The biggest problem with CAES is finding suitable storage caverns. Aquifer storage is a good possibility for CAES (Shepard and van der Linden, 2001).

Superconducting Magnetic Energy Storage (SMES) stores electrical energy in superconducting coils. SMES has the advantage of being able to control both active and reactive power simultaneously. Also, it can charge/discharge large amounts of power quickly.

Hydrogen that can be produced from electrolysis could serve as an energy carrier. During periods of excess demand, it can be used to generate power. The efficiency of electrolysis is about 70%, and the efficiency of power generation is 60%. Thus, the hydrogen storage system, has an overall efficiency of 42%. Hydrogen can be stored in salt caverns, manmade caverns, and depleted oil and gas reservoirs. It costs money to excavate the caverns, and storage of hydrogen in depleted oil and gas reservoirs may contaminate hydrogen. Aquifer storage is the cheapest option. Hydrogen fuel storage will become viable when the appropriate infrastructure for hydrogen production and use comes into existence, and fuel-cell vehicles become popular.

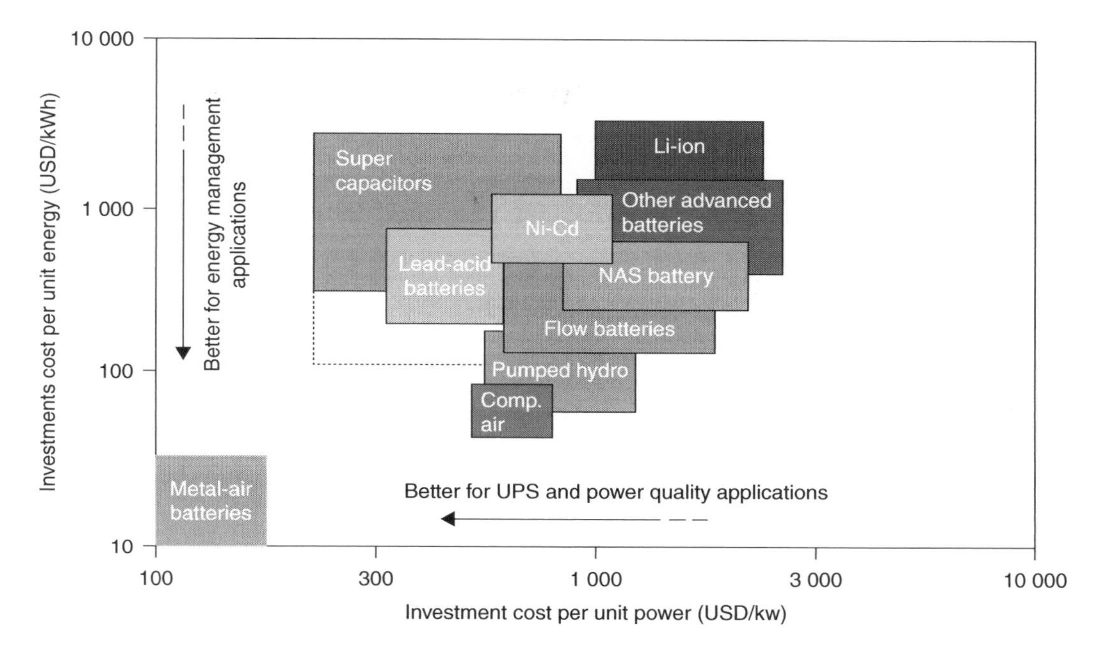

Figure 16.1 Capital cost of different storage options. ETP, 2008. p. 407 © OECD-IEA

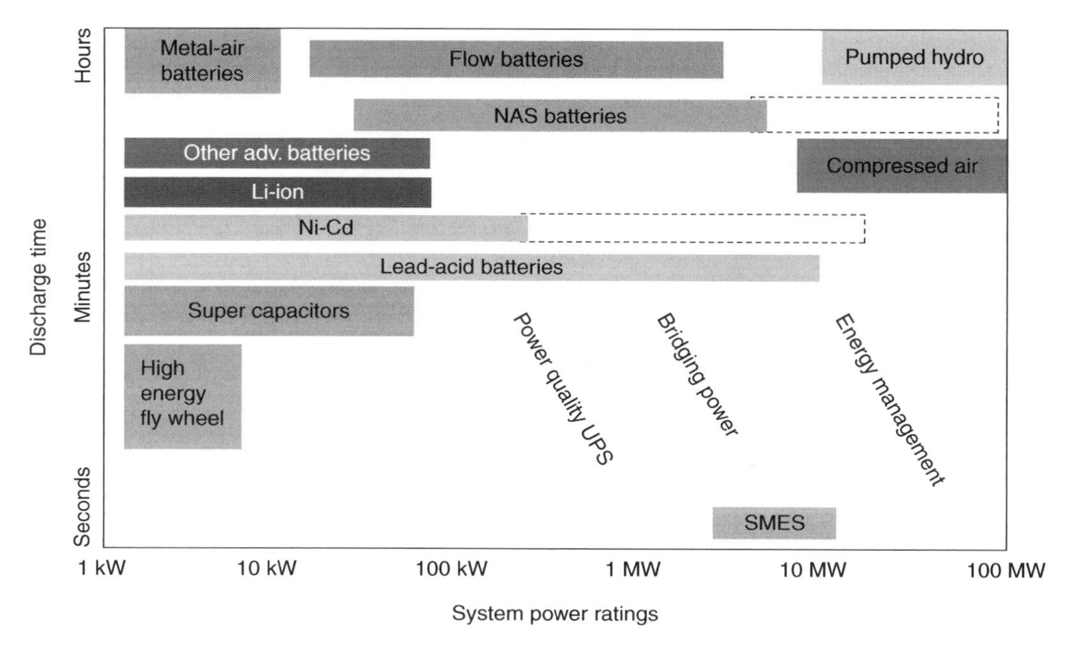

Figure 16.2 Discharge times and System ratings of different storage options. ETP, 2008. p. 408. © OECD-IEA

Table 16.3 Cost comparisons of base-load supply systems

	Investment (USD/MW)	*Fuel (USD/kW/yr)*	*Baseline (USD/yr)*	*CO_2 (t/yr)*	*ACT Map (USD/yr)*	*BLUE Map (USD/yr)*
3 wind turbines + 2 CAES units	4000	0	600	0	600	600
1 wind turbine + 1 NGCC	1500	229	454	2.0	503	848
1 NGCC	500	341	416	2.9	490	1005

Note: Assumes 33% availability of wind turbines, USD 1000/kW for wind turbines, USD 500/kW for CAES, 15% annuity, USD 6.5/GJ gas.

Intermittent supplies, such as wind power, require storage arrangements to ensure uninterrupted supply. Storage is not, however, the least cost option.

Table 16.3 gives cost comparisons of three base-load supply systems (ETP, 2008, p. 410).

The high costs of transmission and distribution can be avoided through decentralized power stations sited close to the demand centers. Small distributed generation units are usually less efficient, and are characterized by higher investment costs per unit of capacity. However, certain renewables (such as, PV, landfill gas, biomass residues) are more suited to decentralized units.

Decentralized generation units are more suitable to places, such as, parts of India and Africa, where grid does not exist or is unreliable. In general, such units are viable in rural and remote areas. However, since it is projected that by 2050, 80% of the world population will live in urban areas, the viability of decentralized power stations will depend upon how they fit into the urban environment.

16.5 DEMAND RESPONSE

Demand Response (DS) seeks to manage customer consumption of electricity in response to the supply conditions (vide Wikipedia: Electricity Distribution). Demand response serves to avoid outages and to help utilities manage daily system peaks. Customarily, the capacity of the electrical systems are sized to correspond to peak demand. So when the demand is less (as in the nights), it amounts to inefficient use of the capital. If peak demand can be lowered through an intelligent management of demand, it would lead to reduction in overall plant and capital cost requirements. Demand response may also be used to increase demand (load) at times of high production and low demand.

Demand Response is different from energy efficiency which means using less power to perform a particular task, whenever that task is performed (in other words, there is no time element built into it).

Most often, consumers pay for electricity tariff at a fixed rate per unit (kWh), independent of the actual cost of production of electricity at the relevant time.

The tariff is fixed by the government or a regulator on a long-term basis. If consumption is made sensitive to the cost of production in the short term, the consumers would (presumably) increase and decrease their use of electricity in reaction to price signals. Since the consumers do not face actual market prices, they have little or no incentive to reduce consumption (or defer consumption to later periods) as there is no benefit to them for doing so.

Whereas nuclear power and thermal power are produced at a constant rate, irrespective the demand, there is intermittency associated with wind power (power is produced when the wind blows), and solar power (which is produced only during day time when the sun shines).

The value of one unit of energy depends upon when it is available, where it is available and how it is available. A unit of energy has more value if it can be made available when needed by the consumer. Thus energy delivered at peak is more valuable than energy delivered off-peak. Also, reductions in energy use are more valuable if they occur at the time of the peak consumption. The capacity value of an energy system is given by the energy that can be reliably delivered at the time of the peak consumption, whereas the energy value of a system is the total amount of energy delivered over the course of a year.

The system of payment to electricity producers is such that it encourages priority usage of lower-cost sources of generation (in terms of marginal cost). In systems where market-based pricing is used, there can be considerable variation in pricing. For example, in Ontario between August and September 2006, wholesale prices paid to producers ranged from a peak of C$318 per MWh to a minimum of negative $C3.10 per MWh (consumers paying real-time pricing may have actually received a rebate for consuming electricity during the latter period). Some times, prices may vary 2 to 5 times in a 24-hour period.

For instance, there is no rigid time when a clothes dryer should be switched on. Using a demand response switch, it can be got switched on during off-peak time, thereby reducing peak demand. As against this, air-conditioning has to be switched on when it is hottest, namely, mid-noon. That is also the time when the solar PV produces peak power.

When an intermittent renewable energy unit like a windmill is combined with a peaking unit such as combustion turbine, and if an analysis of the hybrid system shows it to be the most economic alternative, there is no difficulty in making the choice in favour of the wind mill- turbine unit. Even if the turbine unit alone is found to be cost effective, decision cannot be made in its favour. This is so because the government, as a matter of policy in the context of climate change, is committed to easing out fossil fuel energy generation and promoting renewable energy systems. The turbine unit should therefore be considered as a "necessary evil" in order to make the windmill viable.

16.6 "SMART" GRID APPLICATION

"Smart" grid involves the delivery of electricity from the suppliers to the consumers through the use of digital technology for controlling appliances in consumer's home. This saves energy, reduces the costs to the consumer, and improves reliability and transparency. Fig. 16.3 depicts the general layout of the electricity systems and Fig. 16.4 shows the arrangement of grid.

Future electricity systems are likely to have large component of intermittent power sources such as wind power and solar PV. Under these circumstances, smart grid would

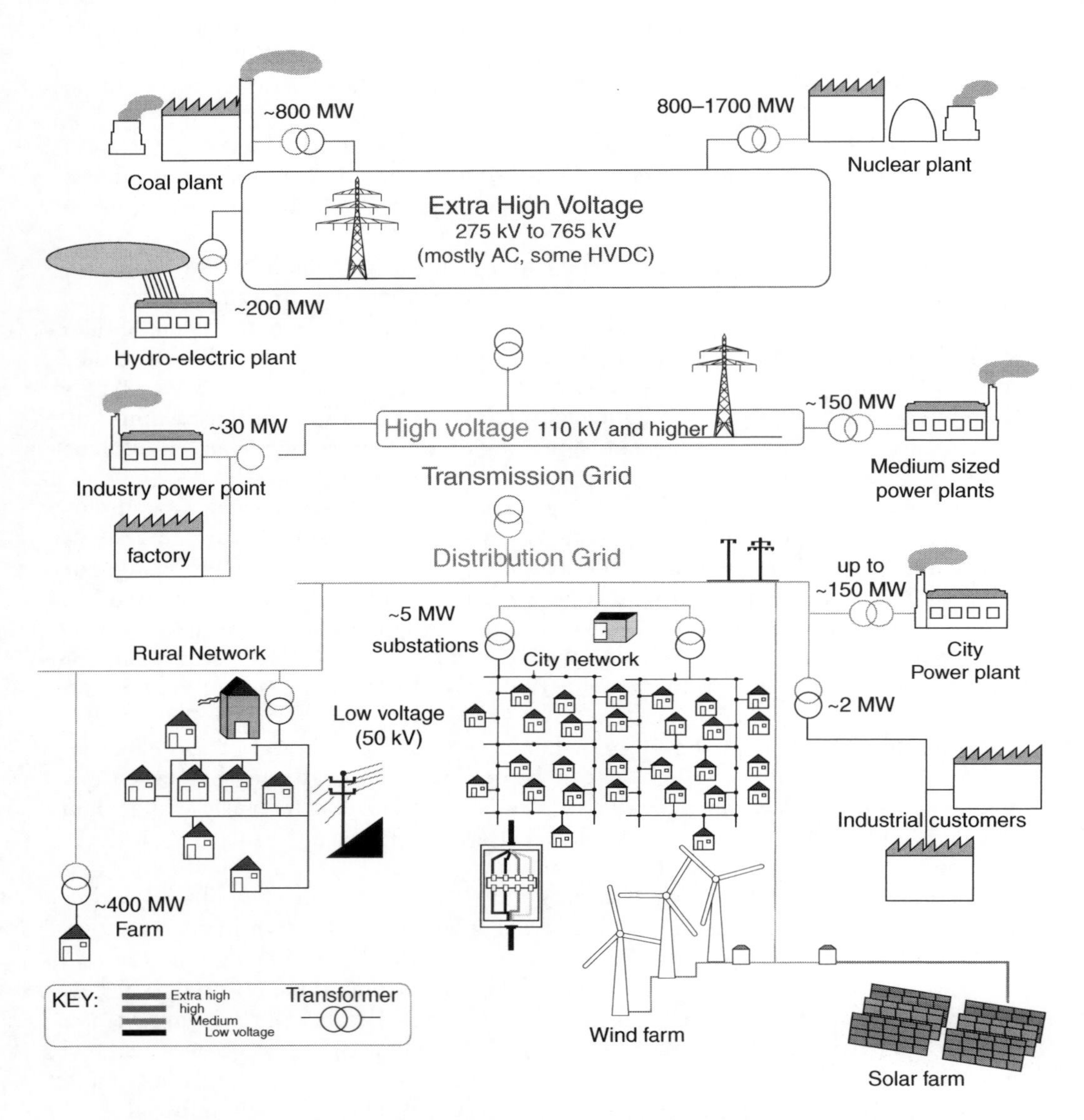

Figure 16.3 General layout of electricity networks. Wikipedia – Electricity Distribution

be an effective way to manage the situation. For instance, in the case of a region heavily dependent on wind power, there are two options for getting over the intermittency problem. One is to build energy storage to deposit excess power. This is an expensive option. A cheaper option will be to use the demand response approach. The excess power instead of being stored, may be used to recharge vehicle batteries during times of excess wind. When the wind dies down, the demand is shed by, say, delaying activation of the refrigeration compressor, or hot water heater coils.

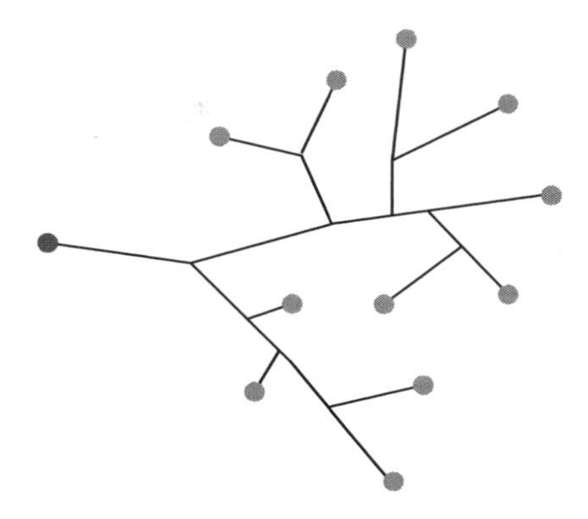

Figure 16.4 Arrangement of grid
Source: Wikipedia – Electricity Distribution

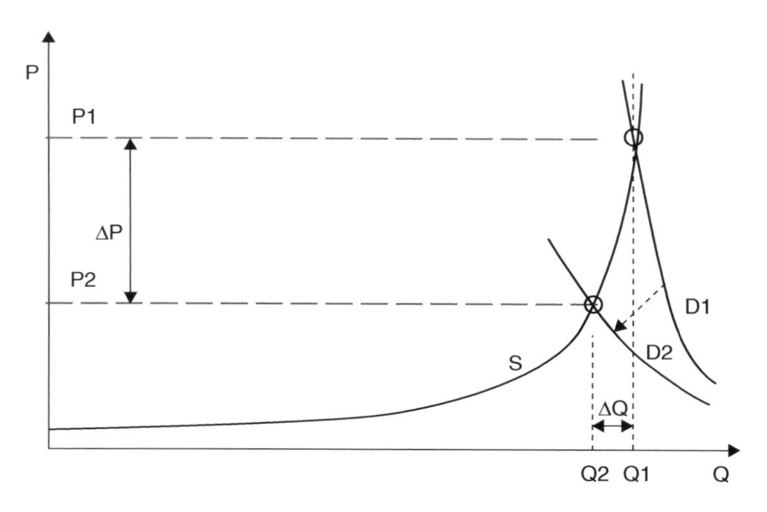

Figure 16.5 Relationship between quantity and power
Source: Wikipedia – Electricity Distribution

16.6.1 Electricity Pricing

The relationship between Quantity and Power under conditions of elastic and inelastic demand is depicted in Fig. 16.5 (source: Wikipedia; Fig. Courtesy: M.G. Tom, 2006).

If the demand is inelastic (curve D1), the price will be high (P1) and it may strain the electricity market. If through the use of demand response measures, the demand could be rendered elastic (curve D2), the price will be lower (P2). A small reduction in quantity (ΔQ) could result in a large reduction in price (ΔP).

It has been found that during the peak hours of the California electricity crisis in 2000/2001, a 5% lowering of demand would result in reduction of the price by 50%. This demonstrates the efficacy of demand response approach.

Carnegie Mellon studies in 2006 found that even small shifts in peak demand would result in large savings to the consumers, while avoiding costs for additional peak capacity demonstrated the profound importance of the demand response in electrical industry: a 1% shift in peak demand would result in savings of 3.9%, which would be in billions of dollars at the system level. A 10% reduction in peak demand could save USD 8 to 28 billion. For this reason, it is worthwhile to make a special effort to improve the elasticity of demand of a system.

A study made by the Brattle Group (USA) in 2007 found that even a 5% drop in the peak demand would bring about an annual savings of USD three billion, by eliminating the need to install and operate 625 infrequently used peaking power plants and the associated delivery infrastructure.

The Independent Electricity System in Ontario, Canada, was built for a peak demand of 25 GW. A maximum demand of 27 GW occurred during only 32 system hours (i.e. less than 0.4% of the time). Thus, by "shaving" the peak demand through appropriate demand response measures, it was possible for the province to reduce the built capacity by about 2000 MW.

16.6.2 Electricity grid and peak demand response

In an electricity grid, it is imperative that electricity production should keep pace with consumption. If this condition is not ensured, there would be instability in the grid with severe voltage fluctuations. Tripping may take place, and this could trigger a chain reaction, with disastrous results.

Governments or electricity corporations optimise the operation of the electricity grid through the following kind of strategy: (i) Total generation capacity is sized slightly in excess of the total peak demand, to take care of unforeseen circumstances, (ii) Least expensive generating capacity (in terms of marginal cost) (say, wind power) is used to the maxiumum extent possible, with expensive source of power (say, nuclear power) being used as demand increases, (iii) The goal of the demand response is to reduce the peak demand in such a manner that there is no risk of voltage fluctuation, while avoiding additional costs for additional plant and infrastructure, and making minimum use of power from more expensive and/or less efficient plants.This will benefit the consumers of electricity through lower prices.

Some types of generating plants, such as, nuclear power plants, must be run at full capacity. Some times there may be enough demand for it. Demand response approach can be used to increase the load during periods of high supply. Pumped (hydroelectric) storage is an economical way to increase the load in order to make use of the excess power. In the province of Ontario, Canada, in September 2006, there was a short period of time when the prices were negative for some category of users, and they had to be paid a rebate. Pumped storage was made use of to get over the problem. Use of demand response to increase load is not common, but may some times has to be resorted to when there is large generating capacity that cannot be cycled down.

In 2006, the province of Ontario, Canada, launched a "Smart Meter" programme to bring the benefits of the demand response to the consumer using the TOU (Time-of-Use) principle. This system has three tiers of pricing: on-peak, mid-peak and off-peak schedules. During winter, on-peak refers to morning and early evening, mid-peak is defined as mid-day to late-afternoon, and off-peak at night time. During the summer, the on-peak covers mid-day to late afternoon, as air conditioning drives the summer demand. In 2007, prices during the on-peak were C$0.097/kWh, i.e. about three times more expensive than off-peak (C$0.034 per kWh). Though the system has not yet gotten into full use, Ontario plans to make TOU metering obligatory by 2010.

16.6.3 Incentives to shed loads

Demand Response incentives to shed loads may take many forms. A utility may pass on to the customers tariff reductions in the price of electricity. During a heat wave, a mandatory cutback may be imposed on high-volume users who are compensated for their participation. Others may receive a rebate for reducing power consumption during the periods of high demand.

Some businesses which have their own captive power stations, usually take steps to stay within that capacity in order to avoid buying power from the grid. Some utilities have framed their tariff structure such that the tariff that a customer has to pay is calculated on the basis of the moment of highest use, i.e. peak demand, in a month. This will serve as incentive to the customer to flatten their energy use, even if that may mean cutting back service temporarily.

Some customers may not be in a position to reduce their demand, or the peak prices may not be that high as to induce the customer to reduce the peak consumption. Automated control systems do exist, but they may not be affordable to some categories of customers.

16.6.4 Technologies for demand reduction

The process of demand response can be automated. Computer systems are available to detect the need for loads shedding, inform the concerned unit how much load need to be shed, implement the directive to shed the load, and communicate compliance to the control unit. Companies such as Ziphany, LLC and Convia have developed the necessary scalable and comprehensive solutions for the purpose.

Electricity demand of a given geographical unit depends upon the size of the population, their life-style, climate, agriculture, industry, tourism, etc. Also, it is highly dependent upon the time of the day (e.g. the demand for air-conditioning is maximum at noontime).Under the provisions of "smart" grid, industrial, residential and commercial users in an area are linked with various power generating units (thermal, wind power, solar PV, nuclear power, etc.) in the area. When it becomes necessary to reduce the peak demand in the area, the central control system may turn down the temperature of heaters, or raise the temperature of some appliances (such as air conditioners and refrigerators) in order to reduce their power consumption. This essentially means delaying the draw marginally. Though the amount of demand involved in this exercise is small, it will have significant financial impact on the system, as electricity systems are sized to take care of extreme peak demands, though such events occur very infrequently.

The city of Toronto is experimenting with a demand response programme (Peaksaver AC) whereby the system operator can automatically regulate air conditioning during the peak demand through allowing the peaking plants time to cycle up. This benefits the grid, and the benefit is passed on to the consumer in the form of lower tariff.

Bonneville Power experimented with such control technologies in residences in the states of Washington and Oregon, and found that the avoided investment justifies the cost of technology. REGEN Energy developed swarm-logic methods to coordinate multiple loads in a facility. New models of appliances such as refrigerators and clothes dryers are being fixed with sensing devices to respond to the directions of grid.

Thus, pricing can be used as an incentive to reduce the great variability in the consumption of electricity in residential and commercial sectors. There are three guiding principles in this regard:

(i) Every effort should be made to make efficient use of a production facility. If a production facility is not used or used insufficiently, it will earn little revenue. It would thus constitute waste of investment.
(ii) Electric systems are sized for peak loads with provision for unforeseen events.
(iii) If peak demand can be reduced by "smoothing", it would mean less investment and more efficient use of generating facility, since significant peak events occur rarely.

16.6.5 "Power Plant in a box"

In San Jose, Calif., on Feb. 24, 2010 (WE), Indian American K.R. Sridhar launched the "Bloom Box" which has the potential to revolutionise electricity production, just as a cell phone did in the case of communications. The "Bloom Box" is a fuel-cell device, consisting of a stack of ceramic disks with secret green and black "inks". These disks are separated by cheap metal plates. The "Bloom Box" can covert air and nearly any renewable and fossil fuel (e.g. natural gas, biogas, coal gas) into electricity by electrochemical process. Since no combustion is involved, there will be no emissions, sound or smell. Unlike solar or wind energy, which are intermittent, Bloom technology would be able to provide electricity 24 × 7. The Bloom Energy Server, a smooth metal box the size of a metal truck, can provide 100 kW of electricity, enough to power 100 American homes or 400 Indian homes. Sridhar says that by 2020, a Bloom Box of 1 kW capacity, costing about USD 3 000, would be available to provide clean, reliable and affordable electricity to individual households.

REFERENCES

Berntsson, T., et al (2007) *Swedish Pulp Mill Biorefineries: A Vision of Future possibilities*. Chalmers Univ. of Technology, Sweden.

Borlée, J. (2007) *Low CO_2 Steels: ULCLOS Project*. Paper presented at the IEA Workshop, Deploying Demand Side Energy Technologies, 8–9 October. OECD-IEA, Paris.

Bowen, C.P. (2006) *Development Trends in Ethylene Crackers: Existing Technologies and RD&D*. Paper presented in IEA/CEFIC Workshop, Feedstock substitutes: Energy Efficient Technology and CO2 Reduction for Petrochemical Products. 12–13 December, OECD/IEA, Paris.

Daggett, D. et al (2006) *Alternative Fuels and their potential impact on aviation*. NASA, Hanover, M.D. http://gltrs.grc.nasa.gov/reports/2006/TM-2006-214365.pdf.

Duleep, K.G. (2007) *Fuel economy of heavy-duty trucks in the USA*. Presented at IEA Workshop on Fuel Efficiency Policies for Heavy Duty Vehicles, June 2007, http://www.iea.org/Textbase/work/2007/vehicle/Duleep.pdf.

ENN (Environmental News Network) (2008) High Speed Rail Advances Globally, Crawls in the US. *Green Energy News*, Feb. 11. v. 12, no.47. www.green-energy-news.com/arch/nrgs2008/20080012.html.

European Commission (2001) *Best Available Techniques in the Non-Ferrous Metal Industry*. IPPC (Integrated Pollution Prevention and Control), EC, Brussels.

European Commission (2007) *Rail Transport and Interoperability*. http//ec.europa.eu/transport/rail/interoperability/tsi_revised_en_htm.

Fargione, J. et al (2008) Land Clearing and Biofuel Carbon Debt. *Science*, v. 319, no. 5867, p. 1235–1238.

Florides, G.A. et al (2002) Measures used to Lower Building Energy Consumption and Their Cost Effectiveness. *Applied Energy*, v.73, no.3, p. 299–328.

Hector, E., and T. Berntsson (2007) *Reduction of greenhouse gases in Integrated Pulp and Paper Mills: Possibilities for CO_2 capture*. Chalmers University of Technology, Sweden.

ICAO (2004) Operational Opportunitiesto Minimize Fuel Use and Reduce Emissions. ICAO Circular 303-AN/176, ICAO, February. Montreal, Quebec.

IEA (2008) *Energy Technology Perspectives*. Paris: OECD-IEA,

IEA (2009) *Transport, Energy and CO_2 :Moving towards sustainability*. Paris: OECD-IEA.

Jakob, M., and R. Madlener (2004) Riding Down the Experience Curve for Energy Efficient Building Envelopes: the Swiss case for 1970–2020. *International J. Energy Technology and Policy*, v. 2, nos. 1–2.

Karagozian, A. et al (2006) *Report on the Technology Option for Improved Air Vehicle Fuel Efficiency: Executive Summary and Annotated Brief*. U.S. Air Force Scientific Advisory Board, Washington, D.C.

Kieran, P. (2003) *World Trends in Shipping and Port Reform*. Port Reform Seminary, South Africa. www.dpe.gov.za/res/peterKieranWorldTrendsinShippingand PortReform.pdf

McNeil, M. et al (2005) *Potential Benefits from Improved Energy Efficiency of Key Electrical Producys: The Case of India*. LBNL, Berkeley, CA.

Passier,G. et al (2007) *Status Overview of Hybrid and Electrical Vehicle Technology, Final Report*, Phase III, Annexe VII, Hybrid Vehicles.IEA Hybrid and Electrical Implementing Agreement.OECD/IEA, Paris. www.ieahev.org/pdfs/annex_7/annex7_hev_Final_rpt_110108. pdf.

Philibert. C. (2006) *Barriers to Technology Diffusion: The Case for Solar Thermal Technologies*.OECD/IEA, Paris.

Rudervaal, R. et al (2000) *High Voltage Direct Current (HVDC) Transmission Systems, Technology*. Review Paper, Energy Week, 2000, Washington, D.C., Mar. 7–8, 2000, www.worldbank.org/html/fpd/em/transmission/technology_abb.pdf

Sachs, H. et al (2004) Emerging energy-saving technologies and practices in the building sector. ACEEE, Washington, D.C.

Seeline Group (2005) *Technology Assessment Study and TRC Analysis*. Seeline Group, Ontario Power authority.

Shepard, S., and S. van der Linden (2001) Compressed Air Energy Storage Adapts Proven Technology to Address Market opportunities. *Power Engineering*, Apr. 2001

Simpson,A. (2006) *Cost-Benefit Analysis of Plug-in Hybrid Electric Vehicle Technology*. Paper presented in the 22nd. International Battery, Hybrid and Fuel Cell Electrical Vehicle Symposium and Exhibition (EVS-22), Yokohoma, Japan. Oct. 23–29, 2006. Conference Paper NREL/ CP-540-40485. www.nrel.gov/vehiclesandfuels/vsa/pdfs/40485.pdf.

SkySails (2006) *SkySails – Turn Wind into Profit. Technology Information,* Hamburg, Germany. http://skysails.info/index.php?id=6&L=1.

Srinivas, S. (2006) *Green Buildings in India: Lessons Learnt.* Indian Green Building Council, Hyderabad.

Thijssen, G. (2002) *Electricity Storage and Renewables. Electricity Storage Association* (ESA), Morgan Hill, CA, and Transmission and Distribution Counselling (KEMA), Arnhem, Netherlands. www.electricitystorage.org/pubs/ 2002/Lisbon_May_2002_KEMA.pdf

Velduis, I.J.S., R.N. Richardson, and H.B.J. Stone (2007) Hydrogen Fuel in a Marine Environment. *Int. J. Hydrogen Energy*, v. 32, no. 13.

Wikipedia – Elecricity Distribution.

World Aluminium (2007) *Electrical Power Used in Primary Aluminium Production.* International Aluminium Institute, London.

Yates, J.R, D. Perkins, and R. Sankarnarayanan (2004) *Cemstar Process and Technology for lowering Greenhouse Gases and Other Emissions While Increasing Cement Production.* Hatch, Canada http://hatch.ca/Environment_Community/Sustainable_Development/Projects/Copy%20of%20Cemstar-Process-final-4-30-03.pdf

Section 5

Making green energy competitive

U. Aswathanarayana

Chapter 17

Roadmaps and phases of development of low-carbon energy technologies

U. Aswathanarayana

17.1 WHY LOW-CARBON ENERGY TECHNOLOGIES?

It is now generally accepted that if we are to escape the catastrophic consequences of global warming, the world must restrict the total carbon emissions to 190 gigatonnes ($Gt = 10^9$ t) by 2050. In 2008, the global carbon emissions were 9 Gt, and the rate of carbon emission is increasing at the rate of 3% per annum. At this rate, the entire carbon budget available to humankind will be reached in 1929 itself. Business-as-usual is hence not a viable option. It is therefore critically important to bring down carbon emissions. The global energy economy needs to be transformed profoundly in the coming decades in terms of ways by which energy is supplied and used. This is to be accomplished through greater energy efficiency, greater use of renewables and nuclear power, CO_2 Capture and Storage (CCS) on a massive scale, and development of carbon-free transport. Among these, improvement in energy efficiency is the least expensive and most effective pathway.

The Developed countries which have 15% of the world's population are responsible for 50% of the CO_2 emissions (Table 17.1).

The CO_2 emissions of the top five emitters in 2005, 2015 and 2030 are given in Table 17.2

Five countries (USA, China, Russia, Japan and India) currently account for 55% of global energy-related CO_2 emissions. The same countries will remain the top CO_2 emitters in 2030, but their relative ranks will change. China has overtaken USA in 2007, and will continue to remain the top emitter. India's rank will jump from fifth to third rank in 2015, and will remain so in 2030.

Table 17.1 Emission of greenhouse gases by some important countries

Country	*CO_2 emissions (million tonnes)*	*Growth rate, 1990–2004*	*CO_2 emissions per Capita (tonnes)*
USA	6046	25	20.6
China	5007	109	3.8
Russia	1524	−23	10.6
India	1342	97	1.2
Japan	1257	17	9.9
Germany	808	−18	9.8
Canada	639	54	20.0
United Kingdom	587	1	9.8
Korea	465	93	9.7
Italy	450	15	7.8
World	28,993	28	4.5

(Source: UN Human Development Report, 2007)

Table 17.2 Top five countries for energy-related CO_2 emissions in the reference scenario

	2005		*2015*		*2030*	
	Gt	*rank*	*Gt*	*rank*	*Gt*	*rank*
USA	5.8	1	6.4	2↓	6.9	2=
China	5.1	2	8.6	1↑	11.4	1=
Russia	1.5	3	1.8	4↓	2.0	4=
Japan	1.2	4	1.3	5↓	1.2	5=
India	1.1	5	1.8	3↑	3.3	3=

Gt = gigatonnes = 10^9 t

(Source: *'World Energy Outlook 2007"*, p. 200)

The International Energy Agency (*Energy Technology Perspectives*, 2008, p.38) came up with two scenarios:

(1) ACT Map scenarios which makes use of technologies that already exist or likely to be available soon, to bring the CO_2 emissions back to current levels by 2050. Emissions need to peak between 2020 and 2030, costing USD 50 per tonne of CO_2 saved when fully commercialized. IEA estimates that to realize the ACT scenario, an additional investment of USD 17 trillion will be needed between now and 2050. This works out to USD 400 billion per year, which is roughly equivalent of GDP of The Netherlands.

(2) BLUE Map scenarios, whereby emissions must be reduced by 50–85% of the current levels by 2050, in order for the global warming to be confined to the IPCC-recommended figure of 2°C–2.4°C. This is more expensive than ACT scenarios as it involves the deployment of technologies which are yet to be developed. Additional investment needed to realize the blue scenario is estimated by IEA to

Table 17.3 Key technologies on the supply-side and demand-side

Supply side	*Demand side*
• CCS fossil-fuel generation • Nuclear power plants • Onshore and offshore wind • Biomass integrated-gasification combined-cycle and co-combustion • Photovoltaic systems • Concentrating solar power • Coal: integrated-gasification combined cycle • Coal: Ultra-supercritical • Second-generation biofuels	• Energy efficiency in buildings and appliances • Heat pumps • Solar space and water heating • Energy efficiency in transport • Electric and plug-in vehicles • H_2 fuel cell vehicles • CCS in industry, H_2 and fuel transformation • Industrial motor systems

(Source: *Energy Technology Perspectives*, 2008, p.46)

be USD 45 trillion between now and 2050. This works out to USD 1.1 trillion per year from now to 2050. This is roughly equivalent of GDP of Italy.

The additional investments needed under ACT and BLUE scenarios appear staggering, but it should be emphasized that these are not net costs. The technology investments in energy efficiency, in many renewables and in nuclear power, have the effect of drastically reducing the fuel requirements.

"In both ACT and Blue scenarios, the estimated total undiscounted fuel cost savings for coal, oil and natural gas over the period to 2050, are greater than the additional investment required" (*Energy Technology Perspectives, 2008, p. 40)*. For ACT scenario, fuel savings exceed additional investments at the discount rate of 3%. The discount rate has, however, to be 10% to fulfill this condition in the case of BLUE scenarios. Thus, the additional investments are cost effective, particularly so in the case of the energy efficiency.

The international Energy Agency, Paris, came up with the following key roadmaps to achieve a sustainable energy future.

17.2 EMISSION REDUCTIONS AND RESEARCH DEVELOPMENT & DEMONSTRATION INVESTMENTS

Abatement of CO_2 emissions in power generation requires significant increase in renewable power generation.

Renewable technologies (wind, solar, biomass, and to a lesser extent, hydro and geothermal generation) figure prominently both in ACT and BLUE Map scenarios (35% and 46% respectively). In the case of nuclear power, Generation III and Generation III+ technologies could deliver the needed outcomes for ACT and BLUE scenarios. Generation IV technologies are, however, needed to reduce costs and minimize nuclear waste and enhance reactor safety.

Table 17.4 is designed to provide the policy makers with information about the technologies available for reducing CO_2 emissions at least cost.

Table 17.4 Emission reduction and RD&D Investment

	CO_2 savings (Gt)		RD&D (USD bn)	
	ACT Map	BLUE Map	ACT Map	BLUE Map
Power generation	**8.96**	**15.13**	**3200–3760**	**3860–4470**
CCS Fossil fuel generation	2.89	4.85	700–800	1300–1500
Nuclear power plants	2.00	2.80	600–750	650–750
Onshore & offshore wind	1.30	2.14	600–700	600–700
BIGCC & Co-combustion	0.22	1.45	100–120	110–130
PV	0.67	1.32	200–240	200–240
CSP	0.56	1.19	300–350	300–350
Coal IGCC Systems	0.66	0.69	350–400	350–400
Coal USCSC	0.66	0.69	350–400	350–400
Buildings	**6.98**	**8.24**	**320–440**	**340–420**
Energy efficiency in buildings & appliances	6.50	7.00	n.a	n.a.
Heat pumps	0.27	0.77	70–100	90–120
Solar space & water heating	0.21	0.47	250–300	250–300
Transport	**8.20**	**12.52**	**260–310**	**7600–9220**
Energy efficiency in transport	5.97	6.57	n.a.	n.a.
Second-generation biofuels	1.77	2.16	90–110	100–120
Electric & plug-in vehicles	0.46	2.00	170–200	4000–4600
Hydrogen fuel-cell vehicles	0.00	1.79	n.a.	3500–4500
Industry	**3.00**	**5.68**	**700–900**	**1400–1700**
CCS industry, H_2 and fuel transformation	200	4.28	700–900	1400–1700
Industrial motor systems	1.00	1.40	n.a.	n.a.
Total	**27.14**	**41.57**	**4480–5370**	**13200–15810**

BIGCC – Biomass Integrated Gasification with Combined Cycle
IGCC – Integrated Gasification fuel-cell Combined Cycle
USCSC – Ultra Supercritical Steam Cycle

(Source: *ETP* 2008, p.132)

17.3 INNOVATION SYSTEMS IN TECHNOLOGY DEVELOPMENT

Technology development goes hand in hand with the innovation process. The framework conditions necessary for the successful prosecution of innovation are: macroeconomic stability, education and skills development, favourable business climate, protection of Intellectual Property (IP) rights, etc. Innovation process is not necessarily linear, and may not proceed smoothly – there can be many impediments en route. RD&D is only part of the innovation scheme – it needs to be adapted depending upon the feedback from the markets and technology users.

The Schematic Working of the Innovation System is depicted in Fig. 17.1 (Source: *ETP* 2008, p.170).

Many projects do not survive the transition from publicly-funded demonstration to commercial viability. Murphy and Edwards (2003) call it a "Valley of Death". At this point, the investment costs and risks may be very high. Neither the public sector nor the private sector considers it their duty to fund this phase. Neither "technology-push" nor the "market pull" may be sufficiently powerful to smoothen the transition. Many

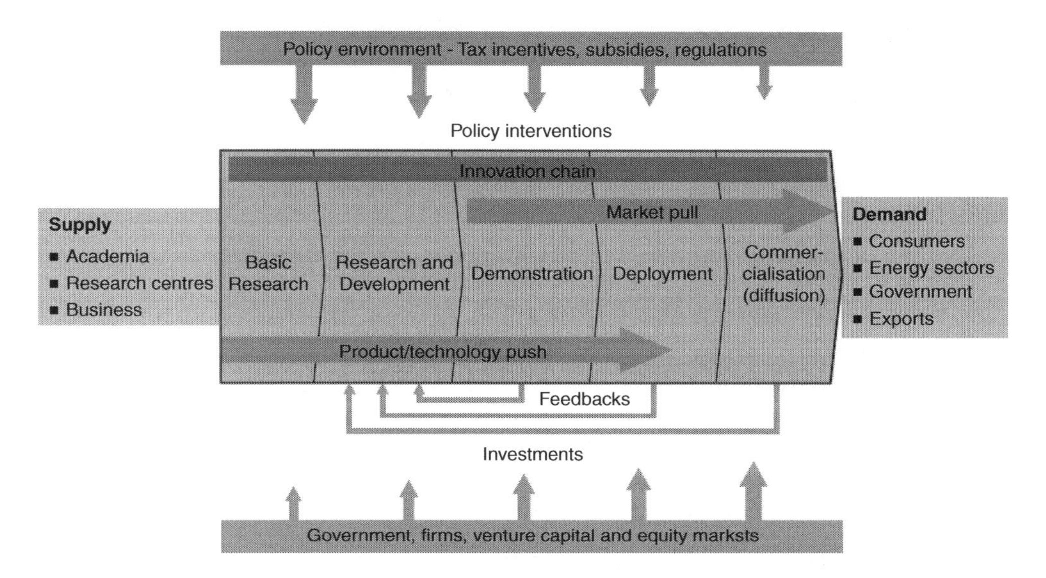

Figure 17.1 Schematic working of the innovation chain
(Source: *Energy Technology Perspectives*, 2008, p. 170, © IEA – OECD)

energy technologies need long lead times and may require extensive applied research and testing, before an invention gets commercialized.

There are a number of ways in which governments could help in navigating the "Valley of death": economic incentives such as tax credits, production subsidies, and guaranteed procurement, and knowledge incentives, such as, codification and diffusion of generated technical knowledge. Public-private research consortia can play an effective role in technology transfer and commercialization. In some countries, technology parks are established to facilitate technology transfer. In these parks, governments give support to individuals or groups of scientists and technologists to perform basic research and applied R&D When some thing viable emerges from this effort, the same group is helped to commercialize the invention. Governments can also create demand for new energy technologies through the promulgation of regulatory requirements. This would induce the supply side to respond to the new regulations.

Some times the same RD&D activities are performed simultaneously in different countries. This redundancy can be avoided by the countries pooling their energy RD&D budgets to perform pre-competitive research that will benefit all. Individual industry players can then draw on this common pool of R&D knowledge to build their own enterprises. Such a process strengthens technology deployment – for instance, the technology strengths of an industrialized country can be combined with the lower manufacturing costs in a developing country.

The private sector investment in RD&D is constrained by the following market realities (Stiglitz and Wallsten, 1999):

- A private company has no incentive in pursuing RD&D which brings about society-wide benefits, since the benefits of RD&D do not uniquely accrue to it, but

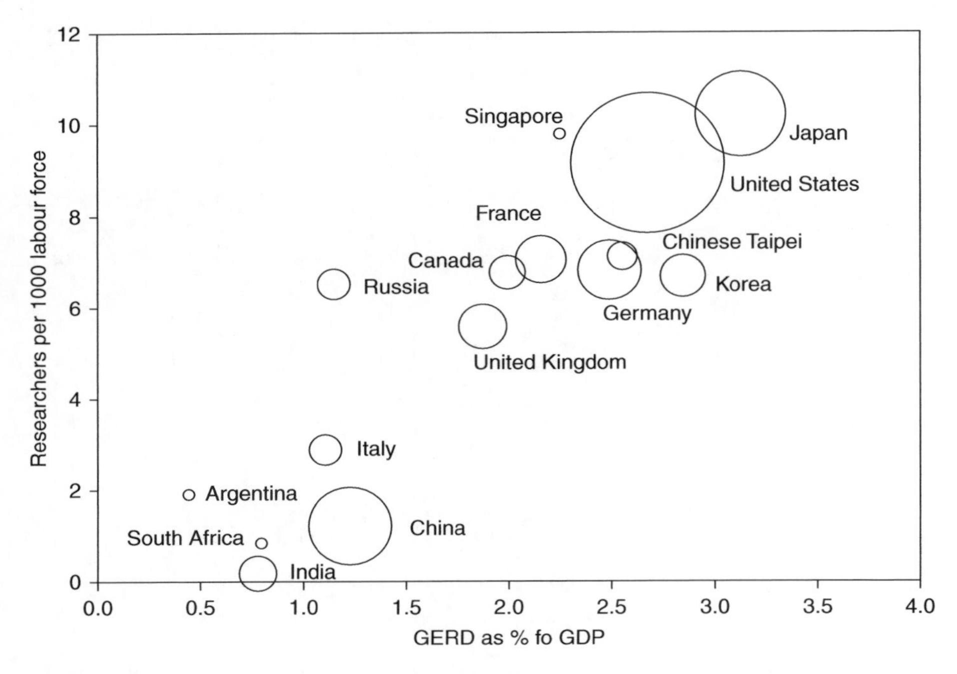

Figure 17.2 Relationship between the government expenditure and the number of researchers (Source: *Energy Technology Perspectives*, 2008, p. 172, © IEA – OECD)

could be made use of by several other companies. For instance, RD&D in regard to artificial photosynthesis (which involves the production of hydrogen, and zero CO_2), has been estimated to cost USD 20–30 billion. Despite the profound importance of this research, no private company would undertake this research. Such research is best sponsored by the governments.

- Private sector would not be interested in innovations that do not bring financial benefits to the company concerned, even though they may bring great environmental benefits to the whole community (for instance, clean air).
- As the performance of companies is judged by the shareholders on a short-term basis, the private sector generally goes in for RD&D that can bring benefits quickly. Thus, areas of RD&D that by their very nature necessarily takes a long time to come to fruition, tend to be ignored by the private sector.

Basic research for achieving public good, (say, carbon dioxide abatement technologies) is best undertaken by public agencies. Governments may directly support the research by private agencies nearer-to-market level, from where the private sector would find it attractive to take it over wholly on their own. Alternately, governments may design a framework to value the public benefits of research funded by private companies.

A number of OECD countries are investing in short-term RD&D in clean coal, energy efficiency and biofuels sectors. Energy RD&D budgets as a percentage of GDP is 0.08% in the case of Japan, and 0.03% in other OECD countries. There is greater

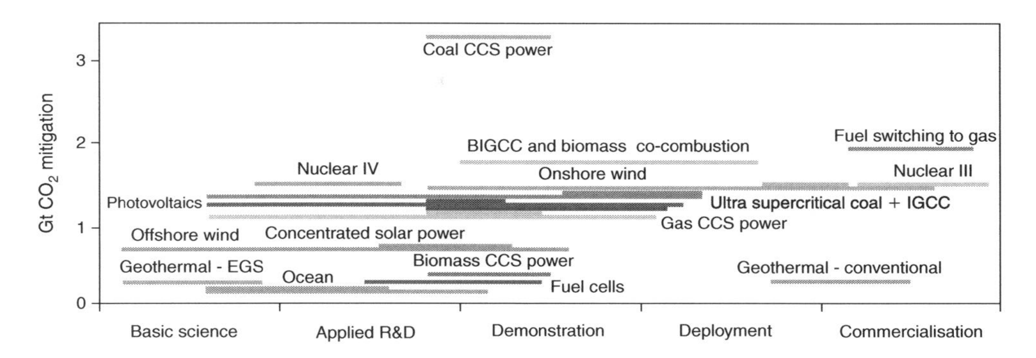

Figure 17.3 CO_2 saving achieved by technology clusters
(Source: *Energy Technology Perspectives*, 2008, p. 181, © IEA – OECD)

emphasis in science and innovation in rapidly industrializing non-OECD countries, like China and India. Good part of energy RD&D in these countries is devoted to adapting and improving technology from the OECD countries. It is, however, a matter of time before they will be able to undertake more sophisticated RD&D activities on their own.

Fig. 17.2 (source: *ETP* 2008, p. 172) is a plot of the relationship between the Government Expenditure in R&D (GERD) as a% of GDP versus the number of researchers per thousand people in the labour force. The circles represent the size of the expenditure in USD billion. The Figure shows that the developing countries need to strengthen their RD&D investments in hardware and human capital.

How the renewable technologies need to be developed in the near-term (i.e. 10–15 years) is shown in Fig. 17.3 (source: *ETP* 2008, p. 181). The X-axis shows the stages in technology development (Basic Science, Applied R&D, Demonstration, Deployment, Commercialization) while the Y-axis shows the CO_2 saving achieved by each technology cluster.

17.4 RESEARCH, DEVELOPMENT AND DEMONSTRATION IN THE ENERGY SECTOR

The decline of RD&D investment in the public sector relative to 1970s and early 1980s is attributed to three factors.

- The oil price shock of 1970s led to a sharp increase in RD&D to find alternatives to oil. When the oil prices collapsed in 1980s and remained low in 1990s, there was no incentive to find alternative fuels.
- As a consequence of the operation of market forces, natural gas technologies for the generation of power and heat got into stream in a big way. No new RD&D investment was needed for this, as it was based on RD&D already made.
- Public concerns over reactor safety, nuclear wastes and nuclear proliferation coupled with cost overruns and production relays, led to reduction in RD&D in nuclear power.

As RD&D is proprietary in the private sector, the actual amounts involved are rarely declared. Some large firms, like Siemens, General Electric and Toshiba, do undertake a vast, multidisciplinary RD&D effort which includes energy RD&D, but the precise breakdown of expenditure for energy RD&D is not known. The RD&D intensity of the total RD&D in the case of power sector is estimated to be 0.5%. This is much less than RD&D intensity of automobile industry (3.3%), electronics industry (8%) and pharmaceutical industry (15%).

Private sector spending in RD&D in energy-related sectors (USD 40–60 billion/year) is four to six times more than governmental spending in the sectors.

Ongoing RD&D activities through IEA and other auspices in regard to key technologies are summarized as follows, technology-wise (source: *ETP* 2008, p. 193–196). These activities listed below include the design, testing, demonstration, evaluation and application of various technological devices.

17.4.1 Renewable Energy Sector

Biomass: black liquor – Production of synthesis gas that can subsequently be converted to a variety of motor fuels; and integration into modern, eco-cyclic, kraft pulp bio-refineries.

Biomass – IGCC – Demonstration of IGCC (Integrated gasification combined cycle) units.

CSP (Concentrated Solar Power) – Solar-driven thermochemical and photochemical processes for production of energy carriers; evaluation of technical and economic viability of emerging solar thermal technologies and their validation.

Geothermal – EGS (Enhanced geothermal systems) – reservoir evaluation and scenario simulation for sustainable technologies; field studies of reservoir performance.

Hydro-small scale – Address technological, organizational and regulatory issues.

Ocean: tidal and wave: Wave and tidal energy converters; develop international standards for wave and tidal energy technology; deployment and commercialization of ocean waves and marine current systems.

Onshore and offshore wind: Benefits, markets and policy instruments related to technology development and deployment; design and operation of power systems with large amounts of wind power; integration of wind and hydropower systems; offshore wind technology development; developing dynamic models of wind farms for power system studies.

Photovoltaics (PV): Design and operational performance of PV power systems building integrated PV, hybrid systems; mini grids, very large scale PV.

17.4.2 Fossil Fuel Power

CCS coal: advanced steam cycle with oxyfueling: Demonstrate the technoeconomic feasibility of CO_2 capture technology for power plants.

Clean coal technologies: Science of coal combustion, conversion and utilization to co-firing and bio-co-processing. Technical, economic and environmental performance of clean coal technologies; opportunities for cross-border technology transfer.

Fuel cells: Development of molten-carbonate, solid oxide and polymer-electrolyte fuel cell systems.

17.4.3 Electricity System

Energy storage: Development of underground thermal storage systems, in the buildings, industry and agriculture sectors; role of electrical storage technologies in optimizing electricity supply and use; role of phase-change materials and thermochemical reactions in the energy systems; ways of facilitating adoption of energy systems.

Transmission and Distribution (T&D) systems: New operating procedures, architectures, methodologies and technologies in electricity T&D networks; Network renewal, integration of renewables, network resilience and distributed generated systems integration. Smart grids.

17.4.4 Industry – Process Innovations

Black-liquor gasification: Black-liquor gasification research on refractory and metallic materials; gas clean-up and black liquor delivery systems; fluid dynamics study of black liquor gasifiers.

Separation technologies, including drying and membrane systems: Advanced separation systems and retrofits; new drying and membrane technologies.

17.4.5 Buildings and Appliances

Heating and cooling: Energy saving potential and environmental benefits of heat pumping technologies; thermally-driven heat pumps; retrofit heat pumps for buildings; ground-source heat pumps.

Passive housing: Solar heating and cooling buildings and community systems; Sustainable solar housing with passive solar design; improved daylighting and natural cooling and solar/glare control; cost optimization of the mix concepts.

Solar heating and cooling: Storage concepts for thermal solar systems; solar heat for industrial processes, and polymeric materials for solar thermal applications.

17.4.6 Transport – Vehicles

Hydrogen fuel cell vehicles: Fuel cells for vehicles, including use as auxiliary power units (APU) and hybridization of the fuel cells with on-board energy devices like batteries and super-capacitors.

Plug-in/electric vehicles: Electrochemical power sources and energy storage systems (batteries, fuel cells, and supercapacitors) for electric and hybrid vehicles.

Hybrid vehicles.

17.4.7 Transport – Fuels

Second-generation liquid biofuels: Liquid biofuels from biomass, especially bio-based ethanol and bio-diesel.

Advanced motor fuels: Synthetic vehicle fuels made by Fischer-Tropsch process, including biomass as possible raw material; production techniques; emissions from production and use, engine performance.

17.4.8 Cross-cutting

Bio-refinery: Co-production of fuels, chemicals, power and materials from biomass for transport, chemical and agricultural sectors; Potential of bio-refineries. Black liquor gasification.

International networks on CCS – Carbon Sequestration Leadership Forum (CSLF); International Network on CO_2 capture; International Network on Biofixation of CO_2 and Greenhouse Gas Abatement with microalgae; Oxy-fuel combustion network; Well Bore Integrity Network.

Nuclear – Generation IV: Advanced R&D.

Hydrogen – International Partnership for Hydrogen Economy (IPHE): International R&D and commercial utilization of hydrogen and fuel-cell technologies; Development of common technical codes and standards for hydrogen economy.

The above list is not exhaustive. Other important technologies with great potential are: thin-film technologies, third generation photovoltaic technologies, deep offshore technologies; plastic recycling and energy recovery, and feedstock substitution such as biopolymer, monomers from biomass and naphtha products from biomass through FT, etc. (ETP, 2008, p. 196).

Organizations, such as the International Energy Agency (IEA) have been promoting international collaboration with focus on technology innovation, through international groupings, such as G-8, G.20, Asian-Pacific Partnership, EU Research Framework, Nordic Energy Research, etc.

Every country needs to identify its energy goals and priorities in the context of their biophysical and socioeconomic situation, map institutions and their RD&D activities in the area of energy technologies, and identify gaps that need to be bridged. By exchanging such studies with other countries, it would be possible to identify the most viable mode of cross-border technology transfer to achieve low-carbon, job-led economic growth.

17.5 RESEARCH, DEVELOPMENT AND DEMONSTRATION POLICIES

Mitigation of climate change is a public good. The goal of the RD&D policy should be to design ways and means by which the value of that public good is built into commercial and innovation systems. The role of the government is most effective when it is able to combine " *supply-push* (i.e. focus on RD&D and technology standards) with *Demand-pull* (i.e. focus on influencing the market through economic incentives such as regulation, taxation or guaranteed purchase agreements)." (*ETP* 2008. p. 184).

Governments and private sector have roles to play in all the five phases of the innovation chain. Generally, governments are expected to play a greater role in the early part of the chain (such as basic science) though some large industrial houses have extensive basic research programmes. Down the line, private sector alone would be involved in the last phase, i.e. commercialization. Roles overlap in the case of Applied R&D, Demonstration and Deployment. Sometimes there may be difficult technical problems that markets fail to address. Through specifying technology standards and participating in full-scale "in the field" demonstration projects, governments may induce private companies to achieve higher technology performance.

Governments could play an effective role in the progress of innovation chain in three ways: (i) direct funding of basic research in universities and national laboratories, (ii) granting protection under IP rights, to enable the innovators to make money from their findings, and (iii) market measures that can indirectly stimulate private sector investment.

While it is generally agreed that the present level of RD&D spending is inadequate, estimates of what constitutes the "right-level" of funding, vary widely – from two to ten times the present level, depending on the methodology adopted. Stern et al (2006) recommended the doubling of the present level of spending. Anderson (2006) suggests that the necessary investment may be estimated as the difference between the average incremental costs of investment in new technologies and that of mature technologies. Another approach is based on the cost of insurance against four types of energy-related risks, namely, oil price shocks, power supply disruptions, local air pollution and climate change.

Innovation becomes progressively more expensive as we move along the chain, from basic research to demonstration. RD&D costs are much lower than deployment costs, and are justified as contributing to cost reduction measures. Also, synergies need to be explored whereby publicly funded RD&D can stimulate privately funded RD&D.

By its very nature, innovation is a risky business. There is no guarantee that higher level of RD&D spending will automatically lead to higher success rates in commercialization of technologies. Some RD&D projects may yield "above-cost" returns, while some may not give any returns. Under the circumstances, a practical approach will be to develop a portfolio of projects in order to hedge risks. Market forces tend to favour least-cost, short-term options. This may result in ignoring some long-term options, which have the potential to deliver huge cost savings and other benefits later. Technologies should be chosen on the basis of their projected outcomes. A portfolio approach needs to be followed in the preparation of technology roadmaps. In sum, an effective portfolio formation and management system is critical for the success of the innovation chain.

Basic research and Applied R&D should proceed in tandem, feeding each other. There has been instances when basic science is available but it has not been commercialized. There are many barriers between the basic research and applied R&D communities in terms of differences in goals, time horizons, organizational structures, intellectual property rights, and so on. Technology transfer from research laboratories to private sector to make useful products, has not worked well. Demonstration projects are useful in transferring cutting edge technologies from the laboratory to the market. Pilot projects enable the identification of real-world performance issues in terms of costs, manpower, market conditions, etc.

Chapter 18

Deployment and role of technology learning

U. Aswathanarayana

18.1 INTRODUCTION

International Energy Agency (IEA), Paris, holds the view that by effectively deploying the energy technologies that are already available or are under development, it is possible for CO_2 emissions in 2050 to be stabilized at today's level. Under the ACT scenario of IEA, deployment costs are estimated at USD 3.2 trillion for power generation, and USD 1.6 trillion for buildings, transport and industry technologies, totaling USD 4.8 trillion. The total learning investment, i.e. additional costs exceeding the presently used fossil fuel technologies, is estimated to be USD 2.8 trillion. The application USD 50/t CO_2 price as economic incentive to reduce CO_2 emissions, would raise the cost of incumbent technologies and bring down the financing needed to bring clean technologies to market (deployment costs) to USD 2.3 trillion. Both OECD countries, and several non-OECD countries, such as China and India, are making new investments in power generation, employing cleaner and more efficient power generation technologies. Investment decisions taken over the next decade will determine the rate of CO_2 emissions during the next 40 to 50 years. Governments have a critical role to play in the diffusion of energy-efficient technologies by enforcing stringent codes and standards.

After Research and Development and Demonstration (RD&D), a technology has to be deployed in the market. Technology learning occurs in the process of deployment, leading to economies of scale, product improvements and cost reductions. This may not go on smoothly, and there may be many impediments in the path.

The stages in the technology deployment are schematically shown as follows (ETP 2008, p. 202):

"**R&D** R&D seeks to overcome technical barriers and reduce costs. Commercial outcomes are highly uncertain, especially in the early stages.

↓

Demonstration	The technology is demonstrated in practice. Costs are high. External (including government) funding may be needed to finance part or all of the costs of demonstration.
↓	
Deployment	Successful technical operation, but possibly in need of support to overcome cost or non-cost barriers. With increasing deployment, technology learning will progressively decrease costs.
↓	
Commercialization (diffusion)	The technology is cost competitive in some or all markets, either on its own terms, or where necessary, supported by government intervention (e.g. to value externalities, such as costs of pollution)."

On the demand side, economically-viable technologies which are capable of delivering two-thirds of the needed reduction in CO_2 emissions, already exist. The commercialization of technologies that are needed for the abatement of the remaining one-third of CO_2 emissions, cannot take place without the support of the government. Though the technologies are cost effective, they have not penetrated the market, as the consumers tend to take a short-term view of costs rather than a long-term life-cycle costs. For example, a filament lamp is cheaper to buy than a fluorescent lamp to start with, but a fluorescent lamp is cheaper in terms of the life-cycle costs because of its lower electricity consumption. Governments may promote the penetration of such technologies through appropriate regulations.

On the supply side, CCS (Carbon dioxide Capture and Storage) and supercritical and ultra-supercritical technologies are expensive. They can become competitive only when a value is attached to the reduction of CO_2 emissions (say, \$ 50/t CO_2). Governments have to identify suitable technological mechanisms and design and implement appropriate policy instruments to remove market and non-market barriers to diffusion.

18.2 TECHNOLOGY LEARNING CURVES

Generally, new energy technologies tend to be more expensive than incumbent technologies. *Technology learning* is the process by which the costs of the new technologies are brought down, through reduction in production costs and improved technical performance. The rate at which consumers switch from old to new technologies will depend upon the relative costs and the value that the consumers attach to the long-term life-cycle costs.

When the private industry finds that a given technological process has a good market potential, they may perform appropriate R&D to make it marketable ("learning-by-searching"), or they may improve the manufacturing process ("learning-by-doing), or the product may be modified on the basis of the feedback from the consumers ("learning-by-using"). The more a technology is adapted, the more will be the improvement in technology.

Technology learning has an important role to play in R&D and investment decisions in respect of emerging technologies. Technology learning curves may be made use of to

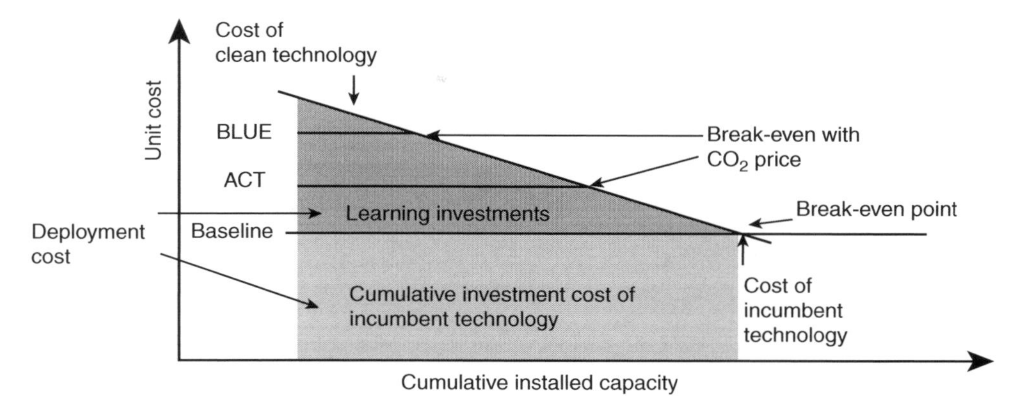

Figure 18.1 Learning curves, deployment costs and learning investments
(Source: *Energy Technology Perspectives*, 2008, p. 204, © IEA – OECD)

estimate the deployment and diffusion costs of new technologies. Governments could make use of this information for decision-making in regard to technology and policy options about new energy systems.

As production doubles, the investment costs decrease. Based on this relationship, it is possible to estimate the deployment costs of the new technologies. In the graph between the cumulative installed capacity and the deployment cost per unit, the blue line (learning curve) depicts the reduction in the cost of new technology as the cumulative capacity increases. The grey line represents the cost of the incumbent fossil fuel technology. The break-even point occurs when the cost of clean (new) technology equals the cost of the incumbent fossil fuel technology. (Fig. 18.1 Schematic representation of learning curves, deployment costs and the learning investments. Source: *Energy Technology Perspectives*, 2008, p. 204).

Deployment costs for making the new technology competitive, are the sum total of incumbent technology costs (yellow rectangle) and the additional costs needed for the new technology to reach the break-even point (orange triangle).

In Fig. 18.1, the line representing ACT map scenario is indicative of the carbon prices of USD 50/t CO_2, and the line representing BLUE map scenario is indicative of the carbon prices at USD 200/t CO_2. Thus, the higher the carbon penalty, the higher would be the cost of the incumbent fossil fuel technology, and the lower would be the learning costs.

Though the learning curves have been constructed for a number of supply-side technologies, demand-side technologies also figure in the learning curves.

The limitations of the learning curves need to be kept in mind when using them to make investment decisions:

- The learning curves are based on price, rather than cost data.
- The factors that will drive the future cost reductions may be different from those of the past.
- The cost of bringing energy-efficient appliances to the market should take into account not only the bottom-up engineering models (which tend to overestimate

Table 18.1 Gives the observed training rates for various electricity supply technologies (the data mostly refers to OECD countries).

Technology	Period	Learning rate (%)	Performance measure
Nuclear	1975–1993	5.8	Electricity production cost (USD/kWh)
Onshore wind	1982–1997	8	Price of the wind turbine (USD/kW)
	1980–1995	18	Electricity production cost (USD / kWh)
Offshore wind	1991–2006	3	Installation cost of wind farms (USD/kW)
Photovoltaics (PV)	1976–1996	21	Price of PV module (USD/W peak)
	1992–2001	22	Price of balance of system costs
Biomass	1980–1995	15	Electricity production cost (USD /kWh)
Combined heat and power (CHP)	1990–2002	9	Electricity production cost (USD/kWh)
CO_2 capture and storage (CCS)		3–5	Electricity production cost (USD/kWh)

(Source: *Energy Technology Perspectives*, 2008, p. 205)

costs as they are based on the higher costs of more efficient components), but also the impact of "learning-by-doing" which tend to reduce the costs.

- Most technologies spill over national boundaries, and hence global learning rates would be more meaningful. Where learning occurs locally (for instance, photovoltaic installations in tropical countries), national learning costs would be more relevant.
- Learning curves may be affected by changes in technology regimes resulting from government regulations, and changes in the design of devices. The learning curve rate may be affected depending upon the starting year from which data has been collected.
- Learning curve rates are also affected by supply-chain effects, such as, shortage of silicon in PV industry, steel for making wind turbines, and reactor vessels in the nuclear industry. This led to innovations, such as Cd-Te/thin-film technologies in PV industry, and 10 MW wind power generators using blades of light-weight materials, and avoiding gear boxes, in the case of wind power installations.

In sum, it is important to remember that the learning curves are not set in stone, but are subject to change as the processes underlying them, change.

18.3 COMMERCIALIZATION OF POWER GENERATION TECHNOLOGIES

Modeling technology deployment costs on the basis of learning rates is not easy – if a low pessimistic learning rate is assumed for a technology, it may be squeezed out by technologies with higher learning rate; if a highly optimistic learning rate is assumed, it may lead to unrealistically high estimates of potential cost reductions.

The International Energy Agency (IEA) camp up with estimated commercialization costs of power generation technologies, based on reasonable learning rates (Table 18.2)

Table 18.2 Applied learning rates for power generation technologies

1	*2*	*3*	*4*	*5*
Onshore wind	1200	7	2020–2025	900
Offshore wind	2600	9	2030–2035	1600
Photovoltaics (PV)	5500	18	2030–2035	1900
Concentrated Solar Thermal	4500	10	Not commercial	1500
BIG/GC*	2500 (2010E)	5	Not commercial	2000
GCC*	1800	3	2030–2035	1400
CCS*	750 (2010E)	3	Post-2050	600
Nuclear III+	2600 (2010 E)	3	2025	2100
Nuclear IV	2500 (2030E)	5	Pst-2050	2000

1. Power Generation Technology; 2. Current investment cost (USD/kW); 3. Learning rate (%); 4. Estimated Commercialization under ACT map; 5. Current target to Commercialization (USD/kW); * BIG/CC – Biomass integrated gasifier/combined cycle; * IGCC – Integrated Gasification Combined Cycle; * CCS – CO_2 capture and storage (CCS)

(Source: ETP, 2008, p. 207)

Table 18.3 Applied learning rates for building, industry and transport technologies

1	*2*	*3*	*4*	*5*	*6*
Fuel cell vehicles	USD/kW	FCV drive system cost	750	22	50
Hybrid vehicles	Car	ICE + electric + battery	3000	20	1500
Lignocellulosic ethanol	USD/Litre	Fuel cost	0.8	10	0.5
FT-biodiesel	USD/litre	Fuel cost	1	10	0.5
Plug-in vehicles	Car	Batteries for plug-ins	9000	20	2000
Gothermal heat pumps	USD/system	Heat pump + installation	15,000	15	7000
Solar heating & cooling	USD/m^2	Panel	630	10	450
Feed stock substitution	Ethylene		1300	10	650
CCS Blast furnace	USD/t CO_2	CCS cost*	150	5	50
CCS Cement kilns	USD/t CO_2	CCS cost*	200	5	75
CCS Black liquor IGCC	USD/kW	Production cost	1600	5	1200

1. Technology; 2. Unit; 3. Boundary; 4. Current cost (USD); 5. Learning rate (%); 6. Cost target to reach commercialization (USD); * cost per tonne of CO_2 captured; FCV = Fuel Cell Vehicle; ICE = Internal Combustion Engine; IGCC = Integrated Gasification Combined Cycle

(Source: ETP 2008, p. 208)

IEA used a discount rate of 10%, and import fuel cost of USD 6.5–7/GJ in the above calculations.

Some of the above technologies are relatively new, and data sets to compute the learning curve rate are not available. In such cases, IEA used its judgment to arrive at reasonable rates.

18.4 DEPLOYMENT COSTS

Though the CO_2 incentive is not in effect yet, it is useful to project what its impact would be on the deployment costs when it is implemented. The carbon incentive would raise the cost of the incumbent fossil technology and would make the new technology competitive at a lower level of deployment. For instance, a USD 50/t CO_2 incentive would lead to 63% reduction in deployment costs for cleaner energy technologies during 2005 to 2050, for buildings, transport and industry (from USD 1.6 trillion to USD 0.6 trillion), and 45% reduction for power generation (from USD 3.2 trillion to USD 1.8 trillion). A USD 200/t CO_2 incentive under the BLUE map scenario has not been analyzed by IEA in detail as it is highly uncertain whether it would be possible to implement it.

It would be instructive to estimate the breakdown of the deployment costs for power generation for Baseline, ACT and BLUE map scenarios for the periods, 2005 to 2030 and from 2030 to 2050. A significantly higher investments are needed for wind, solar thermal, nuclear Generation III and Generation IV and CCS technologies, for ACT scenarios than for Baseline scenarios. The difference between ACT and BLUE map scenarios is minor, and is attributed to higher investment costs for tidal and geothermal technologies under BLUE map scenario.

On the Demand side, hybrid vehicles and solar heating account for the largest share of deployment costs in 2005–2030 period, while the CCS industry is expected to dominate the 2030–2050 period.

18.5 REGIONAL DEPLOYMENT FOR KEY POWER GENERATION TECHNOLOGIES

As should be expected, the projected rate of diffusion of new technologies varies from country to country, depending upon the present position of diffusion and capacity for technology exploitation. The key players are expected to be USA and China.

Onshore wind: Electricity from onshore wind is already competitive with fossil fuel energy at selected sites. It will be competitive globally by about 2020, when the cumulative global capacity reaches 650 GW. Western Europe currently dominates the onshore wind. USA and China will pick up rapidly after 2020. USA is expected to reach a capacity of 200 GW by 2025. China will reach onshore wind power of 250 GW by 2040.

Table 18.4 Regional deployment of power generation technologies

	Wind		Photovoltaics		CCS*		Nuclear		
	2005	2030	2005	2035	2030	2050	2005	2020	2050
OECD North America	13%	24%	27%	25%	35%	25%	34%	31%	27%
OECD Europe	69%	34%	19.5%	25%	35%	16%	32%	25%	15%
OECD Pacific	2%	10%	51.7%	30%	10%	5%	17%	17%	14%
China	3%	21%	0.0%	10%	12%	33%	2%	8%	23%
India	5%	4%	0.2%	5%	3%	10%	1%	3%	7%
Others	6%	7%	1.7%	5%	5%	11%	14%	15%	14%

*CCS – Carbon dioxide Capture and Storage

(Source: ETP 2008, p. 212)

Offshore wind: Western Europe currently accounts for 93% of offshore wind installations in the world. This technology is expected to reach commercialization between 2035 and 2040, when it is expected to reach 250 GW. High costs of offshore wind are a barrier for its spreading.

Photovoltaics (PV): Japan leads the world in PV technology. The PV capacity of Japan is 2.8 GW, which is 47% of the global capacity. Western Europe and USA are the other major centres. It is expected that during 2030–2040, the costs of deployment of PV will become competitive. By 2045, USA will account for 50% of the global capacity of 545 GW.

CCS: A carbon incentive of USD 50/t CO_2 is needed to facilitate the widespread adoption of CCS. Under the ACT Map scenario, CCS deployment is expected to begin in 2020 when USA will have the largest share of CCS deployment. By 2050, China will dominate the CCS field globally, with significant capacities in Canada and India.

Nuclear: Significant deployment of Generation III+ and Generation IV nuclear technologies is expected to take place in Canada and USA, China and India, Russia and western Europe and Japan. High investment costs, concerns about reactor safety, disposal of nuclear wastes and nuclear proliferation, scarcity of highly skilled manpower, are impeding the growth of nuclear power. IEA estimates that Generation III+ technologies will continue to be deployed until 2020 to 2030. After 2030, the focus will be on Generation IV technologies.

18.6 BARRIERS TO TECHNOLOGY DIFFUSION

ETP 2008, p. 215, elucidated different issues involved in technology diffusion.

The rate of technology diffusion depends upon the following market characteristics for individual products: (i) rate of growth of the market, and the rate at which the old capital stock is phased out, (ii) the rate at which new technology can become operational, (iii) the availability of a supporting infrastructure, and (iv) the viability and competitiveness of alternative technologies. Other factors that have a bearing on the rate of diffusion are: government policy in phasing out of constraining standards and regulations, and introduction of new technologies, availability of skilled personnel to produce, install and maintain new equipment, ability of the existing suppliers to market new equipment, dissemination to the consumers of concerned information, and incentives for buying, of new equipment, and extent of compliance with regulations and standards.

Rapid diffusion of technology needs the removal of the following barriers: (i) Investors are not induced to invest due to the non-availability of clear and persuasive information about a product, (ii) Transaction costs (i.e. indirect costs of a decision to purchase and use equipment) are high, (iii) Buyer perceives a risk higher than it actually is, (iv) Costs of alternative technologies are not correctly estimated, and market access to funds is difficult, (v) High sunk costs, and tax rules that favour long depreciation periods, (vi) Excessive/inefficient regulation which does not keep pace with emerging situation, (vii) Inadequate capacity to introduce and manage new technology, and (viii) Non-realisation of the benefits of economy of scale and technology learning.

Technology uptake is faster in rapidly growing markets, such as those of China and India. Technology diffusion is higher for products with shorter life-cycle.

The service life (in years) of important energy-consuming capital goods are: Household appliances: 8–12; automobiles: 10–20; industrial machinery: 10–70; Aircraft: 30–40; Electricity generators: 50–70; Commercial/industrial buildings: 40–80; Residential buildings: 60–100.

Improvement in energy efficiency is an effective pathway to reduce CO_2 emissions.

Governments can promote commercialization of energy-efficient technologies through codes and standards, non-binding guidelines, fiscal and financial incentives, etc.

18.7 STRATEGY FOR ACCELERATING DEPLOYMENT

The choice of industry for being deployed is best left to industry. What the government could do is to remove the barriers that may be impeding the commercialization of improved energy technologies, in such a manner the outcomes that the government is seeking are realized. The development of policy by the government should take into consideration the following criteria: (i) attribution of proper cost to the CO_2 impact of individual technologies, (ii) assurance of policy support to clean technologies, with modifications as the situation on the ground changes, (iii) encourage industry to stand on its own, i.e. without direct support from the government – overgenerous support policies may stifle innovation.

The encouragement of governments to Renewable Energy Technologies (RETs) could take many forms, such as, assured support framework to encourage investment; removal of non-economic barriers, such as beauracracy; a time-frame for declining support in due course; and variable support to different RETs depending upon their maturity. The penetration and deployment of RETs need to be reviewed periodically to ensure that less competitive RET options with high potential for development, are not ignored.

It is expected that OECD countries will embark on clean technologies earlier than non-OECD countries. But as the investments get locked in for 40–50 years, fast-growing non-OECD countries could follow suit, aided by the fact that the costs in non-OECD countries are lower. Also, the non-OECD countries could make use of the opportunity to build new industrial infrastructure. Many developing countries are reluctant to impose tough standards and codes as they fear that this may make the local industries to go out of business. This may lead to commercialization of less-efficient technologies.

18.8 INVESTMENT ISSUES

Investment issues are discussed in terms of three scenarios (Baseline, ACT and BLUE):

Baseline scenarios: Total cumulative investment during 2005 to 2050 in the Baseline scenarios is USD 254 trillion. This looks like a huge sum, but it happens to be only 6% of the cumulative GDP over the period. Demand-side investments involving energy-consuming technologies (USD 226 trillion) constitute the bulk of the investment.

Additional investments needed for the ACT and BLUE Map scenarios (over Baseline scenarios) are USD 17 trillion and USD 45 trillion respectively. Demand-side investments in respect of industry, buildings and transport are higher in ACT and BLUE map scenarios than for Baseline scenarios.

The success of the ACT Map scenario, and more so the BLUE Map scenario, is critically dependent upon the cooperation and coordination between the developed and developing countries in bringing into existence an international framework for incentivising low-carbon technologies and energy efficiency. The World Bank has proposed two new funds, the Clean Energy Financing Vehicle (CEFV) and the Clean Energy Support Fund (CESF). The CEFV will blend public and private sources of funding to promote deployment of clean energy technologies. It involves initial capitalization of USD 10 billion, with annual disbursement of USD 2 billion. The CEFV subsidises the reduction of carbon emissions. Eligible projects will be selected on the basis of the lowest subsidy.

When new technologies are introduced either on the supply-side or demand-side, they face numerous barriers before their full commercial deployment. Financial barriers are far the most important, and are summarized below:

- Investors may perceive a higher risk (in terms of operation and maintenance costs, efficiency and economic life) in the case of new technologies relative to mature technologies,
- Higher initial costs of new technologies may deter investors in the case of immature financial markets,
- Information may not be available to make a comparative study of different investment options, particularly in the absence of knowledge of international standards and codes,
- Small investors may be at a disadvantage as it is more cumbersome to prepare customized financial packages for a larger number of small investors, than for a small number of big investors,
- Unregulated markets may not attach proper value to the environmental benefits of clean technologies,
- Parallel investment has to be made for infrastructure to enable a new technology to take off; alternately, investment in new technology may be made in such a way that it is capable of making use of the existing infrastructure to take off,
- Tax systems generally favour low-investment technologies. New clean technologies with their high initial costs will have to bear a higher tax burden, unless this issue is addressed by the government,
- The perception of an asset owner may be different from that of asset user. For instance, the choice of an owner of an apartment tends to be based on the upfront costs of a device, whereas the tenant living in the apartment would prefer a device that has minimal cost for a life-cycle of energy consumption.

It should be obvious that the above barriers are not just financial alone – they are very much influenced by the behaviour and psychology of the consumer, and the commitment of the governments for the reduction of carbon dioxide emissions, and to minimize the adverse environmental impact of energy technologies.

Chapter 19

Energy efficiency and energy taxation

U. Aswathanarayana

19.1 MATRIX OF ECONOMIC EVALUATION MEASURES

The purpose of a company making an investment to produce a product or provide a service, is always the same any where in the world – it is to make money. Table 19.1 provides the matrix of the investment features and decision criteria concerned. Most of the economic measures are valid for most investments. It is therefore better to compute several of the economic measures to serve as a basis for investment decisions.

In the Table, N means not recommended generally, as it may lead to inappropriate conclusions. It may be noted that several cells are blank – a blank cell signifies that the measure is acceptable. R means Recommended. C denotes a measure which is commonly used to evaluate investments of a specific nature. As no two investments and investors are identical in all respects, the matrix constitutes a quick reference to determine whether or not a more thorough investigation is warranted. A simple analogy is the pathological examination of a patient – to determine the nature of the sickness, and whether more detailed tests are necessary.

The limitations of the matrices should be kept in mind. For instance in the investment decisions matrix, TLCC and RR are not listed as Recommended. Yet the two measures have to be taken into account in cases where a given energy service must be secured whatever the price. These measures are not recommended in general simply because benefits or returns are not taken into consideration in such cases.

Cost-effective alternatives are those with the lowest TLCC, RR, LCOE, SPB and DPB; and the highest NPV, IRR, MIRR, B/C and SIR. It is necessary to keep in mind that when comparing alternatives, different measures may not lead to the same answer (for example, simple versus longer payback periods). Some times, an investment may

Table 19.1 Overview of economic measures related to investment decisions

Investment Features	*NPV*	*TLCC*	*RR*	*LCOE*	*IRR*	*MIRR*	*SPB*	*DPB*	*B/C*	*SIR*
Investment after return					N					
Regulated investment			R							
Financing							N	N		R
Risk							C, R	R		
Social costs	C, R								C, R	
Taxes							N	N		
Combinations Of investments										

A blank cell indicates that the measure is acceptable.

R – Recommended; N – Not recommended; C – Commonly used

Investment Decisions	*NPV*	*TLCC*	*RR*	*LCOE*	*IRR*	*MIRR*	*SPB*	*DPB*	*B/C*	*SIR*
Accept/Reject		N	N		C					
Select from Mutually Exclusive alternatives	R	C		N	N	N	N	N	N	N
Ranking (limited budget)				R	C, N	R	N	N	R	R

Economic Measures
NPV – Net Present Value; TLCC – Total Life-Cycle Cost;
LCOE – Levelized Cost of Energy; RR – Revenue Requirements
IRR – Internal Rate of Return; MIRR – Modified Internal Rate of Return
SPB – Simple Payback period; DPB – Discounted Payback Period
B/C – Benefit-to-cost ratio; SIR – Savings-to- Investment ratio

(Source: "*A Manual for the Economic Evaluation of Energy Efficiency and Renewable Energy Technologies*", p. 36)

involve optimization of two linked parameters, say, an air-conditioner and insulation. The most cost-effective alternative will be a combination of air-conditioner size and amount of installation.

The various economic measures are annotated as follows (source: "*A Manual for the Economic Evaluation of Energy Efficiency and Renewable Energy Technologies*", p. 87–96).

Net Present Value (NPV) – The value in the base year (usually the present year) of all the cash flows associated with a project.

Total Life-cycle cost (TLCC) – The present value over the analysis period of all system resultant costs.

Levelized Cost of Energy (LCOE) – The cost per unit of energy that, if held constant through the analysis period, would provide the same net present revenue value as the net present value of the system.

Revenue Requirement (RR) – The amount of money that must be collected from the customers to compensate a utility for all expenditures associated with an investment.

Internal Rate of Return (IRR) – The discount rate required to equate the net present value of the cash flow stream to zero.

Modified Internal Rate of Return (MIRR) – The discount rate required to equate the future value of all returns to the present value of all investments. MIRR takes into account the reinvestment of cash flows.

Simple Payback Period (SPB) – The payback period computed without accounting for the time value of the money.

Discounted Payback Period (DPB) – The payback period computed that accounts for the time value of the money.

Benefit-to-Cost ratio (B/C) – The ratio of the sum of all discounted benefits accrued from an investment to the sum of all associated discounted costs.

Savings-to-investment Ratio (SIR) – The sum of discounted net savings accruing from an investment to the discounted capital costs (plus replacement costs minus salvage costs).

19.2 TOTAL LIFE-CYCLE COST (TLCC)

TLCC analysis is useful to assess the economic viability of alternative projects. TLCCs are the costs incurred by an investor through the ownership of an asset during the life-time of the asset. These costs are then discounted to the base year using the present value methodology. Renewable energy technologies are characterized by two kinds of costs: investment costs and Operation and Maintenance (O&M) costs, including fuel costs.

In the case of public utilities which do not pay taxes to the government, TLCC can be expressed as TLCC = 1 + PVOM, where

I = Initial investment,
PVOM = Present value of all O&M costs, or

$$\text{PVOM} = \sum_{n=1}^{N} \text{O\&M}_n/(1 + d)^n \qquad (19.1)$$

The TLCC analysis can be illustrated with a simple example (quoted from "*A Manual for the Economic Evaluation of Energy Efficiency and Renewable Energy Technologies*", p. 45)

A five-year life-time of the project and a nominal discount rate of 12% are assumed.

Alternative A: An incandescent light bulb (75 W) costing USD 1 is used every night for 6 hrs. round the year. It needs to be replaced every year, and so during a five-year life-time, five bulbs are required. Electricity costs USD 6 cents/kWh. The bulb is purchased at the beginning of each year, and the electricity is paid at the end of each year. Annual Electricity consumption 164.25 kWh, @ USD 6 cents/kWh, costs $9.86/yr. TLCC for Alternative A works out to $39.56.

Alternative B: A fluorescent lamp (40 kW) costing $15, and has a life-time of 5 years, is used for 6 hrs. every night round the year. It need not be replaced, as it could last during the whole life-time of the project. Annual electricity consumption 87.6 kWh @ USD 6 cents/kWh. TLCC for Alternative B works out to $33.95.

The use of a fluorescent lamp thus saves $5.61.

19.3 LEVELIZED COST OF ENERGY (LCOE)

The calculation of levelized cost of energy (LCOE) enables an investor to decide between different forms of energy generation (say, fossil fuels versus renewable resource) by levelizing different scales of operation, investments, and operating time periods. LCOE can also be employed to evaluate the energy efficiency benefit arising out of an investment (say, incandescent light bulb versus fluorescent bulb).

"The LCOE is that cost that, if assigned to every unit of energy produced (or saved) by the system over the analysis period, will equal the TLCC when discounted back to the base year" (source: "A Manual for the Economic Evaluation of Energy Efficiency and Renewable Energy Technologies", p. 47). LCOE would be inapplicable if the alternatives considered are mutually exclusive (say, large investment vs. small investment).

LCOE can be calculated on the basis of TLCC.

$$\text{LCOE} = (\text{TLCC}/\text{Q})(\text{UCRF}) \tag{19.2}$$

where

TLCC = Total Life-Cycle cost,
Q = Annual energy output or saved,
UCRF = Uniform capital recovery factor, which is equal to

$$= \frac{d(1+d)^N}{(1+d)^N - 1}$$

Assuming d = discount rate as 12%, and N = analysis period as 5 years,

$$\text{UCRF} = [0.12(1+0.12)^5]/[(1+0.12)^5 - 1] = 0.277 \tag{19.3}$$

If a project can be assumed to have not only constant output, but also constant O&M and no financing, LCOE can be calculated from the following formula:

$$\text{LCOE} = \frac{\text{I} \times \text{FCR}}{\text{Q}} + \frac{\text{O\&M}}{\text{Q}} \tag{19.4}$$

Where,

LCOE = Levelized cost of energy
I = Investment
FCR = Fixed charge rate, in this case the before-tax revenues FCR
Q = Annual output
O&M = Annual O&M, and the fuel costs for the plant.

If the purpose of the investment is to improve the energy efficiency, it follows that LCOE has to take into account the energy saved. Thus, instead of calculating TLCCs for different energy-consuming systems, the incremental cost and savings attributable to the energy-efficient system is figured out by levelizing the difference in the non-fuel (electricity) life-cycle costs of the two systems.

We can use the same example as given in 19.2.

The life-time of the project is taken to be five years. A nominal discount rate of 12% is assumed.

Alternative A: An incandescent light bulb (75 W) is used every night for 6 hrs. round the year. Annual Electricity consumption is therefore 164.25 kWh. One bulb costing $1 has to be bought at the beginning of each year for 5 years. The non-fuel cost of this alternative is the discounted cost of buying a new bulb at the beginning of each year, that comes to $(1 + 1/1.12 + [1/1.12^2] + [1/1.12^3] ++ [1/1.12^4] = \$4.03$.

Alternative B: A fluorescent lamp (40 kW) costing $15, and has a life-time of 5 years, is used for 6 hrs. every night round the year. It need not be replaced, as it could last during the whole life-time of the project. The non-fuel cost of alternative B is USD 15. Annual electricity consumption 87.6 kWh.

Investment difference = $15 – $4.03 = $10.97
Energy saving between the alternatives = 164.25 kWh–87.6 kWh = 76.65 kWh
Using UCRF of 0.277, the nominal levelized cost of energy saved, is:
$(10.97/76.65) \times 0.277 = \$0.04/\text{kWh}$.

Thus, the nominal levelized cost per unit of energy saved in the case of Alternative B (USD 4 cents/kWh) is cheaper than for Alternative A (USD 6 cents/kWh). In other words, if we use a fluorescent lamp, it would be as if we get electricity at USD 4 cents/kWh, and is therefore more energy efficient. If nominal cost of electricity drops to less than USD 4 cents/kWh, Alternative A will be the most effective cost instrument. Thus a BPL family which gets electricity free from the government, has no incentive to buy a more efficient but more expensive fluorescent lamp.

Fig. 19.1 (source: "*A Manual for the Economic Evaluation of Energy Efficiency and Renewable Energy Technologies*", p. 51) depicts the costs over the lifetime of the investment and the resulting LCOE. Both the parameters are shown in nominal and real (i.e. constant dollar or inflation-adjusted) terms. The cash flow lines for nominal

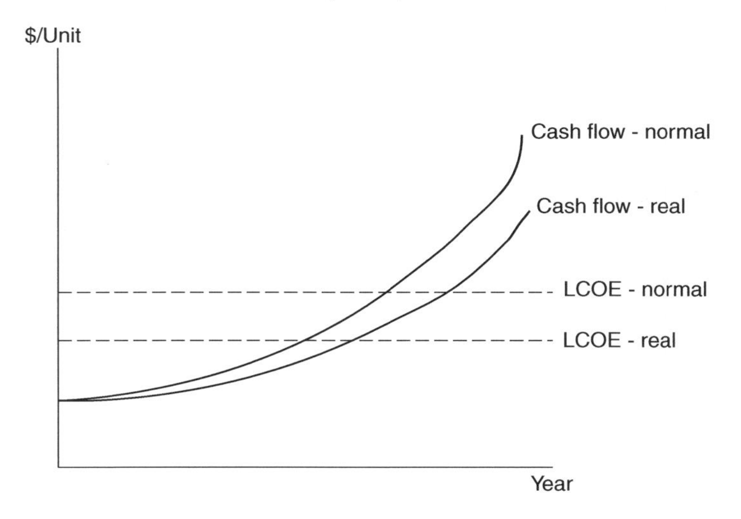

Figure 19.1 Lifetime of the investment and LCOE
(Source: "*A Manual for the Economic Evaluation of the Energy Efficiency and Renewable Energy Technologies*", 2005, p. 51, © University Press of the Pacific)

and real costs are the same in the base year, but the real costs will be less for subsequent years. LCOE (nominal) is higher than LCOE (real). While nominal figures could be used for short-term analysis, the real investment (constant-dollar analysis) would give a clearer picture of actual cost trends. There will be no change in the most efficient option so long as the same method is used.

19.4 ENERGY EFFICIENCY OF RENEWABLE ENERGY SYSTEMS

Apart from the economic measures discussed above, Energy efficiency analysis may require consideration of "system boundaries, optimal sizing, externalities, government investments, backup and hybrid systems, storage, O&M expenses, capacity and energy values, major repairs and replacements, salvage value, unequal lifetimes, retrofits, electricity rates, and programme evaluation" (source: "*A Manual for the Economic Evaluation of Energy Efficiency and Renewable Energy Technologies*", p. 73).

System Boundaries: It may be necessary to extend a system's boundary beyond its direct boundary, for the purpose of evaluating end use markets as well as utility investments. For instance, an electricity grid may involve more than one type of electricity generating system. Pumped storage hydroelectricity can be used to flatten out load variations on the power grid which may be linked to coal-fired plants, nuclear plants or renewable energy power plants. The alternative combinations may be characterized by different time schedules, fuel costs and electricity costs. In such cases, the entire utility system involving extended boundaries, has to be evaluated.

System Sizing: Equipment sizes are determined depending upon a particular technology. For instance, the typical size of a nuclear power plant is 1000 MW, whereas the typical size of biomass power plant is 50 MW. After deciding upon the nature of the power plant, the range of acceptable alternative sizes of the plant are figured out, and their economics are compared. Some times, a backup system is necessary for a particular technology, say, solar technology. The standard methods used in this analysis are the Levelized Cost of Energy (LCOE) and Savings/Investment ratio (SIR).

Externalities: Energy projects have to take into consideration externalities such as air and water pollution, land use, waste disposal, public safety, aesthetics, etc. Examples are: displacement of populations and destruction of fish habitats in the case of hydropower, and noise and visual impacts in the case of wind power. Some of the externalities, such as aesthetics, are notoriously difficult to quantify. Wherever an externality in the form of costs or benefits, is quantifiable, every attempt should be made to do so. A government may seek to penalize a polluter by setting a pollution standard and taxing him if he exceeds that. Alternately, the government may sell pollution permits. Under the circumstances, a company has to decide whether it would be cheaper to modify the process schedule to reduce the pollution within prescribed limits or pay for the pollution. A company may be willing to pay the victim(s) of pollution for the harm/ inconvenience caused to him/ them, but the victim (s) may not be willing to accept payment to incur a cost or forego a benefit.

It is desirable that a sensitivity analysis be made of the measured cost and benefits of the externalities, even though a range of values, rather than firm figures, is available.

Government investments: In the Innovation Chain, Basic Research →Research & Development → Demonstration →Deployment →Commercialization (diffusion),

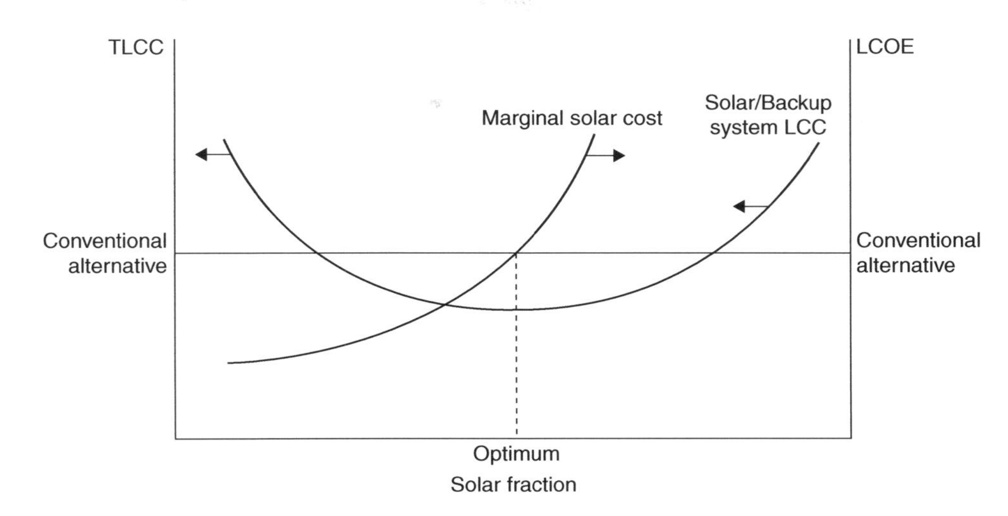

Figure 19.2 Solar fraction optimization
(Source: *"A Manual for the Economic Evaluation of the Energy Efficiency and Renewable Energy Technologies"*, 2005, p. 75, © University Press of the Pacific)

government investments play a major role in the early part of the chain, with the private sector involvement becoming significant in the later part of the chain. The results (say, in photovoltaics) accruing from the government investment in the course of the innovation chain, are available to all. A private investor may make an economic evaluation of these results in order to determine if a particular new technology arising from RD&D, is marketable.

Backups and Hybrid systems: If a renewable energy technology (e.g., solar energy) requires a backup unit (e.g., a fossil fuel system), the cost (capital, operation and maintenance costs, including fuel costs, etc.) of the backup unit, should be included in the analysis. Such a combination of renewable energy and conventional backup system is called a hybrid system.

Fig. 19.2 illustrates the principle of Solar Fraction Optimization (source: "*A Manual for the Economic Evaluation of Energy Efficiency and Renewable Energy Technologies*", p. 75). The Figure shows two curves, one for solar alone and one for solar with backup. The conventional alternative is shown as a straight line. The point of intersection of the solar alone curve with conventional alternative straight line, gives the optimal solar system size which will correspond to minimum life-cycle cost. Also, at this point, the marginal cost per unit of output of the solar energy system equals the marginal cost per unit of the output of the conventional alternative. For solar fractions less than the optimal, the total life-cycle cost (TLCC) of the hybrid system will be higher because of the higher cost of the conventional fuel. For solar fractions higher than the optimal, the increased solar panel cost will make the system's TLCC higher.

Government regulations may sometimes dictate the relative contribution of the two components, in order for the hybrid system to qualify for taxation and other benefits. For instance, Public Utilities Regulatory Policies Act (PURPA) of USA prescribes a

25% limit to the amount of power that could be generated by the fossil fuel, if the hybrid facility is to qualify for federal benefits for renewable energy.

Energy Storage: Energy may be generated and stored during the low-cost, off-peak periods, to be released during high-demand, on-peak periods. There are many ways of storing energy. Pumped storage is a high capacity form of grid energy storage presently available. When there is more generation of electricity than the load available to absorb it (say, during nights), excess generating capacity may be used to pump water to a reservoir at a higher elevation. When the electricity demand is high (as during day time), water is released back into the lower reservoir through the turbine to generate electricity. Thermal energy can be stored in storage systems in buildings, industry and agriculture sectors. Energy can be stored in batteries. Also there can be magnetic storage in superconducting coils.

The high production of solar electricity during summer coincides with the high electricity demand for providing air-conditioning during the daytime.

In the case of renewable energy technologies which are intermittent (such as, wind and solar energy systems), the availability of energy storage will improve the efficiency and economics of a utility system. Storage should not be considered simply as a part of the electricity-generating system – it should be included in the utility system when the economics of a utility system as a whole is evaluated. The economics of different storage technologies may also be compared. The attributes of a storage technology, such as the quantity of electricity stored and discharged, the charge/discharge rate, etc., may be taken into consideration for computing LCOE.

Operation and Maintenance: Operation and maintenance (O&M) costs are of two types: variable costs (e.g. energy) which depend upon the output of a system when it is operating, and fixed costs (e.g. labour) which have to be incurred to keep the system in operable state. For mature technologies, future O&M costs are estimated on the basis of historical performance. For instance, O&M. costs (excluding fuel costs) of fossil fuel plants are projected to be 1–2% of the initial capital cost. The O&M. costs in the case of new renewable technologies, which are in the early stages of technical and market development, are definitely higher than 2%. Some times, companies may have to replace the technology they have been previously using with a more reliable technology which may also be more expensive. Whatever the O & M. costs may be in the first year, it is safe to assume that they will be higher in the coming years, probably rising at the same rate as inflation. This concept is covered in the real-dollar LCOE calculation:

$$\mathrm{LCOE} = \frac{\mathrm{I} \times \mathrm{FCR}}{\mathrm{Q}} + \mathrm{O\&M} \tag{19.5}$$

Where,

LCOE = Levelized cost of energy
I = Investment
FCR = Fixed charge rate, in this case the before-tax revenues FCR
Q = Annual output
O&M = Annual O&M, and the fuel costs for the plant.

Capacity and Energy Value: The value of one unit of energy depends upon when it is available, where it is available and how it is available. A unit of energy has more value if it can be made available when needed by the consumer. Thus energy delivered at peak is more valuable than energy delivered off-peak. Also, reductions in energy use are more valuable if they occur at the time of the peak consumption. The capacity value of an energy system is given by the energy that can be reliably delivered at the time of the peak consumption, whereas the energy value of a system is the total amount of energy delivered over the course of a year.

When an intermittent renewable energy unit like a windmill is combined with a peaking unit such as combustion turbine, and if an analysis of the hybrid system shows it to be the most economic alternative, there is no difficulty in making the choice in favour of the wind mill-turbine unit. Even if the turbine unit alone is found to be cost effective, decision cannot be made in its favour. This is so because the government, as a matter of policy in the context of climate change, is committed to easing out fossil fuel energy generation and promoting renewable energy systems. The turbine unit should therefore be considered as a "necessary evil" in order to make the windmill viable.

Though the availability of wind is generally random, most places have been found to have some time-of-the-day patterns. Such patterns should be taken into account in planning the operational schedule of the backup turbine unit. Since the electricity is supplied from the utility grid, the economic competitiveness of the utility system as a whole needs to be evaluated rather than the evaluation of windmill and combustion turbine unit separately.

Now-a-days, many governments are promoting renewable energy power generation through subsidized loans, guaranteed purchase and other financial instruments. In 1978, the US Government promulgated the Public Utilities Regulatory Policies Act (PURPA) which "requires utilities to purchase power from qualifying facilities (QFs) at a price equal to the specific utility's avoided costs for energy and capacity" ("*A Manual*... p.78). A Qualifying Facility is a power production facility which generates at least 75% of its total energy output from renewable fuels, such as, biomass, waste, geothermal, wind, solar or hydro. A utility may be able to avoid some costs through buying power from a QF. Avoided costs may be in the form of avoided capacity costs (in USD/kW) and avoided energy costs (USD/kWh) or both. A utility has the freedom to negotiate contracts with cogenerators in respect of avoided costs. Avoided costs may include Tansmission and Distribution (T&D) costs, when applicable. T&D benefits may sometimes be large enough to make DSM (Demand Side Management) projects cost-effective.

Major Repairs and Replacements: Every renewable energy system has some components which need to be replaced or repaired. The cost of annual replacements, such as an air filter, should be included in the operating cost estimates. Major repairs may have to be made once or twice during the analysis period, say, at the end of a component's expected life. In such cases, the repair or replacement cost is discounted to its present value and added to the total investment cost, before items such as property taxes, insurance, etc. are added to them. Another approach is to annualize the cost of the replacement and add it to the annual O & M costs. For tax purposes, companies capitalize the repair costs and recover them through depreciation. This approach does not affect a homeowner, who does not depreciate items for tax purposes.

Salvage value: If an investment can be sold or recycled at the end of the analysis period, it is said to have a positive value. On the other hand, if the investment has to be dismantled or destroyed, the salvage value is negative. Generally, salvage value may be considered as the resale value of an investment at the end of an analysis period. For purposes of accounting, salvage value may be treated as a revenue at the end of the evaluation period.

Unequal lifetimes: All the economic measures considered up to this point, assume equal life times for the alternatives considered. Life-cycle cost, required revenues and the internal rate of return are the parameters which are most affected by the length of the lifetime involved. In the case of a long-lived enterprise, investments are summed up over the long period, but the benefits that occur over the long life are not taken into account. In the case of internal rate of return, we do not know what the position would be at the end of the lifetime of a short period investment – for instance, will the same kind of financing and tax depreciation continue to be available as when the investment was first made.

There is, however, ways to compare long-life and short-life investments, based on assuming a number of short-life investments corresponding to the long-life of the instrument, increasing the repair and maintenance costs to cover the longer period, or calculation of the salvage value of the longer term investment.

Another approach is to make use of parameters such as, LCOE and payback, which are independent of the lifetime of the investment.

Retrofits: An economic analysis of retrofits has to address two issues: whether retrofitting is economical, and if so at what point of time would it be most economical to do it. Determining the most appropriate time to do the retrofitting is not an easy task because of the uncertainties involved in predicting the prices of the conventional fuels. A practical way out will be to consider just two alternatives at a time: say, retrofit now or some other base year, and retrofit one year later than the first.

The exercise may be undertaken with the following steps (p.80, "*A Manual* ..."):

1. Assuming the base year to be (say) the present year, compare the economics of the base year with that of next year with and without retrofit. If retrofitting does not lead to any economic benefit, abandon the idea of retrofitting for the present year.
2. Choose another base year, and compare the economics of that year with that of the succeeding year with no retrofit at all. If the next year retrofit is not economical, the base year is optimal.
3. Compare the LCOE of the base year retrofit with that of the next year retrofit. If the base year LCOE is better, that would make it optimal. If not, repeat the exercise with the next year as the base year.

Fig. 19.3 illustrates the Retrofit scenarios (source: "*A Manual for the Economic Evaluation of Energy Efficiency and Renewable Energy Technologies*", p. 81). In case 1, the real conventional O & M costs (including energy costs) are higher than LCOE through out. In case 2, the real conventional O&M costs (including fuel costs) start below the real LCOE of the retrofit, become equal at time t, and rise above it over time. "The optimal retrofit time is the beginning of the year in which the real delivered conventional O&M costs first exceeds the real LCOE of the retrofit". Case 3 refers to retrofit immediately (NPV > 0) or not at all (NPV < 0).

In the above exercise, all costs are assumed to be unchanging over time, but there is always the risk of their changing, thereby modifying the time when the retrofit would be optimal.

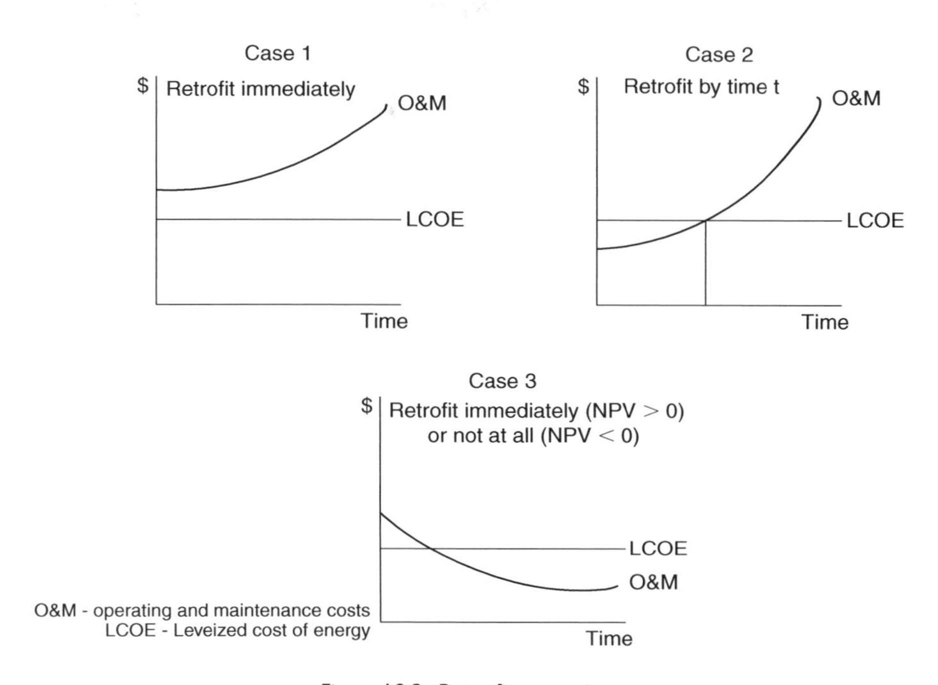

Figure 19.3 Retrofit scenarios

(Source: *"A Manual for the Economic Evaluation of the Energy Efficiency and Renewable Energy Technologies"*, 2005, p. 81, © University Press of the Pacific)

Electricity rates: Electricity rates have a large bearing on the investments and figure prominently in the energy efficiency analysis. Electricity rates (in USD cents/kWh) tend to be different for residential, commercial and industrial customers. Rates may change depending upon the time of use, and size of the load. The relationship between the electricity usage and the energy investment is complicated.

In the developing countries, utility companies tend to be government-owned. While the investments in power generating units and O&M and fuel costs have to be borne by the government willy-nilly, the revenues accruing from the energy supplies tend to be much less than required in order to make the utility economically viable. This is so because there is extensive pilfering of electricity, and populist measures, such as, provision of free electricity to the farmers for pumping irrigation water, and to BPL families in towns.

The aluminium smelter industry is a very heavy consumer of electricity. Consequently, either the industry will have a captive source of cheap power, say, hydropower, or the government provides subsidized electricity at a special cheaper rate in order to promote the industry.

Economic analysis of electricity rates and structure should take into account situations, such as, innovations in technology leading to electricity usage pattern in an industry in terms of change in size and/or timings of peak demands.

19.5 ENERGY TAXATION

Governments obtain revenues by taxing energy. The tax levied on (say) gasoline may be used for a public good, such as the construction of highways. As a part of ameliorating climate change, governments may levy carbon tax to discourage the use of high-carbon fossil fuels. As against this, governments may subsidise generation of energy from renewable energy sources, such as PV, wind and biomass. Differential taxes are some times used as incentives towards a desired purpose. For instance, the European Union levies higher taxes on oil products than on crude oil, to encourage the building of oil refineries in Europe.

Energy and mineral producing states in USA levy a tax called "severance tax" (for "severing" the resource from the earth, i.e. mining). The tax is calculated *ad valorem*. In 1985, when the oil prices were high, severance taxes used to be about 3.3% of the total revenues of the states. Later when the oil prices fell in 1990s, the severance tax revenues came down to 1.3% of the total state revenues. Revenues from oil and gas vary considerably among states. Alaska gets almost half of the revenues from oil and natural gas, whereas Wyoming gets 40% of its revenues from oil, natural gas and coal.

USA has royalty laws for the mineral sector, which are quite different from other countries. The Federal Government owns the energy and mineral resources only in the national forests and the offshore, and may lease it to some other entity to produce a mineral. If a mineral occurs in a private land, the owner of the land is entitled to lease the mining rights.

Around 300 B.C. E, Chanakya wrote the famous "*Arthasastra*", which is a treatise on kingship, state policy, administration, revenues and taxation. According to *Arthsastra*, a farmer has rights only for the top soil in which he grows crops. The mining rights of the sub-soil minerals belong to the king, and hence the mining royalties should go to the treasury. Incidentally, this principle of law holds good to this day in India.

The government share per barrel of oil (1995 figure) varies from about 30% in the case of Ireland to more than 90% in the case of Yemen. Some countries levy lower taxes in order to attract investments in the oil sector, as their reserves are small/oil is of poorer quality/risks are high, etc.

Whatever are the reasons for their adoption, the taxes and subsidies are likely to distort prices and production in energy markets, and decrease efficiency, but they are employed nevertheless as a part of considered policy of the governments.

Though the oil prices in three different markets, say, New York Harbour, Rotterdam and Singapore, are virtually the same, there are vast differences in the gasoline prices at the pump in various countries. This is so because oil products are heavily taxed in almost all countries. Apart from severance taxes, tariffs are levied when energy is traded across borders, and excise taxes are levied when energy is sold to the final consumer.

Conceptually, taxes are no way different from labour and material costs. So all relevant taxes should be included in the economic analysis. As taxes have a profound effect on cash flows, economic analysis of an enterprise should be made on the basis of after-tax cash flows. The point may be illustrated with the comparison of taxation of two utility companies, one using fossil fuel for energy generation and another, the solar power. Depreciation for tax purposes is a major consideration in this analysis. The fossil fuel plant has low capital costs and high fuel costs, and the fuel costs are expended and recovered immediately. As against this, the solar plant has high capital costs and no fuel costs, and the recovery of high capital costs through depreciation take a long time. These considerations apply if private companies are to choose between the two kinds of investments. On the other hand, if the investments for solar power are warranted from societal perspective, government may waive taxes for solar power to make it viable. In such situations, it is better to make the analyses with and without taxes.

Taxes are calculated on the basis of nominal dollars. If real dollars (i.e. constant dollars or inflation-adjusted dollars) are used, the results will be skewed. Also, taxation rates depend upon the type of ownership of a company – sole proprietorship, partnership and corporation. In USA, EE analyses use marginal corporate tax rate of 34%.

19.6 RENEWABLE ENERGY TAX CREDITS

Energy tax credits are provided to renewable energy systems in order to promote them in reference to fossil-fuel technologies. Such tax credits have the effect of enhancing after-tax cash flows and thereby promote investment in these technologies. According to US tax code, when an investor accepts the energy tax credit, the capital investment that is to be recovered through depreciation must be reduced by 50% of the amount of the tax credit. There is also an additional benefit – the depreciation is allowed to be carried through other businesses owned by the investor. This provision has been made because income from renewable energy businesses tends to be meager in the early years. The US Energy Policy Act provides for 10% tax credits to the generation of electricity from solar, geothermal, wind, biomass (including crops specially grown for energy production) sources, at the rate of USD 1.5 cents/kWh.

California allows 30% federal investment tax credit for both residential and commercial solar installations. The State of California is spending USD 3.4 billion to subsidise one million solar roofs which will provide 3 000 MW of solar energy. After his consumption, a homeowner is entitled to sell excess electricity to the state grid.

If taxes owed by a company are less than the amount of tax credit, the unused portion of the tax credit can be carried forward for next year. Suppose a solar company which has made an investment of USD 10 million, owes federal income taxes of USD 750 000/yr. In the first year the company is entitled to receive a tax credit of USD 1 million (i.e. 10% of USD 10 million investment). As the amount of tax credit (USD 1 million) is more than tax owed (USD 750 000), the admissible taxes for next year will be reduced by USD 250 000.

China has emerged as the world's largest market for wind energy. It is now building six giant wind farms with a capacity of 10 000 to 20 000 MW each. For this purpose, wind energy companies get low-interest loans from state-owned banks.

19.7 DEPRECIATION

The capital sum invested in a venture ("depreciable base") is recovered through depreciation by adjusting annually by the amount depreciable ("adjusted base"). Where federal investment tax credits are available, the depreciable base is decreased by half of the tax credit.

In USA, the federal tax rules are implemented through the Modified Accelerated Cost Recovery System (MACRS) using two systems, namely, General Depreciation System (GDS) and Alternative Depreciation System (ADS). MACRS provides the following ways to depreciate property:

- Both the 200% and 150% declining balance (DB) methods over GDS recovery period,
- The 150% DB method over ADS recovery method,
- The straight line (SL) method over an ADS or GDS recovery period.

GDS is generally used for energy efficiency (EE) analyses – it corresponds to shorter recovery period (of, say, 7 years), with greater deduction in the early years. ADS is used for depreciation that is spread over an extended period (of, say, 12 years), when taxable revenues are expected to be greater in the later years than in the early years.

The straight line (SL) annual depreciation is calculated using the following formula:

$$D_n = (C_0 - NSV)/N \tag{19.6}$$

Where

D_n = annual depreciation allowance for year n
C_0 = original cost of the capital investment
NSV = net salvage value (i.e. the estimated salvage value of property, less the cost of removal)
N = depreciation period.

The depreciation allowance for any year n, using the DB method can be computed using the following formula:

$$D_n = B_{n-1} r \tag{19.7}$$

where

D_n = annual depreciation allowance for the year, n
r = annual percentage rate of depreciation applied to the remaining book value (it may be 2 or 1.5, depending upon whether the DB method is 200% or 150%)
n = depreciation period (in years)
B_{n-1} = remaining book value of the asset.

IRS recommends the following depreciation periods for various properties

If a class of property has not been mentioned in IRS, such an item should be depreciated using a 7-year recovery period under GDS or a 12-year recovery period under ADS. Also, some projects, such as a nuclear power project, may include equipment with different depreciation rates.

Type of property	*GDS method (yrs.)*	*ADS method (yrs.)*
Alternative energy property (non-utility generators)	5	12
Nuclear production plant	15	20
Nuclear fuel assemblies	5	5
Hydro production plant	20	50
Steam production plant	20	28
Combustion turbine production plant	15	20
Transmission and Distribution plant	20	20
Non-residential real property	31.5	40

Most investors prefer shorter depreciation periods because an after-tax dollar earned today is worth more than an after-tax dollar earned tomorrow (i.e. time value of money).

Chapter 20

Energy economics and markets

U. Aswathanarayana

20.1 INTRODUCTION

The structure of the energy market changes constantly in response to emerging economic, political, cultural and technological developments. If oil price becomes very high, there will be incentive to look for alternatives such as tar sands, oil shales and bio-diesel. Transportation, which involves moving freight, commuting, recreation, tourism, industry travel, etc. accounts for about 25% of energy consumption in OECD countries. The continuing global economic downturn which started in 2008, meant that people have less money in their hands, and this led to drastic reductions in air travel for holidaying, and automobile travel for socializing and shopping purposes. This development affects the rate of demand for fuel. Information technology is being increasingly employed for streamlining traffic patterns, telecommuting, teleconferencing and e-commerce, thereby reducing the necessity of personal travel and the consequent use of less fuel. Technology may improve the efficiency of various energy uses. For instance, the present-day refrigerators are four times more efficient and cost half, relative to the refrigerators of 1975. The energy that we are saving today by the use of these more efficient refrigerators, is more than all the wind and solar energy produced today. However, as the energy required for a given service becomes less costly because of higher efficiency, the service may be more used. This is a kind of rebound effect.

A number of techniques, such as extrapolation of historical trends, multivariate time series, Bayesian estimations, games theory, etc. can be made use of to predict the amount of energy that need to be produced for various uses, such as, likely price structure, emerging technologies, environmental consequences, and so on. Because of the uncertainties involved, the predictions are projected in terms of bands. Such forecasts

Table 20.1 Per capita GDP, energy use and CO_2 emissions

Country	*Per capita GDP (×2000/capita)*	*Per capita energy use (GJ/capita)*	*Per capita CO_2 emissions (t CO_2/capita)*
India	0.54	22	1.02
China	1.3	52	3.65
France	22.8	186	6.22
Germany	23.7	177	10.29
Canada	24.6	354	17.24
Sweden	29.3	252	5.8
United States	36.4	332	19.73
Japan	38.6	176	9.52
Norway	39.3	253	7.91

(Source: International Energy Agency, "*Key World Energy Statistics, 2006*")

would be useful to industrial establishments to build appropriate capacities, to banks to fund the most suitable technologies, and to governments to design optimal policies. Sometimes the forecasts made by reputable organizations may be so much believed that they may influence the outcome – as a kind of self-fulfilling prophecy. Also, the longer the term of the forecast, the greater will be the uncertainty. For instance, the global economic meltdown in the later part of 2008, and the change of regime in USA in 2009, have given a great fillip to green technologies. This could not have been foreseen.

Thermal energy is produced by the chemical combustion of coal, oil and natural gas. Coal provides 25% of the global energy needs, and accounts for 40% of the world's electricity. Some countries are highly dependent upon coal for their electricity production: Poland – 93%. South Africa – 93%; Australia – 80%, China – 78%, India – 69%, USA – 50%. Oil remains a dominant source of energy in the transportation and the industrial end-use sectors. Energy released from nuclear reaction per unit mass is about a million times more than that of chemical combustion. Hence the quantity of nuclear fuel required to produce (say) one GWe is very small compared to the requirements of fossil fuel (say, coal). Nuclear power currently accounts for about 16% of the electricity production in the world. France leads the world in having 78% of its electricity production from nuclear power. Renewable energy resources include renewable combustibles and wastes, hydro, solar, wind and tide energy, and accounted for 12.7% of the Total Primary Energy Supply. Hydropower is produced from gravitational energy. Presently, hydropower accounts for 90% of the renewable power generation. Brazil currently produces the highest quantity of hydropower in the world (eq. 30 Mtoe).

Globally, nuclear power is enjoying a renaissance in the context of the climate change problems. Though it is not renewable, it is being widely accepted as a "green" power source, as it has no carbon foot-print.

The increase in per capita GDP is generally linked to greater energy use, which entails higher carbon dioxide emissions. However, the example of a number of countries, such as, France, Germany, Japan, Norway and Sweden, shows that it is indeed possible to maintain a high standard of living (as manifested by high GDP) while using comparatively less energy and emitting less carbon dioxide, through the diversification of energy mix (Table 20.1).

Table 20.2 World net electricity generation (in Billion kWh) by type, 2000

Region	*Thermal*	*Hydro*	*Nuclear*	*Geothermal and others*	*Total*
North America	2 997.1	657.6	830.4	99	4 584
Central and South America	204.1	545	10.9	17.4	777.4
Western Europe	1 365.4	557.5	849.4	74.8	2 847.1
EE & FSU*	1 043.7	253.5	265.7	3.9	1 566.9
Middle East	425.3	13.8	0	0	439.1
Africa	333.7	69.8	13	0.4	416.9
Asia & Oceania	2 949.2	528.7	464.7	43.1	3 985.7
World Total	9 318.4	2 625.8	2 434.2	238.71	14617

*Eastern Europe and Former Soviet Union

(Source: International Energy Annual, 2001; EIA/DOE, Feb. 2003)

Table 20.3 Net electricity generation (in Trillion kWh) in the world, by energy resources, 2006–2030

Energy resource	*2006*	*2010*	*2015*	*2020*	*2025*	*2030*	*Annual change (%) during 2006–2030*
Liquids	0.9	0.9	0.9	0.9	0.9	0.9	−0.1
Natural Gas	3.6	4.2	4.9	5.7	6.4	6.8	2.7
Coal	7.4	8.7	9.5	10.4	11.8	13.6	2.5
Nuclear	2.7	2.8	3.0	3.4	3.6	3.8	1.5
Renewables	3.4	4.1	4.9	5.7	6.1	6.7	2.9
Total World	18.0	20.6	23.2	26.0	28.9	31.8	2.4

(Source: "*International Energy Manual*, 2009")

Thermal electricity is dominant in all the regions of the world, except South and Central America in which hydropower is the principal source of energy. The Middle East is almost wholly dependent on thermal power from oil and natural gas. Western Europe has higher proportion of nuclear power than other regions (Table 20.2).

Electricity generation during the period, 2006–2030, is expected to have the following characteristics: (i) Despite its high pollution potential and large carbon foot-print, coal will continue to be the dominant source of thermal electricity, because of the ground realities, (ii) while the use of oil for electricity production will remain unchanged, electricity will increasingly be generated through the use of natural gas, and (iii) Renewables will be increasingly used for electricity generation (Table 20.3).

20.2 MODELING ELECTRICITY MARKETS

Power plants have some special economic characteristics: they need large investments, and they require infrastructure to acquire fuel, and to deliver power to the end-use customers. Where possible, electricity is generated in the pithead thermal power stations and transmitted. In the case of China, there is a mismatch between the natural distribution of coal deposits and the location of coal-using industries. Consequently, about one billion tonnes of coal, which is roughly half of the coal supply, is moved by rail for use in industries (such as, iron and steel, cement, fertilizer, etc). For this purpose, China has to build elaborate infrastructure with dedicated rail links, and trains with payload capacity of 25 000 tonnes.

Capital costs are typically fixed in the short run, as the plants have a long life (say, 40 years). In the long run, however, all costs are variable.

The following equation represents costs of electricity generation (Dahl, 2004, p. 85):

$$TC = FC + VC(Q) \tag{20.1}$$

Where
TC = Total cost,
FC = Fixed cost, also known as sunk cost, which must be borne whether there is electricity production or not. It is composed of the cost of equipment, such as electric generator, or gas scrubbing unit, or the more recent development, namely, equipment for the capture and storage of carbon dioxide. It includes lease of office space, computers, insurance premium, etc.

Q = Quantity of production,
VC = Variable cost
VC (Q) = Variable cost which is a function of Q.

We should also take into account out-of-pocket costs and opportunity costs. For instance, in the case of hydropower, though the out-of-pocket cost of water is zero, there is an opportunity cost, i.e. the foregone value of water in its next best alternative, say, selling the water for irrigation.

20.3 AVERAGE COSTS AND MARGINAL COSTS

There are two kinds of unit production costs – average cost and marginal cost.

Average cost is obtained by dividing the total cost by output.

$$TC/Q = FC/Q + VC(Q)/Q \tag{20.2}$$

Average total cost (TC/Q) equals average fixed cost (FC/Q) plus average variable cost (VC(Q)/Q. Whereas the average fixed cost is constant, the average variable cost may take on a variety of structures. In the case of coal, it may depend on the calorific value and sulphur content of coal, whether the mining is opencast or underground, whether the transportation is by road or canal or rail or sea, and whether there is any subsidy on a particular mode of transportation (say, by rail) and so on.

Marginal cost (dTC/dQ) is computed from the following equation:

$$dTC/dQ = dFC/dQ + dVC(Q)/dQ \tag{20.3}$$

Since FC is constant, dFC/dQ is zero. Thus the marginal cost is a function of change in the variable cost arising from the change in the production system. It may be constant, increase or decrease. For instance, if the rail authorities charge a lower unit cost for large shipments, marginal costs will fall as more electricity is produced.

20.4 LOAD CYCLE

The load cycle depends upon the time of the day, climate, and degree of urbanization and industrialization of an area/community. In the industrialized countries, the load cycle reaches its peak during the day, and off-peak during the night, with shoulder production early in the morning and later in the evening. In countries like USA, there will be peak demand in summer when air-conditioning is used extensively during summer. In countries like Canada where most homes are heated by electricity, the peak demand will be during cold winters.

In order to use electricity as efficiently and cheaply as possible, "smart" grids are being developed. Each energy system has its own set of biophysical and socioeconomic impacts, technological constraints, capital costs, operation and maintenance costs, discount rate, environmental consequences, etc. For instance, nuclear power station, whose capacity is typically in the range of 1 000–1 700 MW, is not suitable for a small town. In the same way, biomass plants whose capacity does not usually exceed 50 MW, are unsuitable for a large cities. Villages in India may sometimes have small plants of less than 100 kW capacity, run on biomass waste, like paddy husk.

Electricity demand of a given geographical unit depends upon the size of the population, their life-style, climate, agriculture, industry, tourism, etc. Also, it is highly dependent upon the time of the day (e.g. the demand for air-conditioning is maximum at noontime). Thermal plants are less able to respond to sudden changes in the electricity demands, and may cause voltage and frequency instabilities. In contrast, pumped storage hydropower plants can respond to load changes almost instantly (within 60 seconds), thereby reducing the need for "peaking" power plants that use costly fuels. Pumped storage (water pumped up into the reservoir during the off-peak period to be released to produce energy during peak times) hydroelectricity can hence be used for load balancing in tandem with large capacity solar plants (which do not generate power in the nights) and wind mills plants (which do not generate power when there is no wind). Intensive efforts are being made to promote zero-carbon road transport by replacing vehicles which burn fossil fuels with cars that run on electricity or other clean energies. In tandem with this, technological improvements in batteries could raise the mileage per charge from the present 145 km to (say) 320 km. The electricity demand at a given time will thus have to take into account millions of electric cars that need to be charged, and millions of appliances that need to be run. Reduction in costs and adopting the products to the markets are only possible through technology learning and deployment in the market place.

The strategy document that will be offered to the customer should contain the following particulars: (i) inventory of the presently operated electricity power sources, in the concerned geographical unit, their capital and operational costs, economics, etc. (ii) Ways of applying information technology to optimize the performance of the present system, and estimating the benefits to the client by this exercise, (iii) projection of electricity demand up to 2030 and 2050, ways of meeting the expected demand, their economics and environmental consequences, and estimating the benefits.

20.5 ENERGY ECONOMICS

There are economies of scale. The largest producer of electricity can offer electricity to the consumer at the cheapest rate, and could drive out of business smaller generating units. A monopoly can then come into existence. A monopolist can utilize his advantage to maximize profits. The profits (π) of a monopolist are total revenues minus total costs:

$$\pi = P(Q) * Q - TC(Q) \tag{20.4}$$

The first order conditions for the maximization of profits is given by the equation:

$$D\pi/dQ = P + (dP/dQ) * Q - dTC/dQ = 0 \tag{20.5}$$

In the interests of social equity, electricity utilities are either owned by the governments or regulated by the governments. Privately-owned utilities cannot raise tariff for electricity at will. When a utility wants a rate increase, it would present its case before the prescribed government authority, and obtain its permission for the same. The price approved in the negotiation will continue until the next case.

Governments optimize social welfare by charging price for a product or service equivalent to marginal cost.

There are three kinds of regulatory practices – rate of return, fully distributed cost and peak load.

Under Rate of Return regulation, a utility should be able to recover the variable costs, plus a fixed rate of return on capital necessary to attract the required amount of capital to the industry. Suppose the firm produces n products ($q_1, q_2, q_3, \ldots q_n$) and charges a price of (p_i) for the "ith" product, its earnings will be $\Sigma_1^n p_i q_i$. The firm is allowed to charge p_i (US cents/kWh) in order to cover its operational expenses, while earning a normal rate of return (s) on its rate base

The basic accounting equation for the purpose is:

$$\Sigma_1^n p_i q_i = \text{expenses} + s(RB) \tag{20.6}$$

Where
p_i = price of the ith service class,
q_i = quantity of the ith service class,
n = number of service classes
s = allowable or "fair" rate of return
RB = rate base (capital stock) of the regulated firm's investment

Many difficulties arise in implementing the Rate of Return regulations.

Let us take into account a coal-fired thermal power station. The utility owning the power station may have a captive coal mine, and may make use of the opportunity to charge higher prices for coal, which will be passed on to the consumer. The capital for the power station may be sought in the form of Bonds, Preferential Stocks and Common Stocks, each of which is characterized by required rate of return. The monetary policy of the national government would affect the rate of return from these investments. Common stocks may be owned individually, and their dividends are dependent upon the economic conditions. We can know the rate of dividends for a given common stock now, but it is not possible to predict what the rate of dividends would be in the future.

A utility may be supplying high-voltage power to industries, low-voltage power for domestic purposes, off-peak power during the night where farmers are provided free electricity for irrigation pumps, and so on. There are several ways of allocating fixed costs among the various classes of consumers who are characterized by different degrees of demand elasticity. For instance, whatever the cost of electricity, a factory should necessarily use a particular fixed amount of power in order to run the machines in the factory – this is a case of inelastic demand. On the other hand, if electricity becomes too expensive, a household will use fewer kitchen gadgets for shorter periods, in order to reduce the electricity expenses. Another way is to allocate fixed costs in the form of fixed charges.

Electricity is charged at higher rate at peak loads than that off-peak loads. Since it is expensive to store electricity, utilities maintain a higher capacity of the plant to cater to the peak loads. This would, however, mean that some capital is sitting idle most of the time. If no peak shifting is possible among the consumer classes, electricity prices may be so fixed that all capital costs are charged to peak users, and only marginal costs to off-peak users.

One way out of the situation is to engage in networking (say) a coal-fired plant with off-peak capacity with (say) a pumped storage plant providing extra electricity needed for the peak periods.

Private companies may generate electricity from coal, natural gas, naphtha, etc. and sell power to the public sector corporations. Nuclear power and hydropower generating facilities are generally government-owned. However, transportation networks and local distribution entities are almost always owned by public corporations in most countries.

In USA, coal-fired thermal power plants produce about half of the electricity production in the country. The share of the capital costs is typically: 44% for electricity production, 22% for transmission, and 34% for local distribution. The shares of operating costs are: 89% production and fuel, 3% transmission and 8% local distribution.

20.6 LEVELIZED COSTS

Levelized costs for electricity are calculated by using discounting to distribute the costs over the production profile. Dahl (2004, p. 323) provides an illustrative example in respect of wind electricity.

Capacity of the wind turbine: 600 kW. Capital cost: Cost of the turbine: \$450 000 + installation cost: \$125 000 = \$575 000. Projected percentage of time that the turbine runs with full capacity: 25%. Lifetime of the turbine: 20 years. Discount (interest) rate: 10%.

Table 20.4 Comparative electricity generating costs (2001 US cents/kWh) including capital

Country	*Nuclear*	*Coal*	*Gas*
France	5.3	6.5	5.8
Russia	5.1	6.0	4.2
Japan	8.7	8.3	9.1
Korea	5.2	4.9	5.1
Spain	7.0	6.0	5.9
USA	4.3	3.8	2.9
Canada	4.3	25.8	3.6
China	4.2	4.3	
	Wind	PV	Fuel cells
USA	5.7	36.9	10.0

*Cost estimates for coal, gas and nuclear are forecasts for 2005. Assumptions are: discount rate of 10%, 40-year plant lifetime, and 75% load factor.

(Source: Dahl, 2004, p. 326)

Electricity is produced, and the revenues collected from it, over time, but the capital costs have to be spent right at the beginning. Hence the discount (interest) rate.

Amount of power that the turbine could generate over its lifetime: 600 kW *24 hrs. *365 days *25% = 1 314 000 kWh

The amount we need to charge per kWh (S_k) in order to recoup the capital costs =

$$\$575\,000 = \sum_{i=0}^{20} S_k * 1\,314\,000/(1+0.10)^i => \tag{20.7}$$

Solving, we get $S_k = \$0.047$ per kWh.

20.7 LIMIT PRICING MODEL

The energy markets are dynamic. For instance, in USA, electricity tariff (cents/kWh) for various modes of generation are: Nuclear: 4.3, Coal: 3.8; Gas: 2.9; wind: 5.7; PV: 36.9; Fuel cell: 10. Though gas source may be the cheapest, enough of that fuel may not be available. Electricity from coal is the next cheapest, and coal also happens to be widely available. But the dominance of coal may not be permanent. When the policy of carbon credits gets implemented, electricity from coal would be costlier. When very large turbines (8 to 10 MW) and large wind farms of several hundred megawatts, become possible, the wind electricity rates will come down and may be able to compete with coal power. This situation is depicted in Fig. 20.1 Limit Pricing Model (source: © Dahl, 2004, p. 276).

In this model, the low-cost fossil fuel is the monopolist, and the higher-cost backstop, such as, wind power, with costs as shown by AC_b is a potential competitor. "Without the backstop, the monopolist should operate where the marginal revenue (MR_m) equals marginal cost (MC_m) charging P_m and producing Q_m. At this price, the backstop would have an incentive to enter" (source: Dahl, 2004, p. 276). The monopolist could

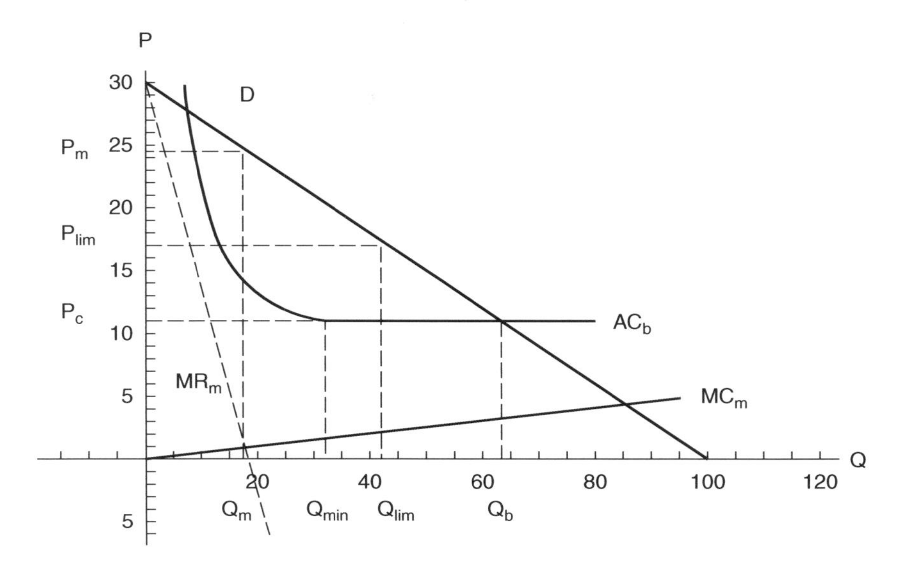

Figure 20.1 Limit pricing model

(Source: Dahl, Carol A. 2004,"*International Energy Markets: Understanding pricing, Policies and Profits"*, p. 276, © Permission by Pennwell Publishing)

discourage the entry of the backstop into the market by charging "a price that is equal to the minimum of the average cost curve for the backstop or slightly below the competitive price (P_c) "(Dahl, 2004, p. 276).

Q_{min} is the smallest output that gives the backstop maximum economies of scale. The backstop would have no incentive to enter the market if the monopolist limits its quantity of production to $Q_{lim} = Q_b - Q_{min}$, and charging the price of P_{lim}.

The options open to the various competitors are very much influenced by the Government policy – for instance, the Obama Government in USA is strongly supporting job-creating green technologies. In his speech in Iowa on the Earth Day (Apr. 22, 2009), President Obama said that US plans to meet 20% of the electricity demand (as against 2% today) through wind power (land and offshore) by 2030, and this would involve the creation of 250 000 new jobs.

20.8 POLLUTION AS A NEGATIVE EXTERNALITY

Pollution is a negative externality, and affects energy policies and prices. It is almost impossible to have a totally pollution-free energy enterprise. A stack gas scrubber in an electric power plant can take out about 96% of sulphur dioxide. To remove the remaining 4% of the gas is technically possible, but prohibitively expensive. So it makes practical sense to learn to live with this small amount of emission. A small amount of pollution may cause no damage to the environment, if the environment is capable of absorbing it. In the graph between pollution (P) vs. costs ($) (Fig. 20.2 Costs and

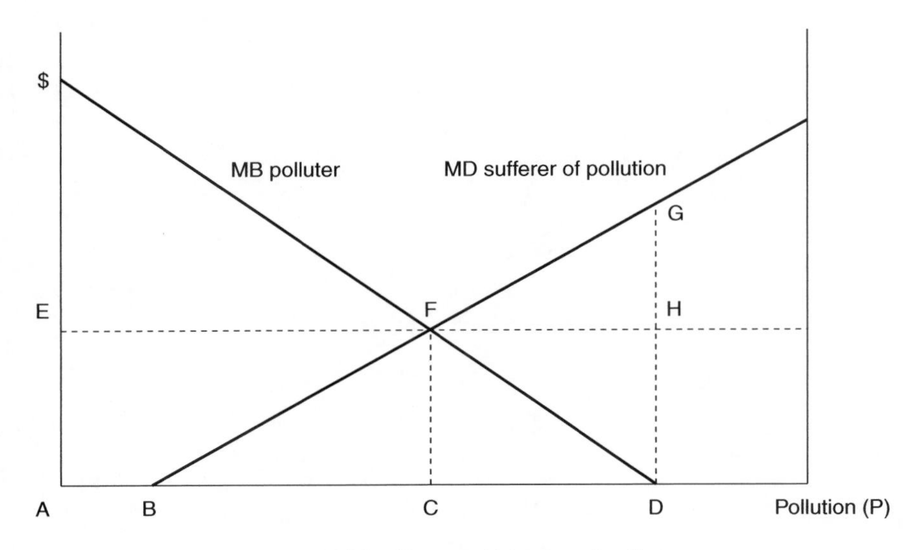

Figure 20.2 Costs and benefits of pollution
(Source: Dahl, Carol A. 2004, "*International Energy Markets: Understanding pricing, Policies and Profits*", p. 204, © Permission by Pennwell Publishing)

Benefits of Pollution – Source: © Dahl, 2004, p. 204), this point of acceptable pollution is represented by B.

If there is no discentive, the polluter would pollute up to point D. The more the departure from B, the more the environment (say, water) will be affected. If the polluter is a coalmine, the effluents from the mine may make a river water more acidic. The population which drinks the water will get sick (health costs), or an industry (say, an oil refinery) in the downstream side, will have to remove the acidity of the water as otherwise the refinery equipment will be corroded (industrial costs). The affected population and the refinery will bring pressure on the government to compel the coalmine to abate the pollution.

In such a situation, the government could take the following steps (Dahl, 2004, p. 206):

(i) Set the pollution standard at C (compliant effluent),
(ii) Set a tax on the pollution of AE,
(iii) Sell pollution permits equal to AC,
(iv) Subsidize cleanup or abatement of AE.

If the government subsidizes the cleanup, the polluter will have no incentive to reduce the pollution. So option (iv) is out. If the effluent is not compliant, the government will penalize the enterprise. If the polluter is allowed to pollute up to D without check, there will be heavy social costs. If the sufferers of pollution make the government compel the polluter to bring down the pollution from D to A, the polluter will go out of business because of excessive costs, and the society will lose the energy generated by the company.

"The polluter pays" principle is now widely accepted. In the context of the carbon dioxide emissions causing the global warming, the US EPA has declared carbon dioxide gas as pollutant. In order to compel the coal-fired thermal power stations to reduce their carbon dioxide emissions (through steps such as supercritical and ultra-supercritical coal combustion, and carbon dioxide capture and storage technologies), the US Government is embarking on the sale of carbon credits to enable an enterprise to pollute beyond B. Since the buying of carbon credits cost money, the polluting enterprise will therefore have the incentive to limit its pollution as close to B as possible.

The point C is therefore a compromise. The company will survive as the costs for polluting up to C will not excessive. The pollution will still be there, but it will not have excessive social costs.

20.9 ENERGY FUTURES AND OPTIONS MARKETS

Energy companies face many risks. The risks include accidents, such as nuclear meltdown, an oil spill, an LNG explosion, or air and water pollution. Crude oil prices and drilling costs may rise; there may be roof collapse or methane explosion in a coalmine; or there may be a fire in a refinery. New governmental regulations about compliant effluents, or carbon credits may raise the process costs steeply.

When the US tanker *Valdez*, owned by Exxon – Mobil, spilled about 260 000 barrels of crude oil in 1989, it caused extensive damage to the Prince William Sound, Alaska. The compensation for the damage were claimed against stocks, bonds and financial agreements which got their value from the assets of Exxon-Mobil, such as oil leases, refineries, tankers, pipelines, and so on.

In 1970s there was much volatility in the energy prices. Taking advantage of the advances in information technology, financial derivative markets came into existence on the basis of financial assets of companies. The *financial derivatives* are in effect financial instruments designed for transferring risk from parties which want less risk to those which are willing to take on the risk – for a price. Most energy derivatives are based on three basic instruments, namely, futures, forwards and options.

"An *energy futures contract* is an agreement to buy or sell a specific energy asset at some future point in time" (Dahl, 2004, p. 273). The contract is arranged through an organized exchange, such as NYMEX (New York Mercantile Exchange) which handles about 80% of the energy futures in the world. It is a standardized contract that can be resold on the exchange. It stipulates the delivery month and delivery price. The buyer pays a small transaction fee, and puts up a margin of typically less than 5% of the value of the contract. Contract gains or losses are settled at the end of each trading day.

A *forward contract* is a bilateral agreement, which can be customized to the needs of the individual customers. It is not a standardized contract and cannot be resold without the agreement of both parties.

An *option* is the right to buy (call) or sell (put) an asset at a specified price called strike price, as a part of the futures contracts. The holder of a call option has the right (but not the obligation) to buy the energy asset at the specified price.

Buying a call option effectively locks in a maximum buying price, and buying a put option effectively locks in a minimum sale price Thus an electricity generating company may reduce its risks by keeping a call option which enables it to buy (say) natural gas at the strike price (say, $2.60 per MMBTU) at a particular time (say, June), and a put option of selling power at (say) $50 per MWh.

The exchanges for the energy futures contracts are located in various countries, and deal with oil, natural gas, coal, and electricity. In 1986, when the crude prices fell from $30 to $10 per barrel, some parties refused to take delivery of oil and pay the higher price. This refusal caused defaults down the rest of the chain. We now have the Brent fifteen-day market. Deliveries are made at Sulom Voe terminal in the North Sea, and 15 days are needed for tankers to pick up the crude. These contracts can be traded on the spot market.

20.10 ENERGY AND INFORMATION TECHNOLOGY

Computers are being increasingly used in design (CAD – Computer-aided Design), engineering (CAE) and manufacturing (CAM). Supervisory control and Data Acquisition systems (SCADA) remotely control processes for pipelines, offshore oil, and gas production, electric utility production and distribution, and so on.

A barrel of oil is a private good which is depletable. When once you use a barrel of oil, nobody else can use it. You can also exclude others who have not paid for it, from using it. On the other hand, information is non-depletable. If you use a megabyte of information, you do not deplete it. Any number of people can use it without depleting it. However, just as the owner of a barrel can exclude you from using it, you can also make the information excludable, i.e., you can get the information only by paying for it. It has been reported that Sun Microsystems paid USD one billion to a Swedish company for MySQL software.

Since information is non-depletable, governments should not charge for online information, as the marginal cost of one more person using the information is zero. For this reason, EIA/DOE post information on the Internet which can be freely downloaded.

In order to be successful, a company should have a clear vision. The vision and mission statement of the California Energy Commission, is a case in point (incidentally, the state of California is the fourth largest economy in the world):

"*It is the vision of the California Energy Commission for Californians to have energy choices that are affordable, reliable, diverse, safe and environmentally acceptable. It is the California Energy Commission's mission to assess, advocate and act through public/private partnerships to improve energy systems that promote a strong economy and healthy environment.*"

On the basis of the mission statements, specific objects, tasks and strategies, are formulated, and implemented.

In his famous book, "*Business at the Speed of Thought*" (1999) Bill Gates argues for an information flow down the chain of command in a company that should empower the employees all along the line to be able to think, act and adapt more quickly.

Moore's law suggests that the computer power doubles every 18 months. Metcalfe's law suggests that the value of a network increases with the square of the number of users. Coase's law suggests that in an organization, the best market governance is that which minimizes transaction costs. Thanks to Internet, we need fewer intermediaries. For instance, a bank transaction costs $1.07 at a bank branch, $0.52 by telephone, $0.27 at ATM, but only $0.01 on the Internet. Digital transactions have the additional advantage of being less liable to distortion, and providing for more interactivity. Goods and services will still be needed to be produced, but they could be produced better, faster and more customized.

Experience has shown that technology changes faster than the ability of new technology users to make the psychological adjustment to the new technology. Hence the continuous upgradation of the knowledge of the employees, is cost effective.

Chapter 21

Renewable energy policy

U. Aswathanarayana

21.1 WHY RENEWABLES?

The green, renewable energy economy is fundamentally different from our 20th. century economy with its overdependence on fossil fuels. The renewable fuels, such as, wind, solar, biomass or geothermal, are entirely indigenous. The fuels themselves are often free. They just need to be captured efficiently and transformed into electricity, hydrogen or clean transportation fuels. In effect, the development of renewal energy invests in people, by substituting labour for fuel. Renewable energy technologies provide an average of four to six times as many jobs for equal investment in fossil fuels. For instance, while natural gas power plant provides one job per MW during construction and ongoing operations and maintenance, equivalent investment in solar photovoltaic power technology would generate seven jobs per MW. The approval of the Renewable Electricity Standard (RES) by USA would involve the construction and maintenance of 18 500 MW/yr of wind, solar, geothermal and biomass plants, and if all the components needed for the project are manufactured in USA, this would generate 850 000 jobs.

If a society decides that climate change should be mitigated, such a policy should be reflected in the choice of technologies made by that society. It should not be forgotten that a desirable outcome, say, the development of renewable energy technologies, does not happen by itself – it should be made to happen through appropriate policy change.

Policy and markets are generally in conflict, as their objectives are different. "A policy is a market intervention intended to accomplish some goal – a goal that presumably would not be met if the policy did not exist" (Komor, 2004, p. 21). The object of the public policy is to promote public good, while the object of the market is to make money as quickly as possible. On the basis of considerations of public good, a government may decide on a policy of promoting renewable energy technologies, but the market would not wish to participate in it unless there is a reasonable prospect of profiting from it.

The trick is in figuring out the "intersection" point, where the two pathways converge – i.e. whereby the market finds that it is possible to make money from an activity that the government wishes to promote. This is easier said than done. The intersection point is not a fixed point – it is a floating point. It is changing all the time in response to changes in technology and market penetration. It follows that the policy makers and private makers have to engage one another in continuous dialogue in order to arrive at the "intersection point" which is acceptable to both the sides.

The power of the technology innovation and market penetration can be illustrated with two recent examples. Apple's iPhone-3GS, costing just $ 200, sold more than one million pieces in three days after it was issued – busy people queued for hours to buy the gadget. Similarly, the demand for Tata's nano, the world's cheapest car at USD 2 000, is in millions – it has already sold 100 000 units.

As Paul Komor (2004, p. 12) puts it succinctly, "If renewables are to succeed, they must succeed in a competitive market". Policy should be aimed at facilitating it.

Energy policy of any country has to have two objectives: job generation as a way of getting out of the recession, and mitigation of climate change impacts through low-carbon technologies. Consequently, governments could consider formulating sustainable energy policy frameworks for their countries based on the following strategy: (i) how to promote greater use of renewable energy for on-grid, large-scale electricity production – this will also help to overcome the intermittency problems of wind power and solar PV, (ii) how to use discentives, such as carbon tax, to phase out fossil fuel use; how to use technology to reduce the carbon footprint and improve efficiency of fossil fuels, where the use of fossil fuels is unavoidable, (iii) how to use market-based strategies, such as green certificates, and how to develop innovative technologies for the production of new kinds of fuels (e.g. algal biofuels), new ways of energy storage, and demand-side management, etc. Energy policy case histories of some countries are analyzed to delineate what works, and what does not work.

Power plants use fuel to generate electricity, which is then transmitted and distributed to the user (domestic, commercial and industrial). It is the generation part of the system that largely determines the cost of the electricity and is responsible for the environmental damage and climate change impact. It is also the part where renewable fuels can play a major role. The chapter seeks to explore the policy framework for promoting greater use of renewable energy for on-grid, large-scale electricity production.

The world energy consumption by source in 2000, was as follows: Oil – 35%; coal – 24%; Natural gas – 21%; Biomass waste – 11%; Nuclear – 7%; Hydropower – 2%; Geothermal, wind, solar – $<1\%$. Thus, fossil fuels account for about 80% of the energy use. The continued use of fossil fuels is not sustainable for the following reasons:

Environmental damage: The burning of fossil fuels (say, coal) in the power plants leads to the production of CO_2 which contributes to global warming and climate change, and sulphur oxides (SO_x) and nitrogen oxides (NO_x) which cause the acid rain.

Table 21.1 Renewable technology summary

Technology	Typical levelized costs (US cents/kWh)*	Advantages	Problems
Wind	4–5	Widespread resource; scalable	Difficult to site; intermittent
Photovoltaics	20–40	Ubiquitous resource; silent, long life times	Very expensive; intermittent
Biomass	4–9	Dispatchable; large resource	Has air emissions; expensive
Hydropower	4	Dispatchable; can be inexpensive	Has land, water and ecological impacts
Geothermal	5–6	Dispatchable; can be inexpensive	Limited resource; depletable

*Levelized cost of energy (LCOE) levelizes different kinds of fuels, scales of operation, investments, and operating time periods.
(Source: Komor, 2004, p. 7)

Fossil fuel resources are finite: As fossil fuel resources get depleted with use, they are bound to get exhausted sooner or later.

Fossil fuel resources are unevenly distributed: About two-thirds of oil resources of the world are concentrated in the Middle East.

Price volatility of oil: Oil prices have been highly volatile. They oscillated from a low of \$10.60 per barrel in Jan. 1999, to a high \$140 per barrel in 2008. This makes energy planning extremely difficult.

The renewables are the fuel of the future, for the following reasons: (i) they have a low environmental impact – there are hardly any direct emissions of CO_2, SO_x, NO_x, particulates and mercury, (ii) they are non-depletable and hence sustainable (geothermal power is finite, and may not strictly qualify as being non-depletable) (iii) they are widely distributed, and (iv) they have wide popular support.

Renewables should not be thought of as the panacea for all our energy woes. They do have problems, however:

(i) Renewables generally cost more. For instance, PV energy (US cents 20–40/kWh) is definitely more expensive. Let us compare power generation with natural gas vis-à-vis wind power. A natural gas turbine costs ~ \$500/kW, which is about half the cost of a wind turbine (~\$1 000/kW). The difference in the initial cost is nullified by the fuel costs. At the natural gas price of ~ \$3/1 000 cu.ft, and since wind is free, wind-produced electricity becomes competitive with natural gas-produced electricity. Whether the two will always be competitive depends upon the future natural gas prices, improvements in the wind turbine construction and time value of money, and so on.

(ii) *Renewable resources are not ubiquitous*: Though the renewable resources are far more evenly spread than the fossil fuel resources, some renewable resources, such as, geothermal resources, are restricted to fault block terrains with Quaternary volcanism, such as, Kenya, Iceland, New Zealand, Italy, USA, etc. Winds are stronger in some areas than others. Insolation (sun light) is less in the Arctic areas.

(iii) *Some renewable resources are intermittent*: Electricity has to be provided on demand. Wind and solar electricity generation is intermittent, being subject to the vagaries of nature, and therefore cannot provide electricity on demand reliably. It is therefore necessary to link them with pumped storage hydroelectricity or fossil fuel combustion turbine. The problem of intermittency of solar PV is sought to be got over by providing systems to store electricity when the demand is less. Wind electricity can be stored. Also, wind farms may be linked together in a grid, so that if winds fail in one place, electricity may be drawn from another farm.

(iv) *Renewables have environmental impacts*: Though the environmental impacts of renewables are nowhere near as serious as those of the fossil fuels, they are not zero. Visual and noise pollution of the wind mills, emission of carbon monoxide and particulates in biomass burning, displacement of a large number of people because of the reservoir construction, etc. are some of the environmental impacts of renewables.

21.2 MARKET-BASED STRATEGIES TO PROMOTE GREEN ENERGIES

The author has a piece of land in his ancestral village in south India, in which his brother grows *casuarina* trees. He is not growing them in order to ameliorate the environment, nor is the government compelling him to do so. He is growing them because there is good money in it. The moral of the story is that it is possible for a community to achieve a desirable environmental objective by creating a situation whereby a desirable environmental objective becomes financially attractive.

Until about 1990, electricity systems in most countries were owned and operated by governmental and quasi-governmental corporations. The system was vertically integrated, i.e. the same company generating, transmitting and distributing electricity, and constituted a "natural monopoly". The objective of these corporations was not to make profit, but to provide a public service. In OECD countries, this system provided dependable, reliable and reasonably priced electricity. There was no incentive for these corporations to introduce new technologies, and bring down costs. In some developing countries, electricity supply was linked to the promotion of social equity – poorer members of the community were provided electricity at subsidized rates or even gratis. No wonder such systems hardly worked well, with frequent brownouts and blackouts.

Two developments (one technological, and one political) since 1990 led to drastic reorganization of the previous "natural monopoly".

Technological advances made it possible for private companies to produce electricity at prices much lower than those being charged by the utility companies. For instance, industry-sized, natural gas – fired power plants were in a position to offer electricity at rates much less than those of the utility companies. Cogeneration technologies, which produce both electricity and heat, became attractive to some industrial users.

Consumers raised the question as to why they need to stick to the default provider, when they could get electricity cheaper elsewhere, or alternatively, generate power on their own.

Right-wing governments in U.K. and elsewhere embarked on privatization of public sector corporations. In 1980s, the British Government privatized British Aerospace, British Telecommunications, British Gas, British Airways, British Steel, British Coal, British Rail, etc. ostensibly to provide more efficient service at cheaper rates, but actually to break the power of the labour unions.

Cap and Trade

Emission trading, also called cap-and-trade, is an administrative arrangement for controlling pollution through providing economic incentives for achieving reductions in the emission of pollutants. A central government authority or an international organization sets a limit or *cap* for the country as a whole. Within the overall cap, individual companies or groups of companies are given allowances or *credits* to emit a specific amount. Companies which need to emit more than the stipulated credit granted to them must buy credits from companies that emit less. This transfer of allowance is called the *trade*. In effect, a company that is polluting more than permissible is thus penalized, and a company which is polluting less than it could, is rewarded. The market forces will compel the polluting company to reduce its pollution. The society therefore achieves pollution reduction at the lowest possible cost.

Let us examine how this system benefits both sellers and buyers, and the society as a whole. Let us take two countries, say, Germany and Sweden. Each can reduce all the required emissions on their own, or they can choose to buy and sell in the market. Let us assume that Germany reduces the emissions more than required, and abate CO_2 emissions at a cheaper cost than Sweden.

Germany sold emissions credit to Sweden at a unit cost, P, while its actual cost is less than P. Sweden bought emissions at unit cost, P. Thus Germany makes a profit for polluting less, while there is no extra burden on Sweden. Thus, the total abatement cost of the two countries together is less than emission trading scenario.

The same principle can be applied by two companies.

Green Certificates

A Green Certificate, also known as Tradable Renewable Certificate (TRC), green tag, and Renewable Obligation Certificate (ROC), is essentially an accounting tool. It monetizes the environmental attributes of renewable-sourced electricity generations. A number of countries have started issuing such certificates.

The Energy Policy of Sweden is based on ensuring high security to energy generation, through obtaining all its energy from renewable sources in the long run. Sweden seeks to generate 12 TWh of renewable electricity during the period, 2007–2016. Producers of green electricity receive a certificate for every MWh of electricity produced from renewable resources. The producer of green electricity could sell such a green certificate and receive an extra income in addition to sale of electricity. This provides an incentive to invest in new renewable electricity. The green certificate is valid for 15 years. Those companies that will enter the market in 2016, will have the benefit till 2030.

21.3 COUNTRY CASE HISTORIES

21.3.1 The Dutch Green Electricity programme

The Dutch green electricity market is the most successful in the world. The success is all the more laudable considering the fact that The Netherlands is a heavily urbanized country, and has few renewable energy resources. Also, the country has access to inexpensive and large Groningen natural gas field. That countries like Holland and Belgium observe a vegetarian day in a week is an indication of the profound feeling they have for all things green. There is little doubt that this psyche contributed in a subtle way to the success of green electricity programme.

By May 2003, 1.8 million households (i.e. 26% of all households) signed up for green electricity. This may be compared to 3 to 6% green penetration in what is considered to be a successful programme component (but not the whole programme) in USA. In 2001, one-fourth of the green electricity sales went to large non-residential green buyers, such as, Dutch Railway, Amsterdam Municipal Water Company and Utrecht City government. Also, the Dutch government purchased power from the neighbouring countries, Germany and France, to avoid carbon emissions.

The are four reasons (in the order listed below) for the success of the The Netherlands green electricity programme (Komor, 2004, p. 110):

(i) *By levying heavy taxes on fossil fuel-based electricity, the price of green electricity was rendered comparable to fossil fuel-based electricity.* This is by far the most important policy action. The *Ecotax* (called REB in Dutch) per kWh is paid directly to the consumers in proportion to their consumption. The Netherlands Ecotax is as high as equivalent of US cents 4.8/kWh in 2001 (compare this with the U.K. climate change levy of US cents 0.6/kWh). An electricity user pays an additional tax of US cents 4.8/kWh, if he opts for non-green electricity. When a consumer is able to get green electricity at the same price as brown electricity, he would natural go in for green electricity, because they get the green attributes at no cost.

    Though the availability of green electricity at the same price as brown electricity is of critical importance, it did not work everywhere. Though Ecotricity products in U.K., and California residential electricity market were offering green electricity at the same price at brown electricity (some times even lower), the kind of market penetration that was achieved in The Netherlands was not achieved in U.K. and California.

(ii) *The green electricity market got into the act early in the game.* Though the residential consumers did not have access to green electricity till 2004, they had access to green products of the competing providers from 2001. The retailers used this window of opportunity to build brand awareness, enlarge their customer base, and promote green market in general. When the residential customers had the option to choose between brown and green electricity in 2004, they opted for green electricity in large numbers.

    In Sept. 1999, the Dutch environmental group WWF mobilized about 2000 volunteers in a marketing campaign with a catchy slogan, "Don't let the North Pole melt ... choose green tariffs". In a spectacular gesture, the volunteers laid a 270 km. long green ribbon along the Dutch coastline. The publicity campaign invoked how the global warming would lead to the melting of Arctic ice, and reduce the habitat of the polar bear, and that the greater use of green electricity is the only way to mitigate this disaster.

(iii) *Dutch companies made use of innovative promotional techniques*: When green choice was introduced from July, 2001, companies undertook various promotional measures. Innovative inducements were offered when a customer signs up for green electricity, such as, Echte Energie providing USD 18 gift vouchers through health food stores, and Caplare providing USD 10 international phone call vouchers to Turkish, Arabic, Vietnamese and Chinese customers. The Greencab company offered taxi service using electric cars powered by green electricity, at no extra cost. Through 700 gasoline retail outlets, Shell provided guarantee of a year of green electricity at a fixed price.

(iv) *Incentives to producers of green electricity*: Apart from the discentive of taxing the producers of fossil fuel electricity, the government provided supply-side incentives to the producers of green electricity. Those investing in green funds were given tax exemption. Some green technologies were allowed accelerated depreciation. Tax credits were provided for some technology investments. Direct support payments were made to renewable generators. A green certificate trading system was brought into existence.

The Dutch model which uses taxes and consumer choice, rather than regulation, to promote renewables, is surely a success since it has been able to achieve market penetration of 26% for green electricity. The Netherlands does not produce enough green electricity, and is therefore compelled to import green electricity from the neighbours. Since a EU-wide green certificate system is not yet operational, this creates administrative problems.

21.3.2 The USA Green Electricity Market

The high per capita CO_2 emissions (19.73 t) of USA are attributable to its high-energy consumption (332 GJ/capita) involving coal and oil. Till recently, USA has been the largest emitter of greenhouse gases (now China has the dubious distinction). Right-wing politicians, industry-funded free-market think-tanks, and contrarian scientists mounted an extensive global warming disinformation campaign to deny that global warming is occurring, let alone that it is caused by greenhouse gas emissions from the industries. Even as recently as November, 2008, the US Chamber of Commerce warned that mandatory CO_2 reductions would have "... a devastating impact on businesses, ... farmers, the fragile economy and job creation". For eight years (2000 – 2008), the Bush White House turned a deaf ear to calls to cut greenhouse gas emissions, and expand renewable energy.

The US Green electricity market has been a mixed bag. As of Dec. 2002, the most successful green power programme signed up 3 to 6 % of the residential customers. The degree of market penetration depended upon how effectively the energy suppliers applied essential marketing principles (such as, branding and market segmentation). The new renewable capacities built and planned in USA as of Dec. 2002, to serve the green market, are given in Table 21.2 (source: Komor, 2004, p. 100).

Table 21.2 New renewable capacities built and planned in USA

Type	*MW installed*	*MW planned*
Wind	913	302
Solar	4.8	1.4
Small hydropower	8.6	2.0
Geothermal	10.5	49.9
Biomass	45.1	76.1
Total	**982**	**431**

The US green power market is relatively new. It was successful where it partnered with environmental groups. Building such partnerships is time-consuming and difficult, but it is well worth the effort.

The whole picture changed profoundly overnight when President Obama came to power in USA in 2009.

According to Sharon Begley (*Newsweek*, Jan. 5, 2009 issue), three big steps are needed to jump-start renewable energy and green technology: (i) Wind and solar sectors need huge upfront capital, and their cash flows are critically dependent upon their ability to borrow large sums of money at low interest rates. To facilitate this, the government should provide loan guarantees for the construction of wind and solar firms, extend tax credits and make them transferable, or even provide direct government loans, (ii) Unequivocal signaling to the industry that CO_2 emissions will cost them, and that they have to cut emissions 80 % by 2050. The industry sees the writing on the wall. Many industries, including the Duke Energy Corporation (which is the third largest emitter of carbon dioxide), ALCOA, Caterpillar, General Electric, BP America, etc. support the mandatory CO_2 cuts, and (iii) Increasing the demand for green energy. If the US Federal Government which has 8 600 buildings, and 213 000 vehicles, switches to green power, that will serve to spur the demand for green energy. The Federal buildings in upstate New York have switched over to wind power, and 110 000 sq. ft. of solar panels have been installed on the complex that houses the mission control for scientific satellites. If the Federal government goes in for solar installations in a big way, that step alone could bring down the cost of solar panels by half, even without needing technological breakthroughs.

The Obama Administration has embarked upon an ambitious, $787 billion stimulus plan. Among the goals of this plan is the reduction of the CO_2 emissions by reducing dependence on fossil fuels and increasing the role of renewables in the energy generation. In his speech in Iowa on Earth Day (Apr. 22, 2009), President Obama said that US plans to meet 20% of its electricity demand (as against 2% today) through wind power (land and offshore) by 2030, and this would involve the creation of 250 000 new jobs. The new Stimulus programme provides $50 billion for energy programmes focused chiefly on energy efficiency and renewable energy. The programme provides funding for "smart" electricity grid; subsidise loans to renewable energy projects; support state energy efficiency and clean energy grants; making federal buildings more energy efficient; grants for research in advanced batteries and electric vehicles; support for basic research in climate science, biofuels, high energy physics; tax incentives for renewable energy; extending tax credit for energy produced from wind, geothermal, hydropower and landfill gas; grants to build renewable energy facilities; tax credit for the purchase of energy-efficient furnaces, windows, doors and insulation; tax credit to families to purchase plug-in hybrid vehicles, and so on.

There is every reason to hope that green energy systems in USA are set to make good progress. Here is an example which demonstrates what a committed government could achieve.

21.3.3 U.K. Green Electricity Market

Komor (2004, p. 82) who made a detailed analysis of the U.K. green electricity market, concluded that it did not work.

U.K. opened its electricity system to retail choices for large industrial houses in 1990, and to residential users in 1998–99. One-third of all the residential users switched providers by late 2001. Green electricity option was available since 1996 when the Renewable Energy Company offered "Ecotricity" (landfill gas-sourced electricity) to commercial electricity users. By 2000, green electricity option became available to all users. If consumers had the choice across other variables, such as greenness of supply, dependability of service, quality of power, etc., there might have been an incentive to go in for green electricity. But cost alone (not greenness or any other factor) became the focus of switching. Many non-switchers were satisfied with their default provider, and thought that switching was not warranted.

The default providers thought that green electricity is a small market and not worth extensive marketing promotion. It turned out to be a self-fulfilling prophecy. There have been some notable exceptions. The RSPB (Royal Society for the Preservation of Birds) which has more than a million members, pitched in for green electricity, and RSPB Energy turned out to be a successful venture.

In early 2000, U.K. Government came up with three pro-renewables policies.

(i) The Climate Change Levy (CCL) is a tax on energy use with the objective of increasing the energy efficiency and reduces carbon emissions. Starting from Apr. 1, 2000, electricity users have to pay an additional tax of 0.43 pence (US cents 0.62 per kWh) on the electricity consumed. This tax is significant, and accounts for 11% of the U.K. industrial electricity price. As renewables are exempt from this, the green electricity is that much cheaper. CCL made green electricity more expensive, because of increased demand. It ended up as a source of uncertainty and risk.

(ii) New Electricity Trading Arrangements (NETA) is a wholesale electricity trading system, by which buyers and sellers agree on prices and make deals. This has created problems for wind-based green electricity providers, since wind power cannot serve as a baseload plant, because of its intermittency. Wind generators saw a 25% drop in prices.

(iii) The Renewables Obligation (RO): The Renewable Obligation requires that companies selling electricity to end users, should ensure that 10% of the electricity they supply comes from renewable sources by 2010. By forcing suppliers to purchase more green electricity, it will increase the demand for green electricity production.

Komor (2004) came to the conclusion that the UK green energy market has not yet taken off.

21.4 LESSONS

A combination of policy incentives and discentives, publicity campaigns and innovative marketing are required in order for the green electricity market to succeed.

(i) *Discentives*: Clear signal by the Government to the industry that emission of CO_2 will cost them, and trade-and-cap regime is unavoidable; Levying heavy taxes on fossil fuel-based electricity to make the price of green electricity comparable to fossil fuel-based electricity.

(ii) *Supply-side incentives to producers of green electricity*: Wind and solar sectors need huge upfront capital, and their cash flows are critically dependent upon their ability to borrow large sums of money at low interest rates. To facilitate this, the government should provide loan guarantees for the construction of wind and solar firms, extend tax credits and make them transferable, or even grant direct government loans, besides allowing accelerated depreciation to green technologies,

(iii) *Increasing the demand for green energy, starting with government establishments.*

(iv) *Public consciousness and Marketing*: Mobilization of the environmental groups for the promotion of "green" consciousness, and partnering with the environmental groups in the identification and implementation of suitable projects; Promotional campaigns by companies.

(v) Research in basic science, applied R&D, Demonstration and Deployment and Technology Learning, to develop ways and means of improving efficiency, bring down capital costs and operation and maintenance costs, create jobs, identify ways of mitigating environmental impacts, of low-carbon energy options.

REFERENCES

Anderson, D. (2006) *Costs and Finances of Abating Carbon Emissions in the Energy Sector*. London: Imperial College.

Aswathanarayana, U. and Rao S. Divi (2009) *Energy Portfolios*, Boca Raton: CRC Press

Boyle, G. (2004) *Renewable Energy*. Oxford: Oxford University Press.

Dahl, Carol A. (2004) "*International Energy Markets: Understanding pricing, Policies and Profits*", Tulsa, Oklohoma: Pennwell Corporation.

Gates, Bill. 1999. *Business at the Speed of Thought*. Newport Beach,CA: Books on Tape

IEA (2007) *World Energy Outlook*, International Energy Agency, Paris.

IEA (2008) *Energy Technology Perspectives*. Paris: International Energy Agency.

Komor, Paul (2004) *Renewable Energy Policy*. Lincoln, NE: iUniverse.

Murphy,L., and P. Edwards (2003) *Bridging the Valley of Death: Transitioning from Public to Private Sector Financing*. Golden, CO: National Renewable Energy Laboratory.

Pearce, D.W. and R. Kerry Turner (1990) *Economics of Natural Resources and the Environment*. Baltimore: The Johns Hopkins University Press.

Stern, N. et al. (2006) "*The Stern Review on the Economics of Climate Change*", UK Treasury, Jan. 2007. Cambridge, U.K.: Cambridge University Press.
Stiglitz, J., and Wallsten, S. (1999) Public – Private Technology Partnership: Promises and Pitfalls. *American Behavioural Scientist*, **43** (**1**), 52–77.
UNDP (2007) *Human Development Report*. 2007. New York: UNDP.
U.S. Department of Energy (2005) "*A Manual for the Economic Evaluation of the Energy Efficiency and Renewable Energy Technologies*", Honolulu, Hawaii: University Press of The Pacific.
Viscusi, W.K. John Vernon and J. Harrington, Jr. (1996) *Economics of Regulation and Anti-trust*. 2nd. Edition. Cambridge, Mass.: MIT Press.

Section 6

A green new deal

K.M. Thayyib Sahini (IAEA, Vienna) (Editor)

Chapter 22

Goals of the green new deal

K.M. Thayyib Sahini, IAEA, Vienna

22.1 INTRODUCTION

It is said that the Chinese term for Crisis consists of two ideograms representing Danger and Opportunity. We have to integrate "green" technologies with economics and policy in order to mitigate the adverse impact of the climate change, and create jobs in the process.

A green new deal is aimed to deliver more service to more people, develop the economy, reduce relative poverty and eradicate extreme poverty, generate more energy, educate more people, utilize resources more effectively and raise the standard of living within the overall frame work of a low carbon future. Energy generation being the reason for two thirds of overall greenhouse gas emissions, this chapter focuses on energy related goals of a green new deal, involving smart electricity grid, zero carbon electricity production, zero carbon transport, zero carbon buildings and zero carbon industry.

22.2 SMART ELECTRICITY GRID

The technical details of smart electricity grid have been given in chap. 16.6.

The function of an electricity grid is to transmit and distribute electricity from the source of generation to its consumers. Even though the technology for different electricity generation methods have advanced with time, the absence of a corresponding advancement in grid technology is the reason for causing problems in electricity transmission such as black outs, brown outs, outage, transmission loss and theft. Both developed and developing countries faces many of these grid related challenges.

As a technological solution to these problems, "Smart Grids" offers a web-based, digitally controlled intelligent delivery system through facilitating bidirectional flow of electricity and data, with multiple benefits. Smart grids combine the principles of decentralization and democratization of energy generation and consumption, improving energy and economic efficiency, better energy and electricity security.

Our low carbon future depends upon the effective utilization of renewables, and the generic nature of current renewable energy technologies is erratic, intermittent, periodic and seasonal. For example, the efficiency of solar panels depends upon the clearness of sky and electricity generation is limited exclusively for daytime. Like wise, another major renewable energy source is wind, its erratic nature makes it an unreliable base load source of power, if the wind doesn't blow at all. Currently an electricity mix of a state is comprised of these many sources, and peak load demand is met through stand-by generators. A smart grid control can resolve this by better digitally controlled demand management. With the use of advanced electronic devices such as Thrysters, HTSC power limiters, capacitors and digital transformers, a smart grid control can reroute power to avoid congestion and deliver both low and high quality power as well as automatic detection of faults. Integrating High Temperature Superconducting Storage (HTSC) transmission wires with Hydrogen pipelines facilitates the transmission of electricity through superconductors. This will prevent transmission loss and improve efficiency as well as the flow of liquid hydrogen, incidentally keeping the superconductor transmission lines from heating. Another promising transmission option is Quantum Nano Transmission wires which is stronger, lighter and has ten times higher electrical conductivity than copper wires. The advantage of this type of transmission wires is that they can be buried underground without shielding or special trenching (Anderson 2004). Most of these technologies need to be developed further to become economically viable alternatives.

Smart Grid not only improves the technical and economic efficiency of transmission lines, distribution networks and control systems, it also offers a better resource management system for the consumers in the form of smart meters, real time pricing solutions, sale of electricity generated from domestically installed solar panels and wind mills. Though Smart Grid concept offers smart solutions compared to the traditional grid systems, smart grid components are much more expensive. Anderson (2004) makes a cost comparison for Smart Grid technologies and found, cost wise most of them are higher by a factor of 10 to 1 000. Such a factual analysis shows that, smart grid technologies need huge investments, and RD&D with combined support from Governments, industry and as well as the general public.

Though challenges remain in the form of cost, investment, and public acceptance, many governments started recognizing Smart Electricity Grid as a reliable option and system to mitigate climate change, and ensure energy security and efficiency. Being one among the largest electricity grids in the world and still based on traditional technology, the US Electricity grid is about to transform itself in to a smart grid. The $3.4 billion grant announced by President Obama is intended to spur the transition to a smart electricity grid. The investment package will be matched by industry funding for a total public-private investment worth over $8 billion (US DOE, 2009).

As a comprehensive smart grid investment package, in addition to the creation of thousands of Jobs, it aims to make the US electricity grid more reliable and to reduce power outages that cost US economy $150 billion a year. It also aims to install 850 censors (Phasor Management Units) which will cover the entire US grid that enables grid operators to monitor grid conditions, there by preventing minor disruptions which otherwise might cause cascading events leading to regional outages or black outs. Installation of 200 000 smart transformers, and almost 700 automated substations will also reduce the number of power outages. On domestic front, with an aim to empower consumers, the package aims to install more than 40 million smart meters, 1 million in-home displays, 170 000 smart thermostats and other 175 000 load control devices which will enable the domestic consumers to realise more economical use of energy. It also aims to reduce the peak electricity demand by more than 1 400 MW (US DOE, 2009).

Like in other parts of the world, the European electricity grid is also designed and developed to distribute electricity generated principally from carbon based generation technologies. In order to meet the future demand for low carbon energy and to enhance energy efficiency, the European Technology Platform (ETP), Smart Grids, was set up by European Commission in 2005 to create a joint vision for the European networks of 2020 and beyond (EC, 2006a). The EU's Strategic Research Agenda (SRA) of the ETP estimates that, "Looking ahead, EU Member States will need to invest in excess of €750 billion in power infrastructure over the next three decades, divided equally between generation and networks" (about €90 billion will be invested in transmission and €300 billion in distribution networks). The SRA has five primary research areas focused on Smart Distribution Infrastructure (Small customers and Network Design), Smart Operation, Energy Flows and Customer Adaptation (Small Customers and Networks), Smart Grid Assets and Assets Management (Transmission and Distribution), European Interoperability of SmartGrids (T&D) and Smart Grids Cross-Cutting Issues and Catalysts (EC, 2006b). In addition to these initiatives, researchers in Europe are also looking outwards for its low carbon future by linking Europe with Africa through the super smart grid (SSG) and thereby utilizing the enormous resource of wind and solar energy available in the deserts of North Africa. The Super Smart Grid would use High Voltage Direct Current technology, which will operate on the top of the current HVAC grid to overcome transmission losses (Battaglini et al., 2009).

The importance of assuring a low carbon future, necessity of improving energy efficiency, and demands from fast growing economy, shaped China's national policy and measures whose aim is to promote and develop a 'strong smart grid' by 2020, which will include distributed power, plug and play and installation of smart meters. The development and construction of Southeast Shanxi-Nanyang-Jingmen ultra-high voltage (1 000 kV UHVAC) transmission line is promising development in this direction. The state grid corporation of China is also building two additional UHVAC transmission lines, each with more that 2000 km in length linking the dams in the southwest to users along the eastern coast of China. Considering the requirement of long distance transmission, the main characteristic of China's smart grid is the UHVAC, which reduces the transmission losses and improves efficiency.

A retrospective analysis of the performance of traditional grid system shows that there was reluctance to push new technologies and systems to modernize the electricity grid. Currently, the world's leading economies are investing billions in Smart grid technologies. Even though there are resource constraints, developing countries can also follow this path of progress and development. As many developing countries still do not have modern grids, this is a good time to install entirely new transmission and distribution systems involving modern and upcoming technologies. From a technological and practical point of view, technological leapfrogging is possible and favors the less developed countries, just like what is happening in telecommunication revolution in the developing countries.

22.3 DECARBONISING ELECTRICITY PRODUCTION

According to IPCC (2007) since 1970, the emissions from the energy supply sector has grown by over 145% and by 2004, CO_2 emissions from power generation represented over 27% of the total anthropogenic CO_2 emissions. In 2007, the global CO_2emissions from electricity generation were about 11 gigatonnes (IEA, 2009).

Electricity, being the lifeblood of modern civilization and a necessity for all economic activities, is a major contributor of carbon emissions. In order to mitigate climate change the world needs zero carbon electricity generation technologies. It is not only the source of electricity generation, but also the technology we use is the reason for carbon emissions. Life cycle emission analysis of various sources of electricity generation shows a range between 10 g/kWh to 900 g/kWh. Wind, Solar, Hydro and Nuclear can generate huge amount of electricity with out carbon emissions. If the whole lifecycle CO_2 emission of each of these generation technologies is analyzed, the emission ranges from 10–30 g/kWh (approx).

Wind: Though intermittent and with a low capacity factor of 20–40 percent, wind is a perfectly renewable energy source with out any CO_2 emissions during generation. The life cycle CO_2 emission analysis of wind energy is 13.5 to 24.5 g CO_2/Kwh (Ackermann, 2005, p. 20). Citing the American Wind Energy Association, Paraschivoiu (2002) estimates, based on an average US power generation mix, that a single 750 kW wind turbine operating at a site for a year with class 4 wind speeds (averaging 12.5–13.4 mph at 10 meters height) is expected to avoid a total of 1 179 tons of CO_2. The levelised cost of electricity produced by Wind turbine averages approximately 5.5–7 euro cents/kWh, and also has a front loaded cost structure, with 65–75% of the investment going for the turbine (Kroch et al., 2009, p. 9). The total wind energy potential is four to five times higher than the current average global power consumption of 15 TW, of which the European Union accounts for half of the generation. Considering this huge potential, wind electricity could be a strong contributor in the energy mix of low carbon economy for all time to come. A combination of government support with corporate investment in the wind energy shows promising growth, making it the fastest growing sector in renewable energy. The growth rate and installed capacities of wind energy in the leading economies, with China 106.5% (12 210 MW), USA 49.7% (25 170 MW), Germany 7.4% (23 902 MW), India 22.1% (9 587 MW), proves this fact (WWEA, 2009).

The story of Muppandal village in South India is an example of renewable energy ushering in economic change. Once an impoverished village, Muppandal and its surroundings lacked economic and job opportunities for the local population. The installation of wind turbines and the availability of electricity created job opportunities for the local population and provided energy for work. Same is the case with Inner Mongolia in China. One notable example for the indirect economic benefit to local economy is "the Huitengxile Wind Farm which became an important attraction of Qahar Youyi Zhongqi, the city where this wind farm is located. Now about half of the local residents' daily income is gained from tourists" (Han et al., 2009). In addition to the generation of zero carbon electricity for the national grid, the US DOE's "Wind Powering America" program aims to create 80 000 jobs and $1.2 billion income for rural landowners and farmers by installing wind turbines in 20 years (US-DOE, 2003). In short, the wind blowing over our head is giving us hope and opportunity towards a low carbon economy; even though siting, land acquisition, installation costs, grid facility and last but not least, the visual disturbance and dislocation of the migratory patterns of birds are posing challenges to policy makers and investors.

Solar: The greatest source of energy, solar power, reaches earth with a maximum density of one kilowatt per square meter and has the potential to generate 1 000 times more energy than the total world energy demand. A zero carbon energy source, the life cycle emissions (CO_2 equivalent) for Photovoltaic (PV) ranges from 17–49 g CO_2/kWh based on US scenarios, which could be reduced even further in future (Fthenakis and Kim, 2007). Currently two generation technologies exist, photovoltaic and concentrated solar power (CSP). According to IEA World Energy Outlook 2009, electricity generation from Photovoltaics will reach almost 280 TWh in 2030, up from just 4 TWh in 2007 and CSP plants from less than 1 TWh to almost 124 TWh in the same time frame. A McKinsey forecast points out that by 2020, the cost of solar electricity would fall from the current 30 US cents/kWh to 10–12 cents at a projected 30–35% annual growth of capacity, from 10 Gw to 200–400 Gw in the same period, which requires a capital investment of more that $500 billion (Lorenz et al., 2008). By 2050, the ACT scenario forecasts a projected growth of solar power to 2 319 TWh/yr and 4 754 TWh/yr in the Blue scenario. The 2008 estimate shows that the world capacity of photovoltaics doubled from 2.65 to 5.8 GW (Euro Barometer, 2009).

Governmental support and investments in many leading economies show the way for an important role for solar energy in the future energy mix and economic development. With an annual production growth rate of 40%, the European Photovoltaic industry's turnover is around 10 Billion Euros. It created approximately 700 000 jobs (EC, 2009).

Germany, being the largest market for PV cells, has a feed-in tariff regime, which requires utilities to pay a guaranteed rate for consumers who feed solar power in to grid. From 850 MW of solar PV in 2006, Germany added another 4 500 MW by 2008 into national grid with a cumulative solar power generation of 5 351 MW. The German spearheaded Desertec project aims at harnessing the solar energy from Sahara to meet ~15% of Europe's electricity consumption with an investment of 400 billion Euros. China aims to generate 20 000 MW of solar power by 2020 and India, through its national solar mission, plans to generate 20 000 MW by 2020 in three phases. The action plan also includes deployment of 20 million solar lighting systems for rural areas by 2022. Sun light is an everlasting source of energy, with a generation system so simple and conveniently modular to any size and space, with instant installation possibilities.

Hydro: Compared to other sources of energy, hydropower is cheap and flexible; even to meet the peak load demand. With a very low life cycle emission of 15 g CO_2 equivalent/kWh, hydropower has a zero carbon emission profile during generation (Gangnon and van de Vate, 1997). Building large dams often create various socio-political and ecological problems, such as relocation of people, submergence of vast areas of vegetation, along with the benefits of power generation, flood control and irrigation. Between 20 and 25% of world's large scale hydro potential has been developed already and generating 15.6% of world's electricity, but still harnessing the 'run of the river' potential is attractive. The potential of such small scale (less than 10 MW) hydropower projects is around 500 GW and only one fifth of such potential has been harnessed so far (Freris and Infield, 2008).

As Europe plans to harness solar energy from Sahara, Africa is looking towards its own hydro electric potential. The grand Inga project on Congo River envisions generating 40 000 MW, enough energy to power several African states. China is planning to raise its hydroelectric capacity up to 300 GW, of which 225 GW is from large and medium scale plants and 75 GW from small scale hydro projects (NRDC, 2007).The small scale hydro projects number 40 000 in China and it electrified 653 rural counties (Huang and Yang, 2009). Though the scope for new large dams is limited, the potential for SHP is still significant in Europe. European Small Hydro Power Association estimates that by 2020, in Europe, the small hydro capacity could reach 16 000 MW. Another great possibility for SHP is in India, which has an estimated SHP potential of 15 000 MW of which less than 20% is being utilized so far.

Harnessing hydroelectricity is a sustainable solution for our carbon worries. A combination of large and small hydro power projects can offer around 7 200 TWh/year (technically feasible hydro potential is 14 370 TWh/year). IEA hydro (2000) gives the breakdown of this economically feasible potential with: Africa-1 000, Asia 3 600, North and Central America-1 000 and South America-1 600 TWh/year respectively.

Nuclear: Among the zero carbon electricity generation sources, nuclear has highest capacity factor, around 90%, and lowest lifecycle emissions, which lie between 2.8–24 g CO_2 eq/kWh e (Weisser, 2007). Nuclear power plants generate electricity with nearly zero carbon emissions. Compared to other zero carbon energy sources, nuclear is free from locational constraints, offers economies of scale and possibility of huge capacity, minimum land requirement, and continuous base load operation, transportability of fuel and stability of fuel price. At the same time it faces the challenges of huge investment, long construction times, philosophical questions of radioactive waste as well as proliferation concerns and ideological opposition.

Nuclear currently generates 370 GW and supplies electricity in more that 30 countries. 56 new units are under construction and around 60 countries showed interest to IAEA in building new nuclear power plants. Basically nuclear reactors are all thermal power houses, and different reactor types offer heat between 300–950 degrees (Kupitz and Podest, 1984). In addition to electricity generation, this wide spectrum of nuclear heat offers many possibilities, especially for sea water desalination, district heating, hydrogen production and as well as for other industrial applications. IPCC's fourth Assessment report (2007) accepts nuclear as a mitigation technology, though it is not yet acknowledged in CDM and JI programmes. Many states in the oil rich Middle East are seriously planning to build nuclear power plants. Generating nuclear power for the primary energy needs instead of burning fossil fuels is an environmentally sound decision taken by those states. UAE already awarded contracts to South Korean consortium led by Korea Electric Power Corporation (KEPCO) to build four NPPs with an investment of $20 billion. India is developing its nuclear energy sector to generate 20 000 MW by 2020. China is investing to generate 60 GW of nuclear power by 2020 and 160 GW by 2030. Like wise, US, UK, France, Russia, and all the leading economies are planning to build new units. Morocco proposes to build two 1 000 MW

NPPs under 'Nationally Appropriate Mitigation Actions' (NAMAs) under the terms of Copenhagen Accord.

A post Chernobyl safety culture, advancement in technology, stable availability of uranium and above all, the low carbon profile favors this technology and a source of a low carbon future, while offering energy security. Nuclear technology and science has a wide range of applications other than energy, such as health, industry, agriculture, water desalination, etc. Thus, the experience of building and operating nuclear power plants tends to develop and nurture a knowledge society capable of assisting in the process of development of a country with the benefits of nuclear science.

22.4 DECARBONISING TRANSPORT

World wide, transport sector alone accounts for 20–25% of total green house gas emissions. As the world becomes more and more integrated due to the forces of globalization, movement of people and goods necessitates voluminous amount of transport. By 2050, as much as 30–50% of GHG emissions are estimated to come from the transport sector (Fuglestvedt, et al. 2008). Absence in technology change and long standing consumer behaviour could increase the emissions from passenger travels between 11 and 18 billion tons CO_2 a year by 2050 (Schaefer, 2009). Present fossil fuel powered transportation system need to be changed drastically in order to achieve a low carbon future. In OECD countries, road transport alone accounts for more than 80% of transport related energy consumption (OECD, 2004). The US passenger and freight transportation release 1 920 million tons of CO_2 and the total worldwide emissions from transport sector are 6 370 Mt CO_2 in 2005 (Schaefer, 2009). Even though Stern review (2007) predicts that the transport sector is likely to remain oil based for several decades to come, increasing efficiency, use of bio-fuels, hybrid, and electrification (generated from low carbon sources) could reduce the total emissions.

Use of hydrogen as fuel for transport is another opportunity to achieve a low carbon future. There are two ways to use hydrogen as a fuel for transport. Hydrogen can be used in (modified) internal combustion engines, which produce a very clean exhaust. The other way is to use hydrogen fuel cells which produce electricity.

The concept of Vehicle to Grid (V2G) is attractive in the sense of maximizing resource utilization. Parked electric car batteries are a good source of power to manage peak load demands. Connected to grid, this distributed source can offer large amount of power collectively. V2G systems facilitate consumers to sell electricity stored in their electric vehicles batteries to national grid through smart grid. To get an idea about V2G, for example, if the 27 million cars in UK are replaced with electric cars having an average battery capacity of 15 kW, this would provide a total capacity of 405 GW (CAT, 2007).

De-carbonization of transport sector needs to be coordinated with electrification (electricity from low carbon sources) and development of mass public transport systems such as developing an extensive network of railways. Expansion of mass transit systems offers immense employment opportunity as well as reducing the use of private vehicles. Mass transit systems employ 367 000 workers in USA, and 900 000 in European Union. In addition to the carbon savings, the 10 year US federal investment program in high speed rail system has an employment potential of 250 000 jobs. South Korea is another example for de-carbonizing the transport sector through public transit system. South Korea which invested $7 billion in mass transit systems including railways, is expecting to create 138 000 jobs, besides meeting its climate change mitigation goals (Barbier, 2009).

Reducing the carbon emissions from transport sector needs a coherent and coordinated action of facilitating policy with strong regulatory mechanism, cutting edge technology driven by research and development, and an economy which pays value for investment. For example, as a policy mechanism, the EU regulation of 130 gm/km and Japanese regulation of 125 g/km for passenger vehicles is a facilitating policy, which promotes research and development of energy efficient and eco friendly engines by automobile companies and as well as developing public consciousness to regulate the emissions which they cause.

22.5 DECARBONISING BUILDINGS

Buildings use energy for heating, cooling and illumination. Besides, there are many appliances and other electric systems such as lifts, elevators, refrigerators, stoves and many other domestic appliances. Apparently, major part of energy is used for heating, cooling and illumination. The world average per-capita energy use is considered as 2,000 W, which varies from 600 W for Bangladesh to 12 000 W for United States. It is not necessary to use 12 000 W of energy to have a life with modern amenities. The Swiss Federal Institute of Technology developed the concept of a 2 000 Watt society. According to this frame work, US energy use has to be drastically cut down by 80 to 83 percent, and Western Europe (W. Europe's per-capita energy use is 6 000 W) by 67 percent (Yudelson, 2009). De-carbonizing the building sector has to be associated with the drastic reduction of per capita energy use in the developed world because, buildings account for 30–40% of the total energy use. In this regard, optimizing building space utilization is equally important since the per-capita energy use depends upon building space available per person. For example, United States has over 850 square feet, EU has 550 to 650 square feet and China has less than 350 square feet of building space per person (Kats, 2009). Optimum utilization of building space has an important role in improving energy efficiency and de-carbonizing building sector.

The energy budget of a building is defined by the climatic and weather conditions as well as its design. A zero carbon building is the one which is carbon neutral. One way to make a building carbon-neutral is to use energy from a low carbon source. However, a life cycle analysis will show that even if a building uses energy from zero carbon source, the materials, and the construction process, maintenance, renovation and even deconstruction might have emitted carbon. To resolve this carbon quotient, the building need to be equipped with renewable energy generation systems such as solar or wind turbines as well as using geothermal energy (connected to a smart grid), thereby generating surplus low carbon energy than is being used.

De-carbonizing the building sector can be approached with tailor made policies aimed at public buildings, corporate buildings, and private buildings such as housing estates. Refurbishment of the existing buildings to make them energy efficient should be combined with the installation of renewable energy generation systems, which will make them carbon neutral. For the new buildings, the low carbon imperative can be integrated right from the selection of location and as well as the design and the installations which use energy.

On a policy level, the European Union directive on energy performance of building (EPBD) has the commendable aim of reducing the energy use in buildings to meet the Kyoto targets. Following this directive the UK government decided that social housing would be carbon free by 2016, the private housing by 2018 and the commercial buildings by 2019 (Yudelson, 2009). The recently announced 'Home Star Energy Efficiency Retrofit Program' by President Obama is aimed to increase the energy efficiency of US homes. It offers subsidies and rebates for home owners to improve the energy efficiency of homes by renovation and upgrading insulation, duct sealing, water heaters, HVAC units, windows, roofing and doors (Whitehouse.gov).

Even though building technology and engineering have advanced considerably during the last 50 years, there is an absence of corresponding improvement in energy efficiency of buildings. Along with the growth of the concept of sustainable development and climate change consciousness, the world began to realize the necessity of such measures resulting in the spread of green building concepts. British Research Establishment Environmental Assessment Method (BREEAM), Green Star (Australia) Comprehensive Assessment System for Building Environmental Efficiency (CASBEE) in Japan, and LEED, (which is a world wide standard) are all mechanisms aimed at standardizing and promoting the sustainability of buildings such as improving energy efficiency and de-carbonizing building sector. Among these, the German 'passivhaus' designation is considered as having the highest efficiency standard for buildings (Kats, 2009). Considering the development in building policy atmosphere across the countries, there is an emerging vision and outlook which realizes that de-carbonizing the building sector is necessary as well as affordable in order to have a low carbon future, and the challenges are investment, technology gap between the developed and developing world, attitudes and mindset of public against change.

22.6 DECARBONISING INDUSTRY

The whole saga of modern economic growth started with the industrial revolution, which is also the cause of earth's changing climatic equilibrium. Anthropogenic emissions due to industrialization fuelled by fossil fuels still continue to be the major reason for Global Warming. They will be continuing their contributing role till 2030, the date by which carbon emissions are to be stabilized, to decline afterwards. Towards such a goal, we need to de-carbonize the industrial sector through the overall improvement of efficiency, technology and utilization.

Decarbonising industry is an all-encompassing term, which includes most of the productive activities. A zero carbon industry will have its end products with minimum emissions, possibly carbon neutral, and offsetting the rest with mitigation actions such as planting trees, or sequestering carbon. An increased energy efficient industry powered by low carbon energy sources could de-carbonize the industrial sector.

The total industrial use of energy in 2006 amounted 156 EJ, which is 32 percent of world energy use, and two fifths of global energy related carbon emissions (UN energy, 2009). "Large primary material industries such as chemicals, petrochemicals, iron and steel, cement, paper pulp and other minerals and metals industries account for more that two third of this amount and substantial reductions are possible (IEA, 2007). Large-scale industries are not the only carbon contributors. The world computing industry is releasing two percent of total emissions which is more that the aviation sector's contribution. According to IEA (2007) estimate, the industrial CO_2 reduction potential amounts 1.9 to 3.2 gigatonnes per year, i.e. about 7 to 12% of today's global CO_2 emissions.

Decarbonising the industrial sector is a long drawn out process, because integrating energy efficiency and carbon saving mechanisms needs additional investment. In this regard the developing economies are in an advantageous situation, because the late bloomer position gives them the opportunity to technological leap-frogging and even technological horse jumping. That's why African countries have the most energy efficient Aluminum smelters and India operating the most energy efficient cement kilns. Initial investment should not be a constraint for improving the energy efficiency, because the returns in long term are high. According to UN-Energy (2009), "investment in energy efficiency of as little as 1.6% of current global fixed capital investment each year to 2020 would produce an average return of 17 percent a year". It is true that industrial activity will cause carbon emissions, but we can aspire and work to increase the energy efficiency and reduce the amount of emissions during the operation by integrating better technologies. This action of technological improvement has to be supplemented with carbon offsetting and contributing for the development and protection of carbon sinks by the industrial sector, according to their emissions.

22.7 CONCLUSION

Different realities such as global warming, rising carbon emissions, daunting demands for energy due to economic and population growth and peaking of fossil fuel resources necessitate alternative sources of energy. Consecutively this alternative energy sources have to be low carbon emitting in order to stabilize the rising global temperature. These demanding factors require an immediate and drastic change of attitude, policy and investment priorities. Activities related to the use and generation of energy is the major area where the world can invest and work for a low carbon future- for securing sufficient energy, better standard of life, achieving the MDGs as well as preserving the planet for the future generation. With an outlook towards such a goal, we need a Green New Deal which will infuse fresh inflow of resource, ideas and capital to develop the basic sectors of economy. With a focus on energy and it related activities, this chapter explored the possibilities, potential and challenges of Smart Electricity Grid, and decarbonising electricity production, transport, housing and industry. There is little doubt that all these goals of our green new deal will necessitate huge investments, and demand political will. They will require public support as well as cooperation of the developed and underdeveloped world, haves and have-nots and corporations and governments. The problems are stupendous, but they will have to be overcome.

Chapter 23

Ways of "greening the economy"

Jayaraj Manepalli, Vienna

23.1 INTRODUCTION

"Green energy" is a term used to describe the energy that is generated from the sources that are known to be non-polluting. Adopting the green energy measures is an environmental-friendly way of generating power—as it is a remedy not only for the harmful effects of pollutants on our environment over the years but also useful to prevent future global warming. Any green energy policy is a result of a particular government's commitment for the environment and promotion of particular forms of technologies, apart from its own business and geo-political interests. Owing to the challenges we face and the growing realisation of the damages caused due to the predominance of fossil fuel based economy, there is an urgent necessity that we take steps towards reducing the carbon emissions and adopting green energy measures.

23.1.1 The challenges in the energy sector

The challenges we face owing to the dependence on fossil fuel based economy are enormous and are growing by the day. Today, the world still relies primarily on fossil fuels as its main source of energy. Fossil fuels are expected to account for 77 percent of the increase in demand between 2007 and 2030 (IEA, 2009). The demand for electricity is projected to grow by 76 percent during the same period, requiring a staggering capacity addition of another 4 800 gigawatts (GW), which is almost five times the existing capacity of the United States. The Report also estimates that still 1.3 billion people will lack access to electricity in 2030, compared to the 1.5 billion people today. Universal electricity access could be achieved with additional power-sector investment of $35 billion per year from 2008 to 2030 and with a modest increase in primary energy demand and related CO_2 emissions.

Another crucial challenge is the concentration of fossil fuel production in a small group of countries. This could lead to a situation where this small group of countries could have an ability to impose higher prices, thereby jacking up the prices worldwide. An example of such a situation is the 1973 oil crisis when the Organization of Arab Petroleum Exporting Countries (OAPEC) announced an 'oil embargo' as a response to the United States policy to aid Israel during the Yom-Kippur war. Price of oil increased four-fold and further crisis could be avoided only because of hectic political and diplomatic efforts and the subsequent end of the war.

The challenge also comes from the developing countries and fast growing economies. The developing countries are expected to contribute 70 percent to the increase in global primary energy use by 2030. As a result, the energy-related GHG (Green House Gas) emissions, mainly from fossil fuel combustion for heat supply, electricity generation and transport account for 70 percent of GHG emissions. Without the near-term introduction of policy actions, energy-related GHG emissions are projected to rise by over 50 percent by 2030 (IEA, 2007). This increases the need to orient the policies towards greening the economy at the earliest.

23.1.2 The Urgency

On a global level, it is estimated that for each year of delay in taking appropriate policy decisions (steps towards green economy) costs an additional 500 billion US dollars, in terms of mitigation costs between now and the year 2030 (IEA, 2009). The proposals by the government to alter any large-scale socio-technical systems to enable sustainable development goals are highly uncertain policy projects. The adoption of such policies might also lead to highly unfavourable domestic political climate—as it means giving preference to long-terms interests over the short-term pain. Some actions may affect the leaders' political survival due to some hard decisions that have to be made. Before those crucial decisions, there is a great need to raise awareness among the citizens on the necessity of taking the long-term interests into consideration. Despite the major risks and pain involved, there is an immediate need for the policy makers to focus on the ways to devise and implement green energy policies. The adoption of such measures has their own challenges as the policies have to come through a complex system of political balancing. Developing countries have priority in social as well as economic development. This needs a fine balancing and integrated policy-making cutting across many areas.

23.1.3 Green Energy

"Green energy" is often understood in the context of renewable energy and sustainable development by most governments and inter-governmental organizations. The concept of a green economy envisages an environmental friendly, green way of producing energy. Green Economy is still an evolving concept, but primarily could be understood as a system where certain sectors like renewable energy contribute in a major way to a country's economy. It should also visualize revolutionary changes without compromising on the people's aspiration for a good life. A green economy should enhance people's quality of life, not just "without compromising" it.

The United States Environment Protection Agency (EPA) describes green energy as "a subset of renewable energy and represents those renewable energy resources and technologies that provide the highest environmental benefit." (US-EPA, 2009). United Kingdom's recently enacted Green Energy (Definition and Promotion) Act-2009 defines green energy as "the generation of electricity or heat from renewable or low-carbon sources by the use of any equipment, the capacity of which to generate electricity or heat does not exceed the capacity specified: i.e., in relation to the generation of electricity, 5 megawatts; in relation to the generation of heat, 5 megawatts thermal and those (adopting) energy efficiency measures."(UK-OPSI, 2009).

EPA defines green power as electricity produced from solar, wind, geothermal, biogas, biomass, and low-impact small hydroelectric sources. All these are known low-carbon sources of producing energy. Apart from the sources of a low carbon, low emission listed above there had been considerable debates on whether the nuclear energy falls under this category too. Nuclear energy production takes a considerable quantity of energy to mine, process and transport enriched uranium fuel. Then there is the issue of hazardous nuclear waste and its disposal and reprocessing issues. However, it is estimated that the cost of making nuclear power, with current legislation, is about the same as making coal power, which is considered very inexpensive. Nuclear energy also does not fall under the purview of carbon taxation, as it does not emit any toxic or greenhouse gases. Hence it is up to the respective states to choose the best possible combination of energy sources.

There are various reasons for countries to make a transition to a low carbon, resource efficient Green Economy, and climate change is an important reason. But energy security cuts in air pollution and diversifying energy sources are also important drivers (UNEP, 2010). OECD's Urban Energy Hand Book list out the three immediate reasons that favour adopting green energy resources and technologies: They make less environmental impact; they do not lead to depletion of resources and importantly, the new system favours decentralization and attempts to seek local solutions to local problems, independent of the national network (OECD, 1995). This enhances the flexibility of the system thereby spreading the economic benefits to the scattered populations, who are often small in number.

23.2 GREENING THE ECONOMY: THE CHALLENGE

The awareness is growing by the day on the harmful effects of the fossil fuel related economy on the health of human beings, thanks to the newer research findings and the theme often promoted in various media and due to the awareness campaigns launched by the NGOs.

While there is such a widespread, growing realization and the eagerness to address the issues, there is still a wide gap between the assumptions and the work actually done in undertaking remedial measures. The reason is far from simple–historically, the usage of fossil fuels has dominated the source of energy needs of human beings. The transition to environmental friendly methods of energy generation or green energy involves a complex process requiring numerous changes–in the way energy is produced, newer technologies are adopted, addressing the effects on the employment opportunities, and most importantly, the transition process itself requires huge investments. Therefore, it affects various vital sectors of the society. From politics, employment, investment policies, research and design, infrastructure to marketing and promotion of the technologies, greening the economy requires suitable responses from various actors. The policymaking in such a situation is an anticipative one, as opposed to the reactive policymaking. This implies that sustainable development favours anticipative policy making, taking the future into consideration, sometimes without any immediate visible benefits.

There is also a huge disparity in the patterns of energy usage. It is a known fact that the advanced and industrialised countries use a lot of energy and as a result, also release carbon in large quantities. Seventy-five percent of the world's energy supply is used by only 25 percent of the people living in wealthy and industralised economies (Fells, 1990). However, it is usually the under-developed or developing countries that face the maximum risk from climate-related threats. Added to this, the under-developed countries' general inability to make huge investments on latest and often expensive technologies to mitigate climate change makes the situation worse.

Therefore, the developing countries' argument is that because of the huge energy needs of the developed and industrialized countries, there is a great responsibility on them to take steps to mitigate the problem. The report by the UK Royal Commission on Environmental Pollution (RCEP, 2000) also echoed the same idea. The report entitled 'Energy-The Changing Climate' suggested that the industrialized world should primarily own up the responsibility for tackling climate change. The right direction for the future was "contraction and convergence" towards equal per-capita emissions across the world.

23.2.1 Carbon Credits: Are these measures enough?

One of the most significant areas where the private sector was involved in the carbon off-shooting measures was through the Carbon Trading scheme (Stern, 2007). This is one of the important methods of taxation aimed at attempting to reduce the growth of carbon in the environment, with the dual objectives of discouraging the obsolete, mainly fossil-fuel based technologies and limiting their use in the future and at the same time, providing incentives for transition towards greener technologies. Carbon Credits or the carbon off-setting measures, i.e. putting a price on carbon are an encouragement and are vital for this transition. While different countries have different pricing mechanisms, they could achieve their stated goals to a certain extent during the last few years.

The European Union's Emissions Trading Scheme (EU-ETS) has been a pioneering contribution of the EU in adopting the Kyoto protocol. The scheme introduced one of the most advanced emissions trading scheme and a directive on renewables. However, even though the Kyoto targets in the context of global warming are modest, they nevertheless, are proving hard for the individual EU member states to comply with (Helm, 2005). Also the EU energy market was highly liberalized over the years. Carbon emissions trading and other market-based environmental governance instruments have also had their share of critics (Newell, 2008). The ETS, despite being considered as a model for similar schemes elsewhere (Skjærseth and Wettestad 2009), had further problems on the reportedly widespread value-added tax fraud (UNEP, 2010). There were also instances of excess allocations by the EU–more permits were issued to emitters than they actually needed. Despite the initial challenges, the ETS scheme has been a pioneer and the centerpiece of the 'new carbon economy'.

Mendonca et al (2007) and others opine that the feed-in tariffs are the most cost-effective policy tools for accelerating the transition to a more sustainable energy system. Feed-in tariffs come with a promise of guaranteed grid access; long-term contracts for the electricity produced and purchase prices that are methodologically based on the cost of renewable energy generation. Despite some progress in most of the countries that adopted these measures, the major challenges still remain–that of poor design and inadequate implementation. Nordhaus (2008) meanwhile suggests a "hybrid" approach–combining the strengths of both quantity and price approaches. An example of a hybrid plan would be a traditional cap-and-trade system combined with a base carbon tax and a safety valve available at a penalty price.

Similar trade and cap schemes have been followed in parts of United States, Australia and New Zealand, which is a promising trend. The focal point in this case that needs further support is towards establishing linkages between carbon trading actors at the regional level. This type of linkage could subsequently evolve towards trading at the national and finally to the international level. The necessity of establishing a global carbon pricing policy has been emphasised at many international forums and is the direction of many nation-states. There are also efforts by countries like the US to have specific agreements (sectorial targets and strategic partnership) with huge emerging economies like China and with other countries like Japan and Canada, (Ladislaw et al, 2009) in terms of reducing the GHG emissions, which is a step in the right direction.

The carbon-trading turnover figures during the recent years show a positive growth. There was an estimated US$64 billion turnover in international carbon markets in 2007, up from US$30 billion the year before (UNEP, 2010). By 2008, the EU ETS, launched in 2005 was worth US$94 billion in terms of revenue (Frost & Sullivan 2009, Capoor and Ambrosi 2008). However much more needs to be done in the area. The Stern Review also reminds us that carbon pricing itself is not enough to mitigate the climate change effects. The carbon market is not expected to bring all aspects of emissions and establishing such a market will also take a long time (Chevalier, 2009). Hence there should be diverse measures all aimed towards the common goal of greening the economy.

23.3 FINANCIAL STIMULI

To a challenge so vast and that is restricted by investment as one of the main factors, financial stimuli is the much needed solution. However, with extensive liberalization of the energy markets (as in the EU zone) and state playing a major role in many resource-rich countries (like Venezuela, Russia, Iran) the promotion of greener technologies is not the prerogative of any single actor but it requires financial stimuli from multiple actors.

The recent economic crisis has affected corporates and governments alike. This trend is witnessed also in the field of energy. According to IEA 2009, the financial crisis has impacted the investment in the field of energy especially in terms of a tougher financing environment, weakening financial demand for energy and falling cash flows. In the oil and gas sector, most companies have announced cutbacks in capital spending by 19 percent when compared to the investments made in 2008. This is expected to lead to further delays in executing the already proposed projects. The report also predicts that the efforts to reduce the CO_2 emissions could fall in 2009 by as much as 3 percent. This decline is much steeper now than at any other time in the last 40 years. The crisis has also led to a deferral of investment in polluting technologies. The challenge now, therefore is to address the losses made owing to the financial crisis during the last years as well as the necessity of catching up with the trajectory of a futuristic low carbon economy.

In a liberalized energy market like the EU, on who does the responsibility of investing in new Research and Development projects that are aimed towards low carbon lie? Is it with the states or with the private actors, in tune with the spirit of market driven principles? Then there are other issues that need to be addressed–like cost-effectiveness, intellectual property, fiscal, regulatory and other issues which fall in the jurisdiction of other institutions. In the case of energy and environment, the governments need to step in with their financial stimuli and also promote newer technologies and not confine themselves to mere regulatory functions in a free market and a liberalized environment. The UNEP 2008 report also admits that a purely market-driven process is unable to address the climate crisis with the necessary scale and speed required and it calls for government action as well as business cooperation to take necessary steps.

Governments usually provide capital subsidies for installation of renewable and green energy systems. Other incentives also exist in the form of tax exemption, credit facilities and third party financing mechanisms. In many developing countries, there is also an incentive-based programme for renewable energy systems. Examples include the World Bank's solar home system project in Indonesia, Sri Lanka, the ESMAP programme in Africa. India and China also promote their own incentive-based projects. Renewable energy is also promoted through micro-credit–this was witnessed in Uganda and Zimbabwe.

The recent financial crisis, even though had reduced the investments in energy sector, had surprisingly led to a sharp increase in energy-efficient measures in public spending by the governments. According to the UNEP report, "leading governments" have committed USD 183 billion to "clean energy" overall in their stimulus packages (UNEP, 2009). The IEA Information paper highlights the measures taken by various governments to promote green energy. Noteworthy among them are Korea's Green Growth Plan, extended into a full five-year plan raising its funding to USD 87 billion, or roughly 2 percent of Korea's GDP. Japan also plans to spend USD 590 million out of the USD 2.18 billion devoted to energy projects funding in the Japanese stimulus package (IEA, 2009a).

The investment in the field of renewable energy is being promoted by China on a massive scale. China today is the largest solar panel manufacturer in the world with more than 40 percent of the market share. China also is in the process of constructing the largest solar photovoltaic power plant in the world (2 000 MW) in the Inner Mongolian desert. This project is expected to provide power for over 3 million homes. China had also set up ambitious plans to tap the benefits of wind-power, and is estimated to surpass United States, the leading country in wind power, within a year or two. A healthy trend that could be emulated by other countries is that 40 percent of the stimulus funds in China are devoted to green energy technologies while in the US, it is 14 percent (Schnoor, 2009).

Despite all these achievements looking very promising, much more needs to be done. Each country should make its own green energy plans taking into consideration the resources available and the respective targets. There is no single pattern for all the countries to follow, however they could formulate their own plans based on the resource availability and demand. There could be a public-private partnership, funding for such projects could be made easier with governmental support, and many policy decisions could be designed according to the specific needs of the particular countries. There should be different levels of funding and also a mechanism that ensures continuity of the funds on a longer term and the implementation of the objectives. Much more spending is needed on energy innovation. Kammen and Nemet opine that the energy innovation programme needs finances on a massive scale–like the Apollo or the Manhattan programmes (Kammen and Nemet, 2005).

The UNEPs Green Economy initiative recommends among other things, that a major portion of the $ 3.1 Trillion in economic stimulus packages be spent on energy efficiency in buildings, renewable energy technologies, sustainable transport technologies, the planets ecosystems and sustainable agriculture (UNEP, 2010).

23.4 RESEARCH AND DEVELOPMENT

Research and Development funding is important because a majority of the world's energy system is based primarily on the utilization of fossil fuels. Newer technologies are needed that are aimed at reducing the carbon emissions, but they are expensive and usually beyond the reach of the under-developed economies. As mentioned earlier, again a complex web of issues confront this area. R&D in technology broadly depends on the financial feasibility of a particular corporation as well as on the successful creation of market demand for that technology or product. The instances where the governments are involved in investing in R&D have seen a gradual decrease over the years.

In case of the US, a considerable decrease in the public sector funding for the Energy Related Research and Development activities was witnessed during the last 30 years. While it was a maximum of $7.3 billion in 1978, the funding fell down to $1.6 billion in 1998. This problem is not confined to the US alone but shared by all other IEA member countries, except Japan. (Runci, 2005).

In the US, federal subsidies granted to energy markets are categorized as R&D expenditures and deployment expenditures, the latter again divisible into direct expenditure, tax-related subsidy and financial support. (IEA, 2008). The largest subsidies for electricity production were directed towards electricity generation from refined coal, followed by wind and solar power. The energy subsidies per million BTUs (unrelated to electricity production) went to bio-fuels, solar energy, and refined coal (in that order). Tax subsidies dominate when looked at the size of the expenditures, but the largest deployment subsidies are not targeted enough to adequately buy down the costs of advanced technologies or help them reach the tipping point to attain widespread diffusion.

One of the reasons for not going green is also because of the high impact on the existing businesses. Midillia et al have showed in their study that all the negative effects on industrial, technological, sectorial and social developments partially and/or completely decrease throughout the transition and utilization of green energy and related technologies when possible sustainable energy strategies are preferred and applied. They had come to this conclusion after taking seven 'green energy strategies' to determine their 'impact ratios' especially in sectoral, technological and application areas (Midillia et al, 2006). Thus, the sustainable energy strategies can make an important contribution to the economies of the countries where green energy is abundantly produced. Therefore, the investment in green energy supply and progress should be encouraged by governments and other authorities for a green energy replacement of fossil fuels for more environmentally benign and sustainable future.

Often, the issue of energy policy suffers because while the green energy policies and technologies are mainly technical issues where the solutions are sought from the technical experts. However, deciding on a policy is more a government's effort and the policies usually have to keep up with the advances in technology. The Socio-technical systems approach of Astrand and Neji (2006) tried to evaluate the green energy technologies while combining the social and technical issues. They highlighted the case of Sweden. The Swedish government, even though having an ambitious plan of expanding its energy production by the use of wind power had a policy that hampered the actual expansion of wind farms. This was because the government insisted on a two-blade design of the wind turbine and limited the supplier choice to only the large national energy providers. This is considered a main cause for slower growth in wind farm expansion when compared to Denmark or Germany.

Harmon and Cowan (2009) analysed the case for green energy from the multiple perspectives framework of Linstone. They concluded that through the interaction of various factors from the technical, organisational and personal framework, the green energy alternative becomes an attractive choice. Their study also indicated that there would be a strong market for green energy over the fossil fuel based energy in the future, especially in the light of current issues like the climate change, increasing economic costs and geopolitical uncertainties. In case of Europe, the heavily fossil-fuel dependent and ageing energy assets have come to the stage where they need replacement. This could be a good starting point for the businesses as well as the respective member states to initiate measures to invest in non-carbon sources of energy.

As mentioned earlier, the new system of transition to green energy measures favours decentralization and attempts to seek local solutions to local problems, independent of the national network. The case of Denmark is a success story in the area of distributive electricity generation. The success was mainly attributed to the political support for such a transition, leading to increase in the energy generation efficiency, reduced carbon emissions and lowered the dependence on the imported fuels. Denmark today has the largest penetration of the Distributive Generation (DG) capacity in the industrialized world. However, the increased DG capacity has led to problems balancing supply and demand. There were instances of coming close to breakdown several times. (Jensen, 2002) The state-owned operator 'energinet' is presently implementing the 'cell' concept, which essentially gives more control to the distribution network operators who are then expected to balance the generation and demand locally as far as possible. A strong support by the government is seen as a vital requirement especially in case of societies that have small number of population scattered over different areas.

Technologies also need robust evaluation and review mechanism in place–comprising of specialists, policy makers and industry. An institutional mechanism that leads to setting up of an 'innovation committee' should be created. Some countries might already have such functioning groups in place. They might need some small change of roles and responsibilities of the members. The Committee could be composed of leading scientists who could review on a continuous basis, the state's RD&D spending programs and budgets, and release public reports about the quality and adequacy of these programs periodically. Other members who maybe included in the committee could be scientists and business representatives who have active role in developing new technologies. Thus, technological advancement could be independently tested and constantly evaluated by this committee before promoting it on a commercial basis.

Most of the socio-technical transitions that had taken place till now, especially in democracies have not been because of the actions of governments alone, because they were a result of the market-based economies. The crucial role of the governments in such a situation is to secure a popular mandate for the transition and also to negotiate and implement that necessary changes required for such a transition.

But when the governments attempted to enforce the transition mechanism, it proved very costly–like the EU CAP, or largely unsuccessful–as in the case of UK where it was planned to supply cheaper electricity by the use of nuclear energy. When we analyze the history of such socio-technical transitions elsewhere, we observe that the transitions are very complex, multi-dimensional and uncertain, and it was possible to a large extent only with the involvement of social groups, processes, technical opportunities and linkages that were not anticipated in advance (Geels, 2002a). Energy systems are also characterized by their complexity, uncertainty and inertia, so many players need to be involved in the transformation and transition processes (Geels, 2002b). Therefore, instead of searching for a "perfect" approach to reducing the GHG emissions that satisfies all the stakeholders, the governments need to design policies with built-in flexibilities, which the Stern Review calls "exit strategies." This gives them enough leverage to quickly switch to newer technologies or stop those policies that could be going in a different direction than that envisaged.

23.5 INFRASTRUCTURE DEVELOPMENT

An innovative mechanism that could be the much-needed solution to the Infrastructure and investment issues is proposed in the 2007 legislation in the US. This proposal envisages the creation of a National Infrastructure Bank, composed of a public–private financing mechanism (Ladislaw et al., 2009). Such a bank is expected to allow the federal government to finance projects of substantial regional or national significance more effectively. Green energy will become increasingly competitive in the market place, more so if the historically high rates of technological improvement continues. However, the challenge is not in improving the technologies but in making them to integrate with the existing systems and infrastructures.

The Stern Review called for suitable exit mechanisms and judicious mix of specific technologies as policy according to the needs of the particular country. The case of

United States is an eye-opener in this regard with reference to policy making in the automobile sector. The federal promotion of alternative fuels for the passenger car market is clearly an example of caution. The federal policy was wavering between many available options. Twenty-five years ago, methanol was vigorously promoted. Then came the phase of electric vehicles, later the hybrid-electrics, and then fuel cells, to ethanol just two years ago, to plug-in hybrids today. This has been an extremely disruptive and wasteful process not only for the automakers but also for the U.S. taxpayers (Ladislaw et al., 2009).

Rather than pushing a single fuel or a particular technology at one point of time, policymakers should set time-bound performance objectives and requirements on a lifecycle basis, and give enough lead-time to ensure quality and reliability. Aggressive policies to transform the system – such as vehicle efficiency standards, low-carbon fuel standards, technology research and development, and incentives to buy more efficient vehicles – should be high priorities. In the recent times, even though the growth rate for renewable energy has been a slight 2.3 percent annually, the geothermal, solar and wind energies are growing the fastest. Renewable energy projects are gaining momentum, although the net growth is very low compared to the immense potential of renewable energy. The global installed wind generation capacity has grown at the rate of 25 per cent per year over the past five years. China, for example installed capacity has nearly doubled every year since the end of 2004 – and the report notes that the wind energy potential under perfect conditions has been estimated at up to 72 000 GW, nearly five times total energy demand (UNEP, 2010).

Since these technologies have fairly small installed bases relative to the world Total Primary Energy (TPE) supply, there is a sign of optimism as these technologies could be used as channels for further growth and acceleration. In such a scenario, as the green energy becomes more widespread and in larger quantities, it becomes more reliable, price will be competitive as it is likely to induce significant competition in the energy markets, ultimately resulting in stabilizing prices and bringing down the prices of all sources of energy.

23.6 EMPLOYMENT GENERATION

A contentious issue when a government attempts an all important policy decision regarding green energy is the issue of employment. According to the IEA (2009) report, the investment in stimulus packages had a great potential of creating new jobs. It has to be understood that creating a new "green job" does not mean shutting down the existing 'polluting' or 'dirty' industries—rather it is aimed towards re-skilling the workers towards the cleaner methods, usually by making slight changes to the existing methods.

The United Nations Environment Programme's (2008) report defines green jobs as

> "Work in agricultural, manufacturing, construction, installation and maintenance, research and development (R&D), technical administrative, and service activities that contribute substantially to preserving or restoring environmental quality. Specifically, but not exclusively, this includes jobs that help to protect and restore ecosystems and biodiversity; reduce energy, materials, and water consumption through high-efficiency and avoidance strategies; de-carbonize the economy; and minimize or altogether avoid generation of all forms of waste and pollution."

The report further argues that the green jobs also should aim to "meet the demands and goals of the labour movement, i.e., adequate wages, safe working conditions, and worker rights, including the right to organize labour unions."(UNEP, 2008) Green jobs ensure that the existing jobs will be suitably oriented towards the goals of pollution mitigation, and increasing the energy efficiency standards. The transition to a green economy needs skilled and trained workers in fields like renewable energy, buildings and construction, transportation, basic industry and recycling, food and agriculture and forestry.

It is estimated that an investment of USD 500 billion in the retrofitting of 50 million buildings would induce the creation of a total of 625 000 direct and indirect jobs (Hendricks et al., 2009). The report also reminds us that the situation does not prevent "well over a million construction workers from sitting idle in a sagging housing market". The reports cautions that the jobs estimated may not be "automatically good jobs", but "they can be made so if adequate attention is given to the development of labour standards, wage classifications, training supports and performance standards".

It is estimated that for every $1 million invested in energy efficiency retrofits generates eight to eleven on-site jobs. White and Jason (2008) and Kammen, Kapadia and Fripp (2004) made a study on the job growth in the clean energy industry across the US. Their findings suggest that on an average, three to five times as many jobs were created by investing in renewable energy when compared to similar investments made in the fossil fuel based energy systems (Kammen et al, 2004). Roger Bezdek opines that the renewable energy and Energy Efficiency sectors generate about 70 percent more jobs than the oil and gas sector–an effective job creation mechanism, generating more than 2.5 times as many jobs per revenue as the oil and gas sector. (Bezdek, 2008) He estimated that a number of 'green jobs' could be created without initiating major policy changes. But when a moderate or major change is made to the existing policies towards green measures, there is a substantial growth in the employment potential. Bezdek estimates that aggressive deployment of renewable energy and energy efficiency measures can net 4.5 million new jobs by 2030–jobs that are widely dispersed throughout the US across all sectors and industries. There is also no danger of 'outsourcing' these jobs as they are on-site based (Bezdek, 2009).

This kind of job creation and the associated technologies could displace approximately 1.2 billion tons of carbon emissions annually by 2030 – the amount scientists believe is necessary to prevent the most dangerous consequences of climate change. In this estimate, approximately 57 percent of carbon emissions reductions would be from energy efficiency while 43 percent would be from renewable energy. The construction industry directly benefits from almost all the growing renewable energy and energy efficiency sectors as well as from improvements in overall economic growth due to energy savings. Farming directly benefits from biomass and biofuel technology growth.

However, a vast majority of jobs 'created' by the renewable energy and energy efficiency industries are in the same types of roles seen in other industries (accountants, factory workers, IT professionals, etc.). These potential jobs are in two categories that every state is eager to attract – college-educated professional workers, with advanced degrees and highly skilled technical workers.

The prudent way to prepare a "green-collar" workforce is to start on the existing foundation of state and local workforce systems. Imparting periodic training in green skills and including curricula to cover new programmes helps the work force to gradually transform towards the requirements of the newer technologies. Green jobs initiatives should mainly address upward mobility of the employees. This would act as an incentive for the employees to move from unemployment or low-wage jobs to jobs that would provide higher wages and benefits. The governments could also consider understanding the targeted green industries at the regional level economies. This could lead to further creation or expansion based on regional networks and partnerships organized by the industry.

In case of Germany, the Government had made ambitious plans to strengthen their laws relating to renewable energy. As a result of some minor changes in their policy, it was estimated that there was an increase from 160 000 jobs to 236 000 jobs between 2004 and 2006. The UK also does not lag behind—it aims at creating 1 million new green jobs, primarily in the field of manufacturing green energy sector over the next ten years. The report on Green Jobs in Australia also estimates that there will be at least 2.7 million new jobs created by 2025, most of them green jobs in Australia if the steps were taken towards making Australia carbon-neutral by 2050 (Australian Conservation Foundation, 2008).

In June 2009, the US House of Representatives passed the Clean Energy and Security Act 2009 which is very comprehensive in addressing various issues relating to transition to a green economy. This is considered a "real attempt" for a national carbon reduction plan. It also address the employment issues by supporting development of a "clean energy" curriculum, additional funding for the worker training programme and climate change worker adjustment assistance to enable smooth transition (Alliance to Save Energy, 2009). The growth of renewable energy and green energy sources in developing countries is also expected to create employment in those countries and elsewhere as the newer and emerging technologies are expected.

23.7 SOCIAL SECURITY

When a decision is taken towards transition to green economy, the issue of social security comes up. Will these new measures render some people to lose their livelihoods, their jobs or – make their lives comparitively more difficult? The implementation of green energy directives entails heavy investments in their new energy-efficient houses? These are a few questions that are often raised by the citizenry. Since the general process of consultation is absent in energy policy making, people tend to have more questions. Therefore, the government has to address the issue in its totality. Any comprehensive energy policy cannot be formulated excluding the other dependant factors. It has to be an integrated approach covering aspects not only of economics or environment but also should incorporate social security, technology promotion, education and awareness.

Some countries, especially the developing countries have a tendency to offer at least one form of energy at a subsidy. This is below the prevailing price in the market. There needs to be a focus on ways to reduce these subsidies. This requires bold political decisions and creating awareness among the people helps to mitigate the hard effects on the political fortunes of the political party in the government. There cannot be a knee-jerk reaction but these subsidies can be gradually minimized according to the respective state's domestic conditions and requirement.

One of the initiatives that gained political support from various countries at the Copenhagen Summit (2009) was the Reduced Emissions from Deforestation and Forest Degradation (REDD) programme. This is a community involvement programme that has a potential to go a long way in not only reducing the climate change effect but also help support local communities. REDD programme involves supporting developing countries to conserve rather than clear tropical forests. This could help in overcoming poverty among the communities by providing them incentives for their greening and conservation efforts. The UNEP Year Book estimates that investing $22 billion to $29 billion in REDD could cut global deforestation by 25 per cent by 2015 (UNEP, 2010).

Much of the social security issues could be properly addressed when there is a common idea driving the policy-mechanism. The issues like climate change, local environmental protection, economic development, health, employment, and energy security all need a comprehensive integration. When the common points are identified, they could be linked up to work towards a common goal. Hence, there needs to be an intensive consultation process with various groups and a successful policy is that which evolves from such an inclusive mechanism, that seeks to minimize the drastic changes that come with such a policy.

23.8 EDUCATION AND OUTREACH

Educating the people about the advantages of switching to green measures or encouraging them to adopt such measures goes a long way in mobilizing the public opinion in favour of positive action. Energy Policy had been exclusively been in the domain of 'technocrats' and 'specialists.' However, bold policies can be taken only when the policy makers enjoy the support of the people. Hence involving more public participation, creating platforms for debates to hear different views and eliciting the opinions of the people are crucial. Any green energy approach that does not have an integral education and training is likely to fail. Therefore education can be considered a prerequisite for the success of a sustainable energy program.

This area had been generally overlooked as it is assumed that the general public is not interested in energy related issues nor has idea about complex technical issues. A study done by Vachon and Menz on the potential influence of a state's particular social, political, and economic interests on its propensity to adopt green electricity policies showed some interesting results. Using an empirical model that combined various social, political and economic indicators as explanatory variables of a state's likelihood to adopt four specific green electricity policies. They concluded that social interests, measured by the level of income, the level of education, and the degree of participation in environmental lobbying groups, were positively linked to the adoption of green electricity policies. Similarly, political interests as measured by the pro-environment voting by states' representatives in the U.S. Congress, also play a positive role in the adoption of such policies (Vachon and Menz, 2006). Therefore education cannot be underplayed in this crucial area.

One of the problems in energy policy making is looking at the issue entirely from an economic perspective. People are seen as a "Demand" while the energy companies are seen as "Supply". The whole energy policy had been built up on this distinction. This could be a good model to evaluate the energy policy in terms of economy. But this also assumes that people are passive and cannot take major decisions on energy saving, which is a wrong conclusion. Therefore, a deliberate push towards measures like collective action–by means of their political participation in the process of energy policy making is required. Publicity campaigns through various media, awareness seminars and other related events diffuse the awareness and knowledge among the wider audience. Usually the states have their public media channels and departments that could be utilized to promote green energy. The business world could also contribute much by promoting such measures as a part of their corporate social responsibility. Integration of the environmental, economic and social dimensions of sustainable development has been a key theme of the International response to the financial, food and energy crises especially during the last few years (UNEP, 2010).

People generally are thought to be not keen in taking an active role in securing low carbon energy supplies. Therefore, the transition has to be both technical as well as social. People also need to be assured that conditions for participation by others will also be created. The Sustainable Development Commission of the UK (2006) opined, "a critical mass of citizens and businesses is ready and waiting to act on the challenge of sustainable consumption. But to act, they need the confidence that they will not be acting alone, against the grain and to no purpose." The UNEP, in February 2010 at its governing council meeting in Bali, Indonesia launched a dedicated website to address issues concerning transition to a low carbon economy. It is a joint project between Low Carbon Economy.com and the United Nation's Climate Neutral Network (CN NET).

The new website seeks to assist knowledge transfer and simplify access to information and tools that could be difficult to trace at a single place. List of every country's carbon policies, commitments, historical performance, future projections and opportunities, as well as country-specific marketplaces, networks, associations and standards, which are relevant to government departments and investment agencies. This information could help diffuse knowledge among a wider area.

23.9 CONCLUSION

This paper draws attention to the urgency and complexity involved in policy-making towards 'greening the economy'. Green energy despite being synonymous with sustainable energy is the one that has a higher relative environmental benefit. The transition towards green policies by itself does not mean shutting down the "dirty" industries but making changes in tune with the requirements of the green energy measures. Adopting such measures means taking some hard decisions duly sensitizing the people on the long-term advantages over the short-term gains.

The transition policies often have to pass through a complex system of political manoeuvring since each political party of a particular country has its own ideological approaches towards a number of issues. There is an urgent need to take measures towards greening the economy and reducing the carbon emissions as the cost of each year of delay is huge. The complexities associated with these transition processes–huge investments in Research and Development, rapidly emerging newer technologies, focus on long-term goals as opposed to short-term gains, addressing infrastructure and employment issues, and addressing the needs of various actors in the whole process.

In the area of carbon trading, the EU-ETS scheme is a pioneer towards a low-carbon economy, which could be emulated at a local, national, regional and ultimately leading towards a global trading mechanism. The focus should be towards establishing linkages between carbon trading actors at various levels. The Green Energy technologies also need robust evaluation and review mechanisms in place. This helps in constant evaluation of the green energy policies. An institutional mechanism that leads to setting up of an 'innovation committee' should also be set up. This helps in keeping up with the advances in technology. There is also a suggestion for a National Infrastructure bank composed of a public–private financing mechanism that allows regional government to finance projects of substantial regional or national significance more effectively. The governments as well as the businesses alike should encourage the financial stimulus and investment towards R&D. Green energy will become increasingly competitive in the market place, more so if the historically high rates of technological improvement continues.

Even though there is no single ideal policy that is suitable for all the countries, individual countries should formulate their own policies keeping in mind their commitment towards the environment and towards international protocols like the Kyoto Agreement and the natural resources and potential energy sources to which the country has access to. Replacing the fossil fuel based assets that have outlived their expected time with green energy measures is a step in the right direction as it is also expected to address the employment related issues. Education and awareness on such transition is also important, as the policy makers require the support of the electorate. This will help people to brace themselves for some hard decisions, but they would be ready to forgo the short-term gains over future benefits.

Chapter 24

Poverty, environment and climate change

K.M. Thayyib Sahini (IAEA, Vienna)

24.1 INTRODUCTION

One sixth of humanity is still living in extreme poverty and struggling to have bare necessities of life . In spite of all the scientific achievements, technological progress and modern economic growth, poverty is still a continuing reality. A common understanding of absolute poverty is deprivation of a person from accessing the basic necessities of life, such as food, clothing and shelter. This is mainly due to the lack of income, even though poverty can be caused by social inequality or social injustice. But the access to basic necessities can't alone assure the happiness and well being of a person that is called relative poverty, which differs across regions and societies. This chapter explores the interrelation between poverty, environment and climate change in the context of energy.

Two hundred years ago, in the wake of industrial revolution and modern economic development, Adam Smith delineated the pain and indignity caused by poverty. Talking about the poor man, Smith (1853, p. 71), in his "Theory of Moral Sentiments" says, "The poor man is ashamed of his poverty; he feels that it either places him out of the sight of mankind, or, that if they take any notice of him, they have, however, scarce, any fellow-feeling with the misery and distress which he suffers". Later, Amartya Sen's studies on famines and poverty exposed the absolute and relative nature of poverty and he characterised poverty as capability deprivation (Sen, 2001, p. 87). Jeffrey Sachs distinguishes poverty in to three degrees, such as extreme or absolute, moderate and relative. By the way of a definition for extreme poverty, Professor Sachs writes on extreme or absolute poverty,

> Extreme poverty means that households cannot meet basic needs for survival. They are chronically hungry, unable to access health care, lack the amenities of safe drinking

> water and sanitation, cannot afford education for some or all of the children, and perhaps lack rudimentary shelter-a roof to keep the rain out of the hut, a chimney to remove the smoke from the cook stove and basic articles of clothing such as shoes (Sachs, 2005, p. 20).

World Bank estimates released in August 2008 shows that, about 1.4 billion people in the developing world (one in four) were living on less than $1.25 a day in 2005, down from 1.9 billion (one in two) in 1981 (see Chen and Ravallion, 2008). These huge groups of population are dispersed in different parts of the world, though a major portion are in sub-Saharan Africa, East and South Asia. Such a reality is posing questions towards the effectiveness of the poverty eradication and development projects. Development theories and economic policies didn't overcame the challenge to eradicate poverty absolutely from the face of earth, but the recent discourses on poverty eradication and economic development changed it as a moral imperative and a tough goal for the able, successful and rich to fulfil in their lifetime. Now there is another challenge looming over the globe, which is the climate change and its related consequences.

24.2 CLIMATE CHANGE CHALLENGE AND POVERTY

Anthropogenic emissions due to industrialization, especially because of the burning of fossil fuels and land use change augment atmospheric temperature, resulting in rise in sea level, receding glaciers and flash floods, frequent droughts, devastating hurricanes and so on. These calamities affect the population globally, but the poor are going to pay the price more because of their vulnerability and lack of choice. The IPCC fourth Assessment report points out that, "As generally known, the impacts of climate change are distributed very unequally across the planet, hurting the vulnerable and poor countries of the tropics much more that the richer countries in the temperate regions" (IPCC, 2007, p. 144). Empowering those bottom billion and vulnerable in order to face the challenge of global warming through adaptation and mitigation is a necessity, which can't wait anymore.

In a recently published report on poverty and climate change, OECD (2008, p. 10) concludes, "many sectors providing basic livelihood services to the poor in developing countries are not able to cope even today's climate variability and stresses. Over 96% of disaster-related deaths in recent years has taken place in developing countries". Thus the relationship between man and nature has a tremendous influence in the socio-economic development. Environmental calamities like flood, drought, erratic rainfall, hurricanes, sea erosion etc destabilizes community life. Such natural disasters normally result in economic hardships and consequently, displacement of people, causing migration, poverty, disease and conflicts.

Common characteristics of the majority of underdeveloped countries are technological backwardness, lack of capital, limited industrial development, poor infrastructure facilities, institutional weakness combined with an overall inadequacy of good social indicators. Presumably this is a cycle of inadequacy, resulting in poverty, disease, distress and political disturbances leading to greater catastrophes. Sadly climate change can induce these phenomena and degrade human life.

24.3 POVERTY AND ENVIRONMENT

There is already a persisting predicament of poverty in many developing countries. Resolving such a development issue itself is challenge for the national governments and the other involved organizations. A natural calamity such as a flood or drought exacerbates poverty and underdevelopment. The prolonged drought in 2005 left many African states is distress and caused an alarming food crisis. The story of such a severe calamity reported in *New York Times* in November 2005 (Wines, 2005) says,

> More than 4.6 million of Malawi's 12 million citizens need donated food to fend off malnutrition until the next harvest begins in April. In Zimbabwe, at least four million more need emergency food aid. Zambia's government has issued an urgent appeal for food, saying 1.7 million are hungry; 850,000 need food in Mozambique, 500,000 in Lesotho and at least 300,000 in Swaziland.

From Africa to Asia, crop failure due to the lack of irrigation and inadequate government support lead to mass suicides of farmers in India. Drought and seasonal variations in rainfall affects farming, leads to crop failure resulting defaults in repayment of loans by farmers in India. Citing India's national crime records bureau, BBC (12 April 2009) reports about 200 000 farmers committed suicide in India since 1997. This number doesn't include number of women farmers who committed suicide because woman farmers are not normally accepted as farmers in India. "By custom, land is almost never in their names. They do the bulk of work in agriculture – but are just "farmers' wives." This classification enables governments to exclude countless women farmer suicides" (Sainath, 2009) which is also exemplifies the gender aspect of poverty.

These narratives of distress points out the fact that, over reliance by a huge number of poor people on traditional farming methods, small land holdings in the absence of large scale mechanized farming exposes those who are dependent on agriculture towards poverty, especially in the wake of environmental imbalances such as drought or flood. The other aspect is the lack of alternatives for income generation; for instance, industries and other services sector. Alternatives to land based occupations such as agriculture and cattle herding could release the pressure on environment.

A case study of Peruvian Brazil nut gatherers shows those who have alternative jobs in the nearby city Puerto Maldonado spend less time in the rainforest than those who don't have such jobs. These alternative salaried and non salaried job opportunities for the Peruvian Brazil nut gatherers stopped them from clearing the forest for crop farming led to the protection of rain forest. Not only crop farming, cattle herding is a threat to rain forests. For example, the Brazilian farmers cleared the rain forest for cattle herding. Swinton et al (2003) finds, "the lack of off-farm employment opportunities is a likely reason that Brazilian rainforest frontier farms are so fixated on clearing land".

A case study on the farming communities of Norte Chico region of Chile shows "how income from nonfarm employment and government credit programs permitted agricultural intensification that allowed environmental recovery of fragile, arid common lands" (Bahamondes, 2003). The Chilean smallholders who worked on commercial grape farms reduced their reliance on extensive goat herding and generated funds for intensive irrigated forage production (Swinton et al, 2003). Considering the fact that 80% of Amazonian deforestation is the result of slash and burn agriculture, indicates the importance of alternative job opportunities for those people.

These above examples show the interrelation of poverty and environment. Persisting poverty could lead people to exploit natural resources around them indiscriminately in an unsustainable manner. This results in deforestation, desertification, soil erosion and flash floods leading to greater hardships for those already suffering. At the same time as we have seen like in Amazonian forest, alternative income generation sources could protect forests and environment. Destruction of environment, especially destruction of tropical rain forest contributes to further changes in climate and induces global warming. Forests are called carbon sinks, because they hold billions of tonnes of carbon, thereby preventing it from being released to atmosphere.

According to a recent study by scientists from University of Leeds (2009), the tropical rain forests remove 4.8 billion tones of carbon emissions from atmosphere every year. The carbon sink (tropical rain forest) in Africa alone absorb 1.2 billion CO_2 each year. The livelihood of those people lives around these forest need to be assured. The Congo basin forest is the second largest carbon sink in capacity after Amazon, which is a home for 24 million people spanning across six countries with a total population of 86 million and covering an area of 4 048 470 km^2. The Congo basin holds an estimated 43 billion tonnes of carbon which shows the importance of this Central African rain forest for the existence of humanity. An alarming reality is, that around 43 0000 square km of such humid forest is wiped out during the period between 1990 and 2005 (Nasi et al. 2008, p. 196–200).

According to Human Development report (2008), 73% of the populations of the sub region are classified as the lowest income countries in the world and poverty is wide spread. The HDR rankings of those Congo Basin countries are Gabon (103), Equatorial Guinea (118) Republic of Congo (136) Cameroon (153), DR Congo (176), and Central African Republic (179) respectively. Being the poorest countries in the world, Cameroon, DR Congo and Central African Republic deserves support to maintain the Congo basin forest.

According to the State of the Forest Report of Congo Basin, "The majority of inhabitants of the sub-region depends on small-scale slash-and-burn shifting agriculture for subsistence- a farming practice which uses the forest as a land reserve for expansion" (Eba'a Atyi et al, 2008, p. 15). Unless there is an alternative source for income generation, the pressure of traditional farming practices will lead to increasing destruction of forest, even though such practices won't help those vulnerable sections to come out of poverty, and resulting natural disasters due to the environmental changes locally and globally.

The US Energy Information Administration (EIA, 2006) data shows net electricity generation of Central African Republic is merely 0.11 Billion kWh, with a world ranking of 185, and total primary energy production is 0.001 and consumption is 0.005 (Quadrillion Btu and ranks 187th), shows that electricity is the biggest infrastructure bottleneck for alternative income generation and industrial development. The total primary energy production and consumption of Cameroon is 0.025 and 0.088 (quadrillion btu) respectively for a population of more that 18 million. These basic facts on energy in these countries confirm the correlation between poverty and the availability of energy and electricity. The less the Total Primary Energy production and consumption, the higher the prevalence of poverty. Incidentally low per-capita energy availability also leads to poor industrialization and heavy dependence on traditional methods of agriculture, and consequent deforestation.

One among the major reasons of rural poverty is lack of alternative income generation opportunities other than agriculture. The mass suicides of Indian farmers point towards this reality. Smallholdings, insufficient or unreliable irrigation, lack of competitiveness, hostile credit atmosphere and absence of adequate government support are the features of agriculture sector in most of the developing countries. Agriculture becomes unprofitable due to these factors. On the other hand, large scale mechanized farming with modern scientific management practices and governmental support assures food security, profit and reliability. Presumably, diversification is the way out of poverty. Developing sustainable industries and services, and also equipping those subsistence agriculturalists with training, capital and infrastructure could release the pressure on environment. Such an alternative approach will empower the rural poor. Being the lifeblood of all productive activity, energy and its uninterrupted availability ensures success of such projects aimed to empower the rural poor.

24.4 ERADICATING POVERTY

Poverty can be eradicated through short-term measures such as economic aid. But in a long term perspective it is only through economic development that poverty can be eradicated. Investments in green energy technologies and projects, and availability of energy itself for those vulnerable sections of society is a long-term measure, which contributes to economic development.

The most comprehensive poverty eradication program is the Millennium development goals adopted by UN General Assembly (2000) during the Millennium Summit 2000. World leaders issued the Millennium Declaration in which nations together agreed for the realization of some time bound targets. The targeted goals concerning development and capability enhancement of poor and vulnerable which came to be known as Millennium Development Goals (MDGs). The declaration set 2015 as target date for achieving most of the development goals. The MDGs aims to achieve a comprehensive development objective which includes, 1) Eradicate extreme poverty and hunger, 2) Achieve universal primary education, 3) Promote gender equality and empower women, 4) Reduce child mortality, 5) Improve maternal health, 6) Combat HIV/AIDS, malaria and other diseases, 7) Ensure environmental sustainability, 8) Develop a global partnership for development. These eight goals are with 18 targets and a series of 48 measurable indicators. Progress of the MDGs are constantly measured and is evaluated, latest one being the MDG report 2009.

According the MDG report (2009) which evaluate progress of the set goals, due to economic crises, there has been a slow down in progress of the goal aimed to reduce the number of people living in extreme poverty. An estimated 55 million to 90 million more people will be living in extreme poverty than anticipated before crises. Likewise, the prevalence of hunger is also on the rise, from 16 percent in 2006 to 17 percent in 2008. The ongoing economic crises may lead to higher global unemployment; rates and could reach 6.1 to 7.0 per cent for men and 6.5 to 7.4 per cent for women in 2009. The recession could prevent an increased aid flow of official development assistance from developed countries, as most of the OECD countries are undergoing economic difficulties. At the same time, during last nine years, remarkable progress has been made in many areas such as overall reduction in extreme poverty, infant mortality, protection of ozone and outstandingly, increase in the number of enrolments in primary education.

In the preface of MDG report (2009), the UN Secretary General notes,

> This report shows that the right policies and actions, backed by adequate funding and strong political commitment, can yield results. Fewer people today are dying of AIDS, and many countries are implementing proven strategies to combat malaria and measles, two major killers of children. The world is edging closer to universal primary education, and we are well on our way to meeting the target for safe drinking water.

Though necessary, successful completion of MDGs depends on many factors. Nine years passed and five more years left for the world to achieve MDGs in the stipulated time frame. Two significant developments wield considerable impact on the development planned in the millennium declaration. The ongoing economic crises and climate change could affect the success in achieving MDGs. As it is clear from MDG report 2009, decrease in flow of ODA (Official development assistance) could affect funding of many development and assistance programmes. Another related aspect is the diversion of funds towards climate change mitigation efforts. The fund diversion, climate change and its related consequences itself could roll back the already achieved standards in poverty eradication and the related development. As mentioned earlier, climate change will impact tropical regions of world, which is also the home for poor and vulnerable.

24.5 ENERGY FOR DEVELOPMENT

MDGs are showing its positive impact on those societies, though extreme poverty and related under development didn't disappear yet. A climate change induced drought, flood, and spread of epidemics combined with food shortage can create havoc. In order to have continuity in the achieved standards and for a further development, the process of economic development has to be continued in those parts of the world. Industry, commerce, agriculture and other services have to be developed and flourish. Communication and transport networks are needed to facilitate development. All kind of economic activity needs to be fuelled by reliable source of energy. This is the missing goal in MDGs. The millennium declaration for development is silent about this basic infrastructure bottleneck which most of the least developed countries are facing. A 2005 study titled "Energy services for millennium development goals" by different UN agencies found that,

> Worldwide, nearly 2.4 billion people use traditional biomass fuels for cooking and nearly 1.6 billion people do not have access to electricity. Without scaling up the availability of affordable and sustainable energy services, not only will the MDGs not be achieved, but by 2030 another 1.4 billion people are at risk of being left without modern energy (Modi et al. 2006, p. 2).

World Bank finds that, though alternatives exist, grid electricity is economical and reliable. Absence of electricity in a household means poor utilization of energy resource. An illuminated home is not a luxury to be desired, it is a basic necessity. It could prevent atmospheric pollution; increase the learning time, there by leading the primary schools children to the secondary level to achieve further progress. Continuous supply of electricity will enable small workshops, industries, and other commercial and business establishments to run without stopping. It will improve the overall socio-economic and cultural well being of people. It will give more employment opportunities and capabilities, which is a way out of poverty and underdevelopment.

According to the UN millennium project (2005), women and young girls are responsible in house holds for collecting water and fuel wood, cooking, and agro-processing, and spend more than 6 hours of their daily life in order to meet these domestic needs. In Sub-Saharan Africa, 90 percent of the population still relies on traditional fuels for cooking, and only 8 percent of the rural population has access to electricity. With electricity, mechanization and modern technology can contribute productive activities. Instead of helping parents, children can go to school; instead of spending long hours for fetching water and firewood, women can contribute for the well being of their family. They can participate in other economic and productive activities using their free time that will empower them and help to achieve gender equality.

The 2009 edition of IEA World Energy Outlook estimates that 1.5 billion people still lack access to electricity. Universal access to electricity could be achieved with an investment of $35 billion per year in 2008–2030 (IEA, 2009). With the availability of electricity, community centers such as schools, health centers, and public offices can function better and deliver the necessary services for people there by lifting the underdeveloped and enhancing capabilities. Traditional methods of farming can be modernized by mechanization with the availability of energy. Instead of depending on seasonal rainfall, motorized water pumps could provide irrigation, resulting in better yields thereby offering financial security to family and food security to community. FAO report on agriculture points outs,

> Fulfilling energy needs of agriculture and rural services is at the core of improving productivity. Land preparation, harvesting, irrigation and processing require different types and levels of energy inputs, both in direct (mechanical, thermal, fossil and electrical energy) and indirect (fertilizer) forms. Without these energy inputs, agricultural productivity remain low and probably well below its full potential (Alexandratos, 1995, p. 386).

As mentioned earlier, the missing development goal in MDGs is energy. Like lifting people out of poverty, or achieving gender equality, connectivity to the grid or electrification of households need to be counted as a necessary goal for development and to be achieved in a time frame of 10–15 years. On the whole achieving the MDGs are also a climate change mitigation option. Empowering those poor people around the carbon sinks will protect the environment. The people around those global commons deserve special support. Channelling finance and technology from the developed world to those vulnerable areas are not only wise but smart as well.

24.6 INTEGRATING POVERTY ERADICATION, ENVIRONMENTAL PROTECTION AND ENERGY SECURITY

Since so many organizations, governments and committed people work for the eradication of poverty, energy planners, technologists and policy makers need not to change their focus. At the same time, this is a perfect opportunity, timing, a critical juncture, which could be used to achieve multiple goals of greening the energy sources and increasing the energy security, while at the same time eradicating poverty by striving towards mitigating climate change.

The goals for poverty eradication, environmental protection and securing energy supplies can be combined. Low carbon energy sources and green energy technologies can contribute for the development and eradication of poverty in many different ways. Investments for implementing the available green energy technologies could provide more jobs, adequate energy to fuel the economic and social activities by reducing carbon emissions, atmospheric pollution and mitigating climate change. A policy atmosphere, which supports or gives priority to such green energy investments could reduce poverty, protect environment and mitigate climate change.

The biggest challenge that poverty eradication projects faced so far has been the availability of sufficient funds, viability of livelihood projects or the longevity of income generation projects. Here we have an assurance or an urge to mobilize funds to mitigate climate change. There is no question of reluctance, as the world saw the heat of debates in Copenhagen. Humanity reached a point of no return when it comes to climate change mitigation. So, funds and political will surely be forthcoming. By integrating the poor into these climate change mitigation and adaptation projects, humanity can come out of still persisting poverty.

The climate change convention (COP15) in Copenhagen witnessed the concern and anticipation of world about global warming. Though the convention couldn't produce a binding agreement over the CO_2 emitters to achieve deep cut in emissions, the summit gave an impression that climate change mitigation and adaptation efforts will be strengthened in the coming years. Through the Copenhagen accord (2009), developed countries agreed to provide $30 billion to developing countries for climate change adaptation and mitigation efforts for the period 2010–12. The result oriented utilization of these funds in the affected/vulnerable countries need to be integrated with development projects. Such an integrated and combined implementation of adaptation and mitigation projects will have multiple benefits. Successive climate change conventions will offer more resources for mitigation and adaptation as the polluters become mature enough to take the complete responsibility in resolving the imminent danger of global warming through the upcoming negotiations. Even though not sufficient, the $100 billion per year commitment by developed countries by 2020 symbolizes a growing effort from the polluters' side. The projected fund of $100 billion for climate change mitigation for developing countries can achieve its goal of mitigation with a beneficial outcome of poverty eradication along with it.

Starting from the bottom, more than 3 billion people including the rural house holds, almost all in low and middle income countries, rely on solid fuels for energy which are the source of atmospheric pollution and causing respiratory diseases including pneumonia, and other acute lower respiratory infections, chronic obstructive pulmonary disease and lung cancer (Bousquet at al., 2007). This is because of the traditional methods of cooking, heating and lighting which uses dung, wood, crop waste or coal in domestic hearths, simple stoves with incomplete combustion. Particularly, women and children are more exposed to indoor air pollution resulting in an estimated 1.5–1.8 million premature deaths a year. "In Africa, approximately 1 million of these deaths occur in children aged under 5 years as a result of acute respiratory infections. 700 000 occur as a result of chronic obstructive pulmonary disease and 120 000 are attributable to cancer in adults particularly in women" (Bousquet et al., 2007, p. 46).

This reliance on solid fuels and its incomplete combustion in inefficient hearths and simple stoves by more than 3 billion people create a strong barrier to achieve the MDGs in its full scale. Because it creates a web of complex realities from which the bottom billions can't escape, i.e., the indoor air pollution that leads to many diseases, and the most affected group is women and children. They suffer in multiple ways, have to spend many hours to collect firewood, and have to spend the almost same amount of time for cooking and other household duties, which expose them to the polluted indoor. This reality has to be changed. The world of poor and vulnerable needs efficient and clean sources of energy. This is an important and urgent necessity. Universal availability of clean sources of energy and energy efficient appliances are part of climate change mitigation and adaptation.

Both grid based and off-grid energy solutions for house hold energy use can transform rural life. Substituting firewood with renewable energy could protect environment and reduce pollution related diseases. Off grid solutions like solar, biogas, wind, small hydro (of course all these small energy generation units can be connected to grid as well) can provide energy as well as jobs for the rural population. Changing the use of solid fuels for house hold energy towards grid electricity, solar, wind, gas and other bio fuels need investment, technology, political will and policy making; but considering the externalities and other health, environmental and economic benefits, change is beneficial in long run. In this regard, Sagar et al, (2009) points out that, "There is a significant gap between existing innovation process and what is needed, especially in developing countries, to meet the range of inter related energy, climate and developmental challenges facing them". In order to bridge this gap of technology and particular nature of local necessities, Sagar et al (2009, p. 283) suggests the establishment of a network of Climate Innovation Centers (CIC) "which uses public-private sector partnerships aimed at developing/adapting technologies and products for climate mitigation and adaptation and overcoming to barriers to market, informed by local needs and contexts, could play an important role". They estimates, each CIC would require an investment of $40 million to $100 million per year and costs a cumulative investment of $1 billion to $2.5 billion in order to establish five such regional CICs for a period of five years, as first phase.

Jeffrey Sachs (2005, p. 41), the architect of MDGs notes in his path breaking work on 'the end of poverty and economic possibilities', "I believe that the single most important reason why prosperity spread, and why it continues to spread, is the transmission of technologies and the ideas underlying them". So let the technology and related know how transmit to the needy under developed regions of the world, let the local ideas to grow and integrate with the mature technologies, thereby contributing local solutions to global challenges. In this context, the idea of establishing Climate Innovation Centers deserves particular consideration.

24.7 CONCLUSION

This chapter tried to explain the intrinsic relation between poverty, environment and climate change in the context of energy generation, consumption and its availability.

Even though poverty is a locally reflected problem, its impacts are global. Presumably poverty in a remote African or south Asian village does affect the people living in developed world by the medium of climate change. It is possible that climate change mitigation and adaptation projects could be combined with poverty eradication and development programmes as well as with green energy and low carbon electricity generation projects.

All the investments in green energy solutions are not necessarily directly related to poverty eradication, but all the green energy solutions are environment friendly, low carbon sources of energy and electricity. Investments for wind and solar energy farms can develop a rural area, and at the same time can give business to some where else, where the components are being manufactured. Same is the case even with off grid solutions like solar lanterns, which illuminates rural households. Biogas plants and heat efficient stoves can take rural households out of indoor pollution and related respiratory diseases.

Natural calamities, especially climate change related ones, throws millions of people out of their livelihoods, thereby inducing the incidence of poverty. While devising energy policy and investment decisions policy makers should give priority to the interest of those vulnerable and deprived sections of people. This can be done by making such group of people the first beneficiaries for tailor made energy solutions specifically designed and planned for particular regions. Taking into consideration geographic and human factors, the poverty ridden underdeveloped parts of the world are also endowed with resources for renewable energy generation. By promoting investment (both public and private) in such energy projects could create hundreds of thousands of jobs directly and indirectly, there by offering opportunity and energy for work and other productive activities for the local population. Illuminating a home means giving light and learning time for the younger generation. Equipping house holds with energy efficient stoves and alternatives to solid fuels provide healthy indoors, which is necessary to empower women and children, leading to greater social development. In short, eradication of poverty leads to protection of environment; investments in green energy solutions create jobs and opportunities; and generating electricity from green sources mitigate climate change and provide energy security, and induce overall economic and social development.

REFERENCES

Ackermann, T (2005) *Wind power in power systems*, John Wiley and Sons, West Sussex.

Alexandratos, N (ed). (1995) *World Agriculture towards 2010: An FAO Study*. FAO and John Wiley, Chichester, 1995.

Alliance to Save Energy website http://ase.org/content/article/detail/6156 Accessed on 29 Jan 2009.

Anderson, R. (2004) *The Distributed Storage*-Generation "Smart" Electric Grid of the *Future*, workshop proceedings, "The 10–50 Solution: Technologies and Policies for a Low-Carbon Future." The Pew Center on Global Climate Change and the National Commission on Energy Policy, Washington, D.C.

Astrand, K and Neji, L. (2006) "An Assessment of Governmental Wind Power Programmes in Sweden–using a Systems Approach" in Energy Policy 34 (2006) pp 277–296.

Australian Conservation Foundation (2008) Growing a Green Collar Economy: Green Jobs Fact Sheet-2008, from http://www.acfonline.org.au/articles/news.asp?news_id=1963 Accessed on 12 Jan 2010.

Bahamondes, M.(2003) Poverty-Environment Patterns in a Growing Economy: Farming Communities in Arid Central Chile, 1991–99, *World Development*, v. 31(11), p. 1947–1957.

Barbier, E.B. (2009), *Rethinking the Economic Recovery: A Global Green New Deal*. UNEP.

Battaglini, A., Lilliestam, J., Haas, A., Patt, A. (2009). Development of SuperSmart Grids for a more efficient utilisation of electricity from renewable sources. *Journal of Cleaner Production*, v. 17, no. 10, July 2009, Pages 911–918.

Bezdek H Roger (2008) Green Collar Jobs in the U.S. and Colorado: Economic Drivers for the 21st Century, Report by American Solar Energy Society, Colorado, US pp 1–84.

Bezdek, H. Roger (2009) Estimating the Jobs Impacts of Tackling Climate Change, Report by American Solar Energy Society, Colorado, US pp 1–20.

Bousquet, J, Khaltaev, N.G, Cruz, A.A, WHO (2007) Global surveillance, prevention and control of chronic respiratory diseases: a comprehensive approach. World Health Organization, Geneva.

Capoor, K. and Ambrosi, P. (2008). State and Trends of the Carbon Market 2008. The World Bank, Washington, D.C.

CAT (2007) Zero Carbon Britain: An Alternative Energy Strategy, Centre for Alternative Technologies, Powys. Accessed from http://www.zerocarbonbritain.com/ (2.03.2010).

Chen, S and Ravallion, M, (2008) *The Developing World Is Poorer Than We Thought, But No Less Successful in the Fight against Poverty*, The World Bank, Washington.

Chevalier, Jean Claude (2009) "Winning the Battle?" in The New Energy Crisis: Climate, Environment and Geopolitics, Palgrave Macmillan, Basingstoke, UK pp 256–280.

Dincer, I. (2000) "Renewable energy and sustainable development: a crucial review" *Renewable and Sustainable Energy Reviews 4, No. 2*, pp 157–175.

Dincer, I., Rosen, M.A. (2005) Thermodynamic aspects of renewables and sustainable development. Renewable and Sustainable Energy Reviews 9, 169–189.

Eba'a Atyi, R, Devers, D, Wasseige, C, and Maisels, F (2008) State of the forests of Central Africa: Regional Synthesis *in* Wasseige, C., Devers D., de Marcken P., Eba'a Atyi R., Nasi R. and Mayaux Ph., (eds) *The Forests of the Congo Basin – State of the Forest 2008*, 2009, Luxembourg: Publications Office of the European Union.

EC (2009) *Photovoltaic Solar Energy: Development and Current Research*, European Commission, Luxembourg.

European Commission (2006a) European SmartGrids Technology Platform: Vision and Strategy for Europe's Electricity Networks of the Future, European Commission, Brussels.

European Commission (2006b) European Technology Platform Smart Grids: Strategic Research Agenda for Europe's Electricity Networks of the Future. European Commission, Brussels.

EU Barometer (2009) Photovoltaic Barometer, March 2009, European Commission, Accessed from http://www.eurobserv-er.org/pdf/baro190.pdf (14.02.10).

Fells I. "The Problem" in: Dunderdale J, (Ed.) (1990) Energy and the environment. UK: Royal Society of Chemistry, London 1990.

Freris, L, and Infield, D (2008) *Renewable energy in power systems*, John Wiley and Sons, West Sussex.

Frost & Sullivan (2009) Asset Management – European Emissions Trading Market Research Report. Frost & Sullivan Consultants, London.

Fthenakis, V.M, and Kim, H.C (2007) Greenhouse-gas emissions from solar electric-nuclear power: A life cycle study, *Energy Policy*, v. 35 (4), pp. 2549–2557.

Fuglestvedt, J, Berntsen, T, Myhre, G, Rypdal, K, and Skeje, R.B, (2008) PNAS, v. 105 no. 2, p. 454–458. Accessed from http://www.pnas.org/content/105/2/454.full (28.02.2010).

Gagnon L., and van de Vate J.F. (1997) Greenhouse gas emissions from hydropower *Fuel and Energy Abstracts*, v 38, no. 4, July 1997, pp. 247–247(1), Elsevier.

Geels, F.W (2002) Technological Transitions as evolutionary Reconfiguration Processes: A multi-level perspective and case study, Research Policy 31, pp 1257–1274.

Geels, F.W (2002) Understanding the dynamics of Technological Transitions: A Coevolutionary and Socio-Technical Analysis, Twente University Press, Twente.

Growing a Green Collar Economy: Green Jobs Fact Sheet-2008, Australian Conservation Foundation from http://www.acfonline.org.au/articles/news.asp?news_id=1963 Accessed on 12 Jan 2010.

Han, J, Mol, A.P.J, Lu, Y, Zhang, L, (2009) Onshore wind power development in China: Challenges behind a successful story, *Energy Policy*, v. 37, 2941–2951.

Helm, D (2005) The Assessment: The New Energy Paradigm, Oxford Review of Economic Policy 21, pp 1–18.

Hendricks B., Goldstein B., Detchon R. and Shickman K., (2009) Rebuilding America: A National Policy Framework for Investment in Energy Efficiency Retrofits, Center for American Progress and Energy Future Coalition.

Huang, H, and Yang, Z, (2009) Present situation and future prospect of hydropower in China, *Renewable and Sustainable Energy Reviews, v. 13, nos. 6–7*, p. 1652–1656.

IEA (2007) World Energy Outlook, International Energy Agency, Paris

IEA (2008) World Energy Outlook, International Energy Agency, Paris

IEA (2009) World Energy Outlook, International Energy Agency, Paris

IEA (2009) Philippine de T'Serclaes, Emilien Gasc, and Aurélien Saussay (2009) "Financial Crisis and Energy Efficiency" Information Paper, International Energy Agency (IEA).

IEA Hydro (2000) *Hydro Power and World's Energy Future*: The role of hydropower in bringing clean, renewable, energy to the world, IEA Hydro, Accessed from: http://www.ieahydro.org/reports/Hydrofut.pdf (02.03.10).

IEA (2007) *Tracking Industrial Energy Efficiency and CO2 Emissions*, International Energy Agency, Paris.

IPCC (2007) Climate change 2007: mitigation of climate change: contribution of Working Group III to the Fourth Assessment Report of the Intergovernmental Panel on Climate Change, Cambridge University Press, Cambridge.

IPCC, (2007) Climate change 2007: Synthesis Report-An assessment of the Intergovernmental Panel on Climate Change, IPCC, [Online] Accessed from, http://www.ipcc.ch/pdf/assessment-report/ar4/syr/ar4_syr.pdf [23.07.09].

Jenson, J (2002) Integrating CHP and Wind Power-How Western Denmark is leading the way, Cogeneration and on-site Power Production 3, 55–62.

Kammen, D. M. and G. F. Nemet (2005) "Reversing the Incredible Shrinking Energy R&D Budget" Issues in Science and Technology, 22: 84–88.

Kammen, D. M., Kapadia, K. and Fripp, M. (2004) Putting Renewables to Work: How Many Jobs Can the Clean Energy Industry Generate? A Report of the Renewable and Appropriate Energy Laboratory, University of California, Berkeley.

Kats, G, (2009) *Greening Our Built World: Costs, Benefits, and Strategies*, Island Press, Washington DC.

Krohn, S, Morthost, P-E, Awerbuch, S, (2009) *The Economics of Wind Energy*: A report by the European Wind energy Association, Brussels.

Kupitz, J, and Podest, M, (1984) *Nuclear heat applications: World overview, IAEA Bulletin*, v. 26, No. 4.

Ladislaw, Sarah and others (2009) A Roadmap for a Secure, Low Carbon Energy Economy: Balancing Energy Security and Climate Change, Report of Centre for Strategic and International Studies and World Resources Institute, Washington D.C. pp 1–36.

Lorenz, P, Pinner, D, and Seitz, T. (2008) The Economics of Solar Power, *The McKinsey Quarterly*, June 2008.

Mendonça, Miguel (2007) "Feed-in Tariffs: Accelerating the Deployment of Renewable Energy" in Powering the Green Economy, Earth Scan Publishers, London.

Midillia, A, Dincer I and Aya, M (2006) 'Green energy strategies for sustainable development' in Energy Policy 34 (2006) pp. 3623–3633.

Modi, V., S. McDade, D. Lallement, and J. Saghir. 2006. Energy and the Millennium Development Goals. New York: Energy Sector Management Assistance Programme, United Nations Development Programme, UN Millennium Project, and World Bank. Washington DC.

Nasi, R, Mayaux, P, Devers, D, Bayol, N, Eba'a Atyi, R, Mugnier, A, Cassagne, B, Billand, A and Sonwa, D (2009) A First Look at Carbon Stocks and their Variation in Congo Basin Forests *in* Wasseige C., Devers D., de Marcken P., Eba'a Atyi R., Nasi R. and Mayaux Ph., (eds) *The Forests of the Congo Basin – State of the Forest 2008*, Luxembourg: Publications Office of the European Union, Brussels.

Newell, P. (2008). Civil Society, Corporate Accountability and the Politics of Climate Change. Global Environmental Politics, 8(3), pp. 122–153.

Nordhaus, William (2008) A Question of Balance:Weighing the Options on Global Warming Policies, Yale University Press, New Haven.

NRDC, The National Development and Reform Committee and the People's Republic of China. (2007) The Medium and Long-term Plan of Renewable Energy Source Development. Beijing, China.

OECD (1995) Anonymous. Urban Energy Handbook, Organization for Economic Co-Operation and Development, Paris.

OECD (2003) Poverty and Climate Change: Reducing the Vulnerability of the Poor through Adaptation. http://www.oecd.org/dataoecd/60/27/2502872.pdf (Accessed 18.12.2009).

OECD, (2004) *OECD environmental strategy: 2004 review of progress*, OECD Publishing, 2004, Paris.

Paraschivoiu, P. (2002) *Wind turbine design: with emphasis on Darrieus concept*, Presses inter Polytechnique, Montreal.

RCEP (2000) Report 'Energy-The Changing Climate', The UK Royal Commission for Environmental Pollution.

Robert R Harmon and Kelly R Cowan "A Multiple Perspectives View of the Market Case for Green Energy" in Technological Forecasting and Social Change 76 (2009) pp 204–213.

Runci, Paul (2005) Energy R&D Investment Patterns in IEA Countries: An Update Research Paper PNWD-3581, Joint Global Research Institute, University of Maryland.

Sachs, J.D. (2005) *The end of Poverty: Economic Possibilities for our Time*, The Penguin Press, 2005, New York.

Sagar, D.A., Bremmer, C. and Grubb, M (2009) *Climate Innovation Centres: A partnership Approach to meeting energy and Climate Challenges*, Natural Resource Forum 33 (2009) 274–284.

Sainath, P (2009) Neo-Liberal Terrorism in India: The Largest Wave of Suicides in History, Counter Punch.Org http://www.counterpunch.org/sainath02122009.html (Accessed 13.12.2009).

Sarah, White & Walsh, Jason (2008) Greener Pathways – Jobs and Workforce Development in the Clean Energy Economy, Report by Centre on Wisconsin Strategy, University of Wisconsin-Madison. http://www.cows.org/pdf/rp-greenerpathways.pdf Accessed on 18 Jan 2009.

Schaefer, A. Heywood J.B, Jacoby, H.D, Waitz, I.A, (2009) *Transportation in a Climate-Constrained World*, MIT Press, Cambridge, MA.

Schnoor, L. Jerald (2009) "Jobs, Jobs and Green Jobs" in Environmental Science and Technology 2009, 43 (23), p 8706.

Sen, A.K. (2001) *Development as Freedom*, Oxford University Press, Oxford.

Skjærseth, J.B. and Wettestad, J. (2009). The Origin, Evolution and Consequences of the EU Emissions Trading System. Global Environmental Politics, 9(2), pp. 101–122.

Smith, A. (1853) *The Theory of Moral Sentiments*, Henry G. Bohn, London.

Stern, N.H, (2007) *The Economics of Climate Change: The Stern review*, Cambridge University Press, Cambridge.

Swinton, S. M, Escobar, G, and Reardon, T (2003) Poverty and Environment in Latin America: Concepts, Evidence and Policy Implications. *World Development, 31 (11)*, p. 1865–1872.

The White House (2010) Fact Sheet: Homestar Energy Efficiency Retrofit Program http://www.whitehouse.gov/the-press-office/fact-sheet-homestar-energy-efficiency-retrofit-program (Accessed 03.03.10).

UN-Energy (2009) *Policies and Measures to realize Industrial Energy Efficiency and Mitigate Climate Change*, United Nations.

UNEP (2008) Renner, Michael, Sweeny, Sean and Kubit, Jill "Green Jobs: Towards decent work in a sustainable, low-carbon world, Report of United Nations Environment Programme (UNEP/ILO/IOE/ITUC), From http://www.ilo.org/wcmsp5/groups/public/—dgreports/—dcomm/documents/publication/wcms_098503.pdf Accessed on 28 Jan 2009.

UNEP (2010) Year Book, New Science and Developments in our Changing Environment, United Nations Environment Programme.

University of Leeds (2009, February 19). One-fifth Of Fossil-fuel Emissions Absorbed By Threatened Forests. *Science Daily.*, from http://www.sciencedaily.com/releases/2009/02/090218135031.htm Accessed 14.01.2010.

UK-OPSI Office of the Public Sector Information, United Kingdom website http://www.opsi.gov.uk/acts/acts2009/ukpga_20090019_en_1#l1g1 Accessed on 20 Dec 2009.

US Dept. of Energy (2003) Wind Energy for rural Economic Development, USDOE- Energy Efficiency and Renewable Energy, Accessed from http://www.nrel.gov/docs/fy04osti/33590.pdf (14.01.10).

US DOE (2009) President Obama Announces $3.4 Billion Investment to Spur Transition to Smart Energy Grid, US department of energy, October 27, 2009 http://www.energy.gov/8216.htm.

US-EPA United States Environment Protection Agency website http://www.epa.gov/greenpower/gpmarket/index.htm Accessed on 12 Dec 2009.

Weisser, D (2007), A guide to life-cycle greenhouse gas (GHG) emissions from electric supply technologies, *Energy* 32 (9), pp. 1543–1559.

Wines, M. (2005) Drought Deepens Poverty, Starving More Africans, New York Times, November 2, 2005, http://www.nytimes.com/2005/11/02/international/africa/02malawi.html?_r=1&pagewanted=2 (Accessed 14.12.2009).

WWEA (2009) *World Wind Energy Report 2008*, World Wind Energy Association, Bonn.

Yudelson, J, (2009) *Green Building Trends: Europe*, Island Press, Washington DC.

Section 7

Overview and integration

U. Aswathanarayana (India)

The book deals with five themes.

Theme 1: Renewable Energy Technologies (RETs)

There is little doubt that the Renewables are the energy resources of the future, for the simple reason that they are not only "green" but most of them do not get depleted when used. The BLUE Map scenario envisages a strong growth of renewables (reaching about 20 000 TWh/yr by 2050) to achieve the IPCC target of 450 ppm of CO_2.

Wind energy is believed to be the most advanced of the "new" renewable energy technologies. Wind power (2 016 GW) is expected to provide 12% of the global electricity by 2050, thereby avoiding annually 2.8 gigatonnes of emissions of CO_2 equivalent. Wind power sector would need an investment of USD 3.2 trillion during 2010–2050. The lifecycle cost is projected to be USD 70–130/MWh for onshore wind, and USD 110–131/MWh for offshore wind. The life cycle costs of wind energy are sought to be reduced through resource studies, technology (e.g. larger rotors, greater heights, deep water foundations for offshore turbines), supply chains, and mitigation of environmental impacts.

Solar PV will work wherever the sun shines. Its levelized cost (US cents 20–40/kWh) is several times more than electricity from fossil fuels (US cents 3 to 5/kWh). Solar energy is expected to grow thousand-fold between now and 2050. Technical advances in thin-film production and "building-integrated PVs" (BIPV) as well as massive application are bringing down the costs rapidly, however.

Second-generation biofuels, produced by enzymatic hydrolysis of cellulosic feedstock, and gasification of a variety of biomass material, have a great future. Single-cell algae are being used to produce a chemical "mix" that is chemically identical to petroleum crude, which is also carbon neutral and sulphur-free. Small ($\sim$100 kW) power units, which burn biomass wastes like paddy husk, are very useful to villages, which are not connected to grid.

Presently hydropower accounts for 90% of the renewable power generation in the world. Though hydropower is the cheapest way to produce electricity, it has become controversial because of human and ecosystem problems of large dams. Pumped storage is the highest capacity form of energy storage.

Geothermal energy is confined to areas of high heat flow. It is non-polluting and can be generated round the clock. High temperature geothermal sources can be used to generate electricity.

The use of tidal energy to generate power is similar to that of hydroelectric plant. The estimated global potential of wave electricity is 300 TWh/yr. The 740-m long Rance Barrage in France which produces 480 GWh of electricity, is one of the few operating tidal energy plants in the world. Considerable R&D effort is needed to ensure the commercial viability of ocean energy.

Countries have to decide upon the actual optimal mix of RETS, and timing of the policy incentives, depending upon their biophysical and socioeconomic situations. The level of competitiveness will depend upon the evolving prices of competitive technologies. The deployment of Renewable Energy Technologies (RETs) has two concurrent goals: (i) exploit the "low-hanging" fruit of abundant RETs which are closest to market competitiveness, and (ii) developing cost-effective ways for a low-carbon future in the long term. Highest priority should be given for the removal of non-economic barriers.

The transition to mass market integration of renewables requires some policy interventions. Such interventions should be able to lead to a future energy system in which RETs should be able to compete with other energy technologies on a level playing field. When once this is achieved, RETs would need no or few incentives for market penetration, and their deployment would be accelerated by consumer demand and general market forces. Technology-specific support schemes need to be fashioned depending upon the level of maturity of a given RET at a given time, employing a range of policy instruments, including price-based, quantity-based, R&D support and regulatory mechanisms.

All RETs are evolving rapidly, in response to technology improvements and market penetration. Renewable Energy policy frameworks should be so structured as to facilitate technological RD&D and market development concurrently, within and across technology families.

Theme 2: How to make Green Energy competitive

The renewable fuels, such as wind, solar, biomass, tides, and geothermal, are inexhaustible, indigenous and are often free at source. They just need to be captured efficiently and transformed into electricity, hydrogen or clean transportation fuels. In effect, the development of renewal energy invests in people, by substituting labour for fuel – renewable energy technologies provide an average of four to six times as many jobs for equal investment in fossil fuels. That said, the most important challenge facing the renewables is to achieve market penetration. This section is addressed to possible ways of making the renewables competitive.

Public – private partnership is the most effective way to achieve success in the operation of the Innovation Chain: Basic Research → Research & Development → Demonstration → Deployment → Commercialization (diffusion). The goal of the RD&D policy should be to design ways and means by which the value of a public good (say, climate change) is built into commercial and innovation systems. The role of the government is most effective when it is able to combine *supply-push* (i.e. focus on RD&D and technology standards) with *Demand-pull* (i.e. focus on influencing the market through economic incentives such as regulation, taxation or guaranteed purchase agreements). Generally, renewable energy technologies tend to be more expensive than incumbent technologies, like fossil fuels. *Technology Learning* can be made use of to bring down the costs of the green technologies, through reduction in production costs and improved technical performance.

The Levelized Cost of Energy (LCOE) which is calculated by levelizing the different scales of operation, investments and operating time periods between various forms of energy generation, can be made use of to make investment choices and evaluate the efficiency benefit arising from an investment. The value of one unit of energy depends upon when, where and how it is available. The capacity value of an energy system is given by the energy that can be reliably delivered at the time of the peak consumption, whereas the energy value of a system is the total amount of energy delivered over the course of a year. The efficiency and economics of a renewable energy facility are optimized on this basis.

A combination of policy incentives and discentives (such as, Cap-and-Trade regimes, Green Certificates, loans at low interest rates, tax credits, accelerated depreciation), enhancing the demand for green energy starting with government establishments, publicity campaigns and innovative marketing, are required in order for the green energy to achieve high market penetration.

Theme 3: How to reduce CO_2 emissions and improve efficiency and employment potential of Supply-side Energy Technologies

For the mitigation of climate change, CCS (Carbon dioxide Capture and Storage) is a technology option that would allow the continued use of fossil fuels. Pulverised coal combustion (PCC) accounts for 97% of the coal-fired capacity. Supercritical steam plants and Integrated coal gasification combined cycle (IGCC) plants have been able to achieve high thermal efficiencies of 42 to 45%. Post-combustion capture of CO_2, followed by geological storage, is nearest to commercialization. CCS has considerable flexibility in technological improvement, such as the use of new absorbers. Flue gas scrubbing with amines is the most promising for plants of 500 MW or higher capacity. Transport of CO_2 is the key for CCS deployment. Pressure vessels can be used to transport CO_2 in the liquid form. Pipeline transport may be used for supercritical CO_2 above the critical point (31.1°C; 73.9 bars). Sleipner-type offshore storage of CO_2 has technical, social, political and economic advantages.

Present global nuclear share of electricity amounts to ~15% (370 GWe), and it helps reducing the global emission by ~3 giga tonnes of CO_2. Nuclear power is slated grow to 473–748 GWe by 2030. It is seen as a renewable energy, considering its vast energy potential. Current nuclear reactor technologies are based on the utilization of thermal or slow neutrons to fission low-enriched uranium (3–5% ^{235}U) using light water as moderator and coolant. Most of the reactors currently operating utilize less than 1% of the fissionable content of the fuel. The rest is treated as waste in once-through fuel cycles. In a closed fuel cycle the unutilized materials are recycled. In fast breeder reactors the energy utilization of the fuel is multiplied by almost 100 times, as the reactor "breeds" more fissile material than it consumes. India has come up with the design of a low-enriched uranium – thorium fuelled, heavy water reactor (AHWR-LEU), whose fuel cycle is proliferation-resistant, and which can be deployed without ant safety and security prescriptions. Such a technological innovation allows the bypassing of proliferation and security legal frameworks which are difficult to enforce.

Next Generation Green Technologies, which are in the process of development, may have a potential comparable to other renewable energies. Biomass gasification is potentially more efficient than the direct combustion of the original fuel. In the town of Güssing in Austria, a plant supplies 2 MW of electricity and 4 MW of heat, generated from wood chips, since 2003.

The global marine energy resource is estimated to be the order of 200 GW for osmotic energy; 1 TW for ocean thermal energy; 90 GW for tidal current energy; and 1–9 TW for wave energy. The total worldwide power in ocean currents has been estimated to be about 5 000 GW. It is estimated that capturing just 1/1 000 of the available energy from the Gulf Stream, would supply Florida with 35% of its electrical needs. On an average day, 60 million km^2 of tropical seas absorb an amount of solar radiation equal in heat content to about 250 billion barrels of oil. Ocean thermal energy conversion (OTEC) uses the temperature difference that exists between deep and shallow waters to run a heat engine. The osmotic pressure difference between fresh water and seawater is equivalent to 240 m of hydraulic head. Theoretically a stream flowing at 1 m^3/s could produce 1 MW of electricity. The worldwide fresh to seawater salinity resource is estimated at 2.6 TW.

Tidal power has potential for future electricity generation. Tides are more predictable than wind energy and solar power. Wave energy can be considered as a concentrated form of solar energy. Useful worldwide resource has been estimated at >2 TW. Low head hydropower applications use river current and tidal flows to produce energy. If the viable river and estuary turbine locations of the US are made into hydroelectric power sites it is estimated that up to 130 000 gigawatt-hours per year could be produced.

Enhanced Geothermal System (EGS) technologies "enhance" geothermal resources in hot dry rock (HDR) through hydraulic stimulation. It is reported that in the United States the total EGS resources from 3–10 km of depth is over 13 000 zetta joules. Out of this over 200 ZJ would be extractable, with the potential to increase this to over 2 000 ZJ with technology improvements.

Algae can be used to produce not only several kinds of fuel end products, but also byproducts which have wide ranging applications in chemical and pharmaceutical industries. They can be grown using land and water unsuitable for plant and food production. They are energy-efficient. They consume carbon dioxide. They can be mass produced. Algae may be cultivated in photobioreactors, and harvested using rotary screening methods. Expeller press and ultrasonic assisted extraction technologies may be used to produce energy products, such as, biodiesel, ethanol, methane, hydrogen, etc.

Theme 4: How to reduce CO_2 emissions and improve efficiency and employment potential of Demand-side Energy Technologies

2050 is only 40 years away. During the next five to ten years, it is imperative that we shift to long-term trajectories while meeting the interim targets in respect of Industry, Buildings and Appliances and Transport. This would involve undertaking the required RD&D programmes, improving efficiencies, achieving increased market penetration, making appropriate investments, changing of the policies, and so on.

Industry: Industry-caused CO_2 emissions (6.7 Gt in 2005) constitute about 25% of the total worldwide emissions. Iron and steel industry accounts for about 30% of the CO_2 emissions, followed by 27% from non-metallic minerals (mainly cement), and 16% from chemicals and petrochemicals production. If the Best Available Technologies (BATs) are applied worldwide, current CO_2 emissions can be reduced by about 19% to 32%. Improvements in steam supply systems and motor systems have the potential to raise efficiencies from 15% to 30%. If CHP is included in the process designs, it will reduce heat demand per unit of output. There are three main ways in which CO_2 emissions from the industries can be reduced: (i) through improvements in efficiency that can be brought about through the recycling of waste materials, and changes in the product design, (ii) feedstock substitution, such as the greater use of biomass, and (iii) CO_2 capture and storage (CCS).

Buildings and Appliances: The buildings sector employs a variety of technologies for various segments, such as building envelope and its insulation, space heating and cooling, water heating systems, lighting, appliances and consumer products. Local climates and cultures profoundly affect energy consumption, apart from the life styles of individual users. Buildings are large consumers of energy – in 2005, they consumed 2 914 Mtoe of energy. The residential and service sectors account for two-thirds and one-third of the energy use respectively. About 25% of the energy consumed is in the form of electricity. Thus the buildings constitute the largest user of electricity. Globally, space and water heating account for two-thirds of the final energy use. About 10–13% of the energy is used in cooking. Rest of the energy is used for lighting, cooling and appliances. The end-uses dominated by electricity consumption are important from CO_2 abatement perspective, in the context of the CO_2 emissions related to electricity production. CO_2 emissions can be reduced significantly through the use of Best Available Technologies in the building envelope, HVAC (heating, ventilation and air conditioning), lighting, appliances and cooking. Heat pumps and solar heating are the key technologies to reduce emissions from space and water heating.

Transport: Presently, transport accounts for about 19% of global energy use and 23% of energy-related carbon dioxide emissions. High growth rates are forecast for surface, air and marine transport for decades to come. Hence the transport energy use and CO_2 emissions are projected to increase by nearly 50% by 2030 and more than 80% by 2050.

This future is not sustainable. In order to achieve a low-carbon, sustainable future, it is necessary for the governments to embark on two pathways simultaneously: Firstly, through appropriate regulatory mechanisms, governments should strive to improve the efficiency of today's vehicles and for the deployment of transition technologies such as plug-in hybrids. Secondly, RD&D should be promoted for the development and deployment of long-term technologies, such as, biofuels, electric and fuel cell vehicles. Governments should make investments in infrastructure such as efficient, affordable and dependable public transportation (as in Singapore), and providing incentives for making rail travel preferable to air travel for journeys around 600 kms. International cooperation is essential to reach these goals.

The world average of Transmission and Distribution losses is 14.3% of the gross electricity production. It is high in India (31.9%) and low in Japan (8.7%). High voltage D.C. transmission is coming into vogue, as it is more economical than A.C. transmission for long distances (>500 km). Electricity can be stored, but only through other forms of energy. Pumped storage is the preferred option. It has an efficiency ranging from 55 to 90%, system rating of about 100 MW, and discharge times of hours. Pumped storage plants can respond to load changes almost instantly (less than 60 seconds). Compressed air energy systems (CAES) have efficiencies of about 70%. The biggest problem with CAES is finding suitable storage caverns. Aquifer storage is a good possibility for CAES Superconducting Magnetic Energy Storage (SMES) stores electrical energy in superconducting coils. SMES has the advantage of being able to control both active and reactive power simultaneously. Also, it can charge/discharge large amounts of power quickly.

Electricity demand of a given geographical unit depends upon the size of the population, their life-style, climate, agriculture, industry, tourism, etc. Also, it is highly dependent upon the time of the day (e.g. the demand for air-conditioning is maximum at noontime). Under the provisions of "smart" grid, industrial, residential and commercial users in an area are linked with various power generating units (thermal, wind power, solar PV, nuclear power, etc.) in the area. When it becomes necessary to reduce the peak demand in the area, the central control system may turn down the temperature of heaters, or raise the temperature of some appliances (such as air conditioners and refrigerators) to reduce their power consumption. This essentially means delaying the draw marginally. Though the amount of demand involved in this exercise is small, it will have significant financial impact on the system, as electricity systems are sized to take care of extreme peak demands, though such events occur very infrequently.

"Bloom Box", recently unveiled by K.R. Sridhar, has the potential to revolutionize electricity production. It is a fuel cell device consisting of a stack of ceramic disks coated with secret green and black "inks". It can convert any renewable and fossil fuel (e.g. natural gas, biogas, coal gas, etc.) into electricity, 24 × 7. Since no combustion is involved, there would be no noise, smell or emissions.

Theme 5: A Green New Deal

The present fossil fuel dependent energy generation and consumption is the largest contributor of carbon emissions, and being the cause of global warming ought to be replaced with low carbon sources of energy, such as renewables and nuclear. Low carbon and renewable energy generation by wind, solar, hydro and nuclear (uranium can be extracted from sea water) could resolve the energy and climate challenges of humanity. Investing in these technologies and sources will generate millions of jobs, revive the economy and bring overall socio-economic development. In order to establish this change, we also need transmission and distribution networks based on the smart and super smart grid technologies, which will facilitate distributed electricity generation, improve efficiency in transmission, resolve peak load challenges and provide options for the users to control their consumption patterns. These green goals have to be supplemented with qualitative improvement in the user end, such as de-carbonizing the transport, building and industrial sectors of the economy. By analyzing the low carbon possibilities of energy generation, transmission and consumption/use, this chapter shows how a green new deal with integrated energy goals for a low carbon future and sustainable development can be promoted.

There is urgent need to take measures towards greening the economy and reducing the carbon emissions to keep up with the emission targets for the future well being of the planet. The necessary policy making is this area is complicated by a variety of factors—huge investments needed in Research and Development, rapidly emerging newer technologies, focus on long-term goals as opposed to short-term gains, addressing infrastructure and employment issues, and addressing the needs of various actors in the whole process. This chapter while highlighting the urgency of the issue makes an analysis of the challenges and prospects based on experiences from select countries across the world. There is also a necessity for the suitable integration of the green energy policies across various sectors to derive maximum benefit.

There are ways of addressing poverty in the context of environment and climate change. Lack of economic opportunities forces people to exploit the natural resources around them unsustainably, leading to destruction of forests and other carbon sinks.

Income poverty leads to inefficient use of energy in house holds, and the absence of electricity and clean fuels for the household use drastically reduces the productivity and socio-economic well being of the people. This can be overcome by combining poverty eradication and development projects with climate change mitigation programs. Investment in energy and electricity generation projects, especially renewables, will induce economic and social development and provide energy security. Eradicating poverty can be made a part of climate mitigation action. Climate change induced natural calamities affects the tropical countries which are mostly underdeveloped and vulnerable. Looking at the climate change mitigation discourse, it is clear that the world is getting serious about the role of socio economic development in protecting the environment and mitigating climate change and also realizing that the investments can be combined.

Quo vadis?

Paul Krugman, Nobel laureate in Economics, wrote a highly thought-provoking article entitled, " Building a Green Economy", in the *New York Times* of Apr. 5, 2010. Climate Change is a classical case of "negative externality" – economic actors (say, a coal-fired thermal power station and the user of electricity) impose costs on others, without paying a price for their actions. Two approaches have been attempted to limit the negative externality – pollution tax and cap-and-trade. Acid rain is caused by the emissions of sulphur dioxide from power plants. It was controlled by the government prescribing compliant effluents, and taxing the power plants that were emitting beyond the permissible limits. Pollution tax is a disincentive. A company avoids pollution tax by reducing its pollution to be within compliant limits. Recently, US EPA formally declared CO_2 as a pollutant attracting pollution tax. Pollution tax in respect of CO_2 is vigorously opposed by the coal industry. In the case of cap-and-trade, the government issues a limited number of licenses to pollute. Companies which need to pollute more need to buy licenses from those which have pollution to spare. The incentive here is for a company to reduce pollution to avoid paying for extra pollution.

Pollution tax accrues to the government. In the case of cap-and-trade, the potential revenue goes to the industry instead of the government.

According to Nicholas Stern, author of the famous Stern Report, the emission of carbon dioxide and other greenhouse gases, is the "biggest market failure the world has ever seen". Markets cannot be trusted to set the matters right, and government interventions cannot be avoided. Many of the costs of climate change arise from skewed climate patterns, with more droughts, floods and severe storms. It should not be forgotten that London is at the same latitude as Labrador. Without the Gulf Stream, western Europe would be barely habitable. Poor countries are most vulnerable. Industrialized countries have certain amount of flexibility, but they cannot altogether escape the adverse effects. For instance, the American Southwest could become a dust bowl, if mitigating steps are not taken.

Climate change by 2100 will lower the gross world product by 5%, while stopping it would cost 2%. Then why not go for stopping it? This is because we have to spend the money now in order to protect the future generations, and many are reluctant to do so. The last time when the earth experienced warming at about the present level was during the Palaeocene- Eocene thermal maximum about 55 million years ago when temperatures rose by 11 degrees Fahrenheit. As is well known, this is the time of mass extinctions! Martin Weitzman argues that our policy analysis should be based on the non-negligible probability of utter disaster, forget about the uncertainty of Climate models and difference of opinion of ways of mitigating the adverse consequences of the climate change. Krugman supports Weitzman, and aggressive action to reduce emissions.

Author index

Subject index